CW01176852

DISLOCATING THE ORIENT

DISLOCATING THE ORIENT

BRITISH MAPS AND THE MAKING OF
THE MIDDLE EAST, 1854–1921

Daniel Foliard

THE UNIVERSITY OF CHICAGO PRESS
CHICAGO AND LONDON

The University of Chicago Press, Chicago 60637
The University of Chicago Press, Ltd., London
© 2017 by The University of Chicago
All rights reserved. No part of this book may be used or reproduced in any manner whatsoever without written permission, except in the case of brief quotations in critical articles and reviews. For more information, contact the University of Chicago Press, 1427 E. 60th St., Chicago, IL 60637.
Published 2017
Printed in the United States of America

26 25 24 23 22 21 20 19 18 17   1 2 3 4 5

ISBN-13: 978-0-226-45133-6 (cloth)
ISBN-13: 978-0-226-45147-3 (e-book)
DOI: 10.7208/chicago/9780226451473.001.0001

Publié avec le concours de l'Université Paris Ouest Nanterre La Défense

Library of Congress Cataloging-in-Publication Data
Names: Foliard, Daniel, author.
Title: Dislocating the Orient : British maps and the making of the Middle East, 1854–1921 / Daniel Foliard.
Description: Chicago : The University of Chicago Press, 2017. | Includes bibliographical references and index.
Identifiers: LCCN 2016035742 | ISBN 9780226451336 (cloth : alk. paper) | ISBN 9780226451473 (e-book)
Subjects: LCSH: Great Britain—Relations—Middle East. | Middle East—Relations—Great Britain. | Middle East—Maps—History—19th century. | Middle East—Maps—History—20th century. | Cartography—Great Britain—History—19th century. | Cartography—Great Britain—History—20th century. | Cartography—Political aspects—Great Britain. | Imperialism—History—19th century. | Middle East—Name.
Classification: LCC DA47.9.M628 F65 2017 | DDC 327.4105609/034—dc23 LC record available at https://lccn.loc.gov/2016035742

♾ This paper meets the requirements of ANSI/NISO Z39.48-1992 (Permanence of Paper).

CONTENTS

Introduction · 1

PART I. FROM SEBASTOPOL TO SUEZ (1854–1869)
1. The Mid-Victorian Perspective:
   A Fragmented East · 15
2. Labeling the East · 52
3. Maps for the Masses? · 67

PART II. A SHIFTING EAST IN THE AGE OF HIGH IMPERIALISM (1870–1895)
4. Oriental Designs · 95
5. Virtual Travel in the Age of High
   Imperialism · 139

PART III. THE FABRICATION OF THE MIDDLE EAST (1895–1921)
6. Seeing Red? · 169
7. Enter Middle East · 207
8. Falling Into Places · 233

General Conclusion · 266

Acknowledgments · 275   *Notes* · 277
Notes on Methodology and Select Bibliography · 319
Index · 327

For high-resolution images, raw data, and a bibliography, see www.dislocatingtheorient.com.

INTRODUCTION

Divisions of the globe are evolving constructs.[1] Their denominations sometimes tell us more about the imaginations and ideologies of those who devise and demarcate them than about the territories and populations they encompass. Some regional entities are more problematic than others. They raise more questions, trigger more debates. One in particular, the Middle East, became instrumental in how people in Europe and America viewed the world in the 20th and 21st centuries.[2] This book investigates the origins of this specific geographical entity.[3]

The "Middle East" was initially conceived as a way of delineating the transitional space between Europe and India, a region that the originators of the concept wished to see under the influence of the British Empire. Though the dismantling of that empire after the Second World War should have led to the concept's obsolescence, it acquired a life of its own during the Cold War. While these 20th-century conflicting visions and uses of the "Middle East" are well documented, the genesis of the term has attracted much less attention. Studies consistently devote their attention to the evolution of the concept after the First World War rather than considering its earlier origins.[4] This volume sets out to explore the genealogy and prehistory of this geographical topos. It is an attempt at what Marc Bloch called retrogressive history.[5] The "Middle East" has to be seen as an emerging notion, the culmination of, rather than the starting point in, a process of conceptual coalescence.

One of the greatest potential defects of retrospective analysis is determinism. It is fatally easy to imagine a coherence when there was none, or induce a linear scientific progress where what was actually at stake was a messy process.[6] I therefore choose deliberately not to use the expression "Middle East" in sections dealing with pre-1900 documentation. In analyzing the origins of the use of that term it is crucially important to contextualize it alongside other usages rather than anachronistically concentrating on the one phrase which happens to interest us. That means following the grain of 19th-century conceptual divisions rather than resisting them. There were, of course, precursors of what would become referred to as the "Middle East" in the early 20th century. Nineteenth-century cartographers and geographers felt a need for subdivisions of Asia which would take into account the area located between the Mediterranean and India. James Wyld published, for instance, a remarkable map of the *Countries lying between Turkey & Birmah*, which placed Persia and the Arabian Sea at the very heart of a growing imperial network of communications.[7] Older maps, such as Jean-Baptiste Bourguignon d'Anville's *Première Partie de la Carte d'Asie*, already prefigured a transitional area between the Levant and India in the 18th century.[8] Yet, while the "Middle East" was not invented ex nihilo, to impose it as a post facto analytical schema upon 19th-century documentation would ineluctably distort the analysis of it.

I will therefore refer to an ill-defined East and no less approximate Orient in the early chapters of this book. Semantic vagueness is intentional in this instance. It is not at odds with 19th-century usages of both concepts. Their meaning was often context specific and, as Martin Lewis and Karen Wigen have shown, the scope of the "Orient" or the "East" varied over time.[9] The malleable nature of these categories is illustrated by the widely varying definitions adopted by writers in the 19th century. The "East" often referred to the eastern Mediterranean and its adjoining territories. The education of the Victorian elite rested a good deal on Edward Gibbon's *Decline and Fall of the Roman Empire*.[10] That work conditioned its 19th-century readers to see the "East" as the eastern Roman Empire and its immediate neighbors. So, for example, John Carne's use of the term in his 1830 travelogue included Syria, Palestine, Arabia, and Egypt, and that denomination was a commonplace among his contemporaries.[11] This minimalist East was also, among other things, the geographical reflection of the Grand Tour. The 1862 Prince of Wales's educational journey to the "East" was in fact a trip to Cairo and Constantinople, not to India.[12] Decades later, the East could refer to Russia and Turkey in Europe for Henry de Worms, who became undersecretary of state for the colonies in 1888.[13] For Valentine Baker, one of the Great Game heroes, the East went as far as Central Asia.[14] George Curzon eventually subdivided the area into three sections in 1894: the Far East, the Central East, and the Near East.[15] The same volatility characterized uses of the "Orient," which, as noted by Lewis and Wigen, had a "cultural connotation" in contrast with a more geopolitical "East."[16] Indeed, "East" and "Orient" were often used interchangeably. Both terms should be viewed in this book through this 19th-century prism, since it will use "Orient" and "East" in their various fluid Victorian denotations, to mean the region from the eastern Mediterranean to the confines of Persia and from the Arabian Sea to the Black Sea.[17]

I have selected various maps to introduce each chapter of this book. They serve two purposes. First, they offer a guide to the reader as to the evolving conceptual historical geography of this region. Second, they should be approached as test cases for understanding how maps were, from their genesis to their usage, a set of "unfolding practices."[18] These are microstudies, revealing the larger forces at stake in the exploration and construction of the area. They add up to an informal atlas, one that focuses on some of the more significant features of the British relationship with the area. These interconnected case studies and the title of the book indicate that the focus of this work is on one category of documentation in particular: maps. Even if the Middle East is often regarded as a political entity rather than a truly geographical one, maps represent points of entry through which to investigate these cultural constructs. Their elaboration, compilation, semiotics, publication, and reception furnish a set of clues by which we can set about exploring the various social groups, networks, and British imaginations whose preconceptions they document. Besides, maps do not exist in a vacuum. They are the product of an "assemblage of people and of discursive practices."[19] Studying how they interact with other forms of knowledge as well as how they articulate with textual material is central to understanding them. The book therefore focuses on the actual processes of cartography, that is, the manner in which the British visualized the Middle East through literal mapping and how maps were made. But, in so doing, it is also about the mapping going on in the mind.[20]

Three analytical themes weave together the wide array of maps I consider in this work. The book first considers geographical and cartographic imaginations. It draws on a body of literature which considers how people view and categorize the world around them.[21] From Edward Said's study of imaginative geographies, to the work of Derek Gregory, who underlined how mapping was always "situated, embodied, partial," a prolific field of study considers maps and geographical material from a critical perspective.[22] Others, notably Brian Harley and Felix Driver, have insisted on the intrinsically social nature of the production of geography and cartography.[23] Maps are not only forms of knowledge, but also processes as well as performances, which thereby reveal assumptions and, in turn, shape reality.[24] They can be understood in terms of narratives of socially constructed meanings. Building on these analyses, this work isolates the contrasting clusters of cultural representations, ideologies, and stereotypes which framed the East from a British perspective. By geographical imaginations, I am referring to a mode of interpreting reality which stems from the accretion of social and collective representations of spatiality. These specific forms of imaginations give meaning to the world, in terms of space, distance, or territory. They consolidate the social and cultural links within a given group, while never being entirely independent of external influences, comparisons, and circulations. Institutions tend to regulate them, to a greater or lesser extent. Maps convey, reflect, and shape these social representations. They thus entertain a dual relationship with imaginations, representing what is a distant reality. In that sense, they claim to reflect the world for the map reader, while also providing a frame and a specific discourse through which the reality might in its turn be modified. Travelers looking at a map to plan a journey, diplomats using cartography as a cognitive instrument to solve a crisis—these are some of the various instances of this dialectic process. This critical perspective on imaginations necessarily emphasizes the many-sidedness and discursive nature of the cultural representations of place. Drawing on Felix Driver's work on "the heterogeneity of geographical knowledges," the book therefore distances itself from the essentialism that characterized Edward Said's approach of Orientalism.[25]

A large body a scholarship is already devoted to British mapping in the 19th century. Matthew Edney's research on the Survey of India, Haim Goren's interest in the British expeditions in Mesopotamia and Palestine, Graham Burnett's analysis of the geographical construction of British Guiana, and John Moscrop's volume on the Palestine Exploration Fund are all crucial points of reference for this book.[26] Yet most of these works focus either on the physical activity of topography, on the one hand, or on the evolving British geographical imagination, on the other. This study's integrated approach to the construction of a regional entity offers a new perspective on the mapping of Asiatic and European Turkey, Palestine, Syria, Egypt, and Persia. It implies taking into account the entire process, from the actual survey in the field to the subsequent uses to which maps were put, their impact and the controversies which they provoked. I document the accretion of multiple layers of geographical meaning to form fundamental assumptions, ones which then are adopted by the topographer and inculcated in the schoolroom, in order to tie up the production, the content, and the reception of cartographic material. Though necessarily patchy, this longitudinal section of the British geographical imagination in the 19th and early 20th centuries is a principal focus of this book.

The second area of concern is the production of cartographic knowledge and its economics. I want to give due weight to the importance of the infrastructure that partly determined the cultural superstructure. How maps of the East were made, printed, sold, distributed, handled, and studied is fundamental to understanding this kind of documentation's cultural impact. Mapping the East, and maps of the East, were part and parcel of a much wider discourse. That discourse embraces the imperatives of empire and exploration, not to mention missionary religious endeavor. But behind it lay the technical constraints of the mapping process, as well as the need for local assistance. So the subject of this book is a set of multifaceted and contested mapping processes taking place at multiple sites within an evolving chronology.

The interaction between documents and reality is the third theme of this book. Did modern maps and geographical discourses transform the East? The answer to that question has to start with the central relationship between cartography and empire.[27] It is sometimes the accepted view that 19th- and 20th-century maps were instruments of British imperial control. Commenting on British surveys in Palestine in the 1860s, Kathleen Stewart Howe sees maps as means by "which imperialism manifests its central act."[28] The assumption that cartographic and geographical enterprises in the 19th century were acts of subjugation, figuring empty territories to be reclaimed and selling imperial values at home, needs nuancing.[29] This is not to argue that an enveloping culture of supremacy did not pervade these discourses. As noted by Matthew Edney for India, "imperialism and mapmaking intersect at the most basic level."[30] Yet, so far as the area under study here is concerned, it is doubtful whether British agents even felt the need to justify their actions through the articulation of a coherent ideological system before the First World War. Too overarching a description of what the development of British world power implied for the area would thus fail to reveal the reality for what it was: an intricate network rather than a raw projection of superiority. John Darwin rightly emphasizes "the chaotic pluralism of British interests" that presided over imperial expansion.[31] This was specifically true of the territories on the High Road to India, from Egypt to Persia. They were progressively drawn into a global system of influences and interests, dominated by Britain but without being an integral part of the formal empire: "almost an empire, in all but name."[32] Even though I will use the term "imperialism" in a broad sense, inclusive of the informal dimensions of British dominance in the East, I shall necessarily be taking into account the mutations of its various expressions during the 19th and early 20th centuries. Ideologies and less systematized aggregates of representations considerably evolved over time. Alfred Milner's constructive imperialism in the 1900s had little to do with Palmerston's liberal empire. The pluralistic nature of British projections in the East calls for context-specific analyses. I therefore apply a historicized approach to the various shades of imperialism to be found in the mapping of the Orient.[33]

This book traces the origin story of the "Middle East." It first sets out to analyze what this genealogy reveals about Britain itself. The argument, drawing on recent scholarship on the reciprocal influences between the expansion of Europe's global influence and metropolitan cultures, will underline how multilayered and contentious worldviews were within Britain.[34] A great variety of interests, forces, and emotions characterized what might too simplistically be seen as the single course of late 19th-century imperial culture and ideology. Part of the mapping of the East in the 19th century reflects the visions of those who saw themselves as agents of progress (chaps. 4, 6, and 7). They aimed at removing the blanks on the map. They located resources, opportunities, and potential routes. Some even cartographically legitimated imperial expansion in the area as early as the 1870s. Others, remarkably so in the Orient, mapped a world still untouched by modernity. Their explorations, as well as their publications, reflected the powerful attraction that only such a history-saturated area could create. This Byronic romance of the unspoiled East existed in juxtaposition with a more biblical perspective (chaps. 1 and 2). Large segments of British imaginations were deeply shaped by a strong Victorian vernacular biblicism. This religious component, coupled with Renaissance-rooted antiquarianism and Romantic exoticism, competed with more secular and commercial forces of modernity's sycophants. It is one of the primary aims of the books to demonstrate how education, cartography, and geography reflected these multifaceted Orientalist encounters and their imprint on British culture.

My emphasis on the role of knowledge (and ignorance) in British global expansion, and on the complex ways in which this power was constructed, contested, and received both in the empire and domestically also aims to contribute to the growing literature on culture and empire. Catherine Hall has already made plain that the cultural consequences of Britain's overseas expansion for English imaginations were never uncomplicated. Her dialectical study of the changing perceptions of race between Birmingham and Jamaica is a masterly demonstration of how crucial it is to understand British history in relation to the United Kingdom's global experience in the 19th century. Others, in particular John Mackenzie and Andrew Thompson, have pointed at the pervading presence of the empire in metropolitan culture, as well as the necessary breakdown of the phenomenon along social lines.[35] This book owes a great deal to these perspectives on imperial histories. It sets out to show that not only did the empire play a key role in shaping cultural practices, but also the metropole and the British world system as a whole were ineluctably intertwined (chaps. 3 and 5). The Orient was not, as we shall discover, painted red on the imperial map. But there is no doubt that it ineluctably played a crucial role in the first globalization of cultural circulation.

This work also highlights the articulation between cartographic knowledge and power. It emphasizes how effective cartography and exploration were to the wider project of opening the East to trade. Explorations of the Euphrates with a view to opening steamer routes (chap. 1), geological surveys in Egypt to locate resources (chap. 6), and early 20th-century oil prospecting in Persia (chap. 7) are all illustrations of this. Military mapping (chaps. 1 and 8) is another manifestation of the power of the map. This book also examines how cartographic and geographical knowledge helped bring the area into a standardized legal framework by way of boundary commissions and cadastral registers (chaps. 4 and 6). While this work does not downplay the power of maps and geographical knowledge, it shows that overly simplistic binaries do not mirror the articulation between cartographic intelligence, imaginaries and geopolitical power with sufficient accuracy. Christopher Bayly, in a groundbreaking work on imperial intelligence in India, analyzed how East India Company officials were often hindered by their ignorance and reliance on local information. They failed to understand India and the challenges posed by native economies of information as they turned to the "routinized and abstract information" of surveys and statistics in the second half of the 19th century.[36] A similar turn characterized the cartographic and geographical approach of the East in the late 19th and early 20th centuries. The romantic polymath of the mid-Victorian age gave way to supposedly more expert explorers. Thematic cartography, be it ethnographic, geological, or economic, depicted an area that could be rationally understood and developed. Maps and geographical knowledge were increasingly used to legitimize Britain's growing interference in the late 19th and early 20th centuries in the area. Cartographic propaganda during the First World War and the uses of maps during the post-1918 peace conferences strengthened the impression of systematization and coherence. The area, its inhabitants, and its resources were supposedly mapped out, described under the rational gaze of experts and European decision makers (chap. 8). Decades of accumulated knowledge apparently corroborated British and French designs. Yet this was mostly a reconstruction a posteriori of a decades-old process that was full of twists and turns. The long-term cultural processes that presided over the invention of the "Middle East" as a representational category have a chaotic history. I shall pay particular attention to the influence of cartographic discourses on policies and their implementation in order to trace these evolutions.[37] The wealth of archives examined here puts Victorian and Edwardian narratives of the progress of exploration and mapping, as well as the semblance of coherence displayed by the promoters of the creation of a British empire in the Middle East after 1918, into perspective.

This is a book about the idea of the "Middle East," but it is also a study of the Middle East. In that sense, it is more than just a case study on maps. It tells us about the consequences of the intrusion of Western knowledge and technology in the region. Cartography and geography are often described as typical instruments of the West's quest for categorization in the imperial age. Chapters 4 and 8, for instance, both underline the part played by British boundary-making experts in drawing the borders of many Eastern states. Yet, on several occasions, this book shows how reciprocal influences, countermapping, and recycling of Western cartographic rhetoric by local powers deeply determined the piecemeal ordering of the Orient by maps and mapping. The study of the interaction between maps and hegemony also raises the issue of the imperfections, contradictions, and discontinuities of European knowledge during the colonial period.[38] Maps and mapping were quickly appropriated and manipulated by local interests in order to resist discourses created from outside the area. This book will examine these signal distortions and how they participated in the elaboration of an East that was the product of an asymmetrical dialogue, and never a unilateral construction. It also throws the spotlight on how crucial Western cartographic discourses were to the "invention" of Eastern communities (to use Benedict Anderson's phrase). Those discourses became instrumental by their being adapted, as well as contradicted, by those communities.[39] The part played by British explorers in locating Eastern Christians (chap. 1), the systematic recording of tribal lineages in the Gulf and their subsequent legitimation (chap. 6), and the Westernization of Ottoman mapping (chap. 4) are all manifestations of that process. The 1919 Paris Peace Conference (chap. 8) witnessed a wide array of participants, be they Kurds, Zionists, or exiled Armenians, attempting to shape the negotiations thanks to cartography. Cartographic and geographical arguments fostered non-European contestation during the negotiations. This subversion of Western discourse substantiated emerging national feelings, potential states, and projected borders. This study therefore shows that the expansion of British influence in the area and its consequent intellectual mastery was never left unanswered.

Three preliminary questions need to be addressed. First, the subtitle of this book suggests that there was, indeed, an entity that one can delineate as "British maps of the East." But was that actually the case? As with other forms of scientific and cultural productions, maps are the expression of a multitude of circulations. Surveyors in the field collected information from local inhabitants; cartographers compiled data from a variety of works, which in some cases were not actually "British." London-based mapmakers and publishers did not hesitate to copy the works of their Continental colleagues, who often did the same in turn. While a full comparatist perspective has not been adopted in this already substantial work, considerable attention is devoted to the unraveling of the pluralistic elaboration of the sample of British knowledge under study here.[40] Not only was British cartography inherently international and interconnected with its French, German, and even Ottoman counterparts; it was also characterized by the intrinsic plurality of the four nations. A distinct Scottish cartographic production developed in Edinburgh in the 19th century, whereas London was a dynamic mapmaking center.[41] Be they English, Welsh, or Scottish, a variety of vernacular Orientalisms, largely conditioned by national and local cultural specificities, shaped "British" imaginations in relation to the East.

Second, what was the cumulative weight of the gravitational pull which the Orient exerted on European imaginations? The wealth of an archive is no safeguard against what Pierre Bourdieu called the "epistemocentric error."[42] The researcher's theoretical perspective, unless it is deconstructed, distorts the object of inquiry and overestimates its significance. Before we undertake this tour of the Orient together, we need to consider whether the "transitional East" was a relevant cultural entity of real significance from a British 19th-century standpoint.[43] Having no direct means to answer this question, I resorted to some indirect measures that offer some telling evidence. I first analyzed fifty years of the annual publications of the Royal Geographical Society (RGS; fig. i.1). The RGS was created in 1830 when the preexisting African Association and the Palestine Association were merged into the new structure. The society aimed to promote exploration and also to publish every form of geographical information. It was instrumental in developing a specifically British geographical and cartographic grasp of the world. The statistical analysis of the publications of the Royal Geographical Society provides some surprising

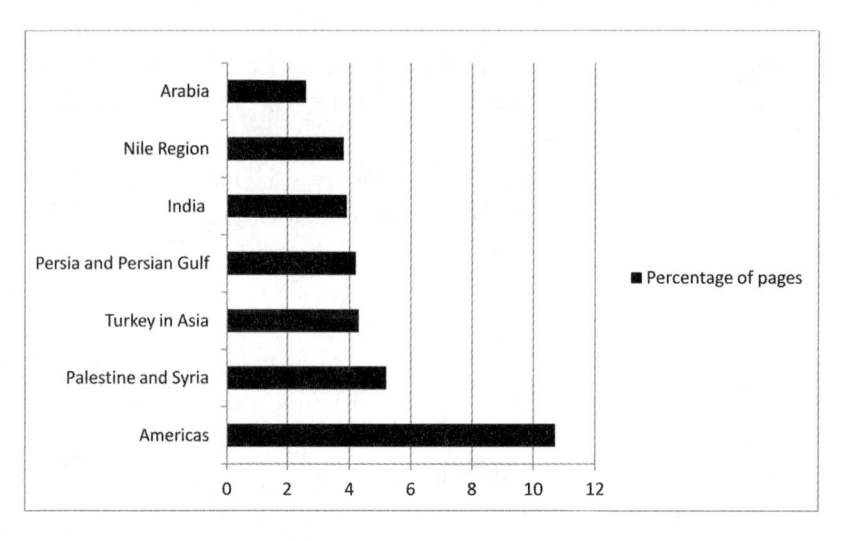

FIGURE I.1. The world according to the publications of the Royal Geographical Society (1831–1880): Proportion of pages per subregion showing the relative importance of areas in the East. © Daniel Foliard.

results.[44] Persia and the Persian Gulf accounts for 4.2% of the total: more than India (3.9%). If we add to this the articles devoted to Turkey in Asia, Persia, the Persian Gulf, Arabia, and Palestine/Syria, we begin to take stock of the significance of these areas from the RGS's standpoint. They account for more than a fifth of the total number of pages. The Americas, by comparison, only account for 10.7%. Interestingly, this characteristic of the RGS's preoccupations remained constant. Mesopotamia, Syria, and Arabia attracted almost unvarying proportions of the attentions of British geographers over the years, such that the evolutions of the Eastern Question remain a constant predominant focus, barely reflecting the trends in exploration in other lands. Another set of data derived from the archives of the Foreign Office's map library from 1850 to 1919 confirms that impression (fig. i.2).[45] While limited in scope, this graph provides a visualization of map uses in a crucial governmental institution, the British government's Foreign Office. It shows the relative importance of the areas eventually united under the label "Middle East" in the Foreign Office's outlook on the world. Quite predictably, the proportions of maps of the region rose during the Crimean War, the Russo-Turkish war, and in the aftermath of the 1882 intervention in Egypt. Maps became increasingly common tools of government after the late 1870s, with an even more consistent rise from the early 20th century until the 1914–18 peak. The East remained an object of

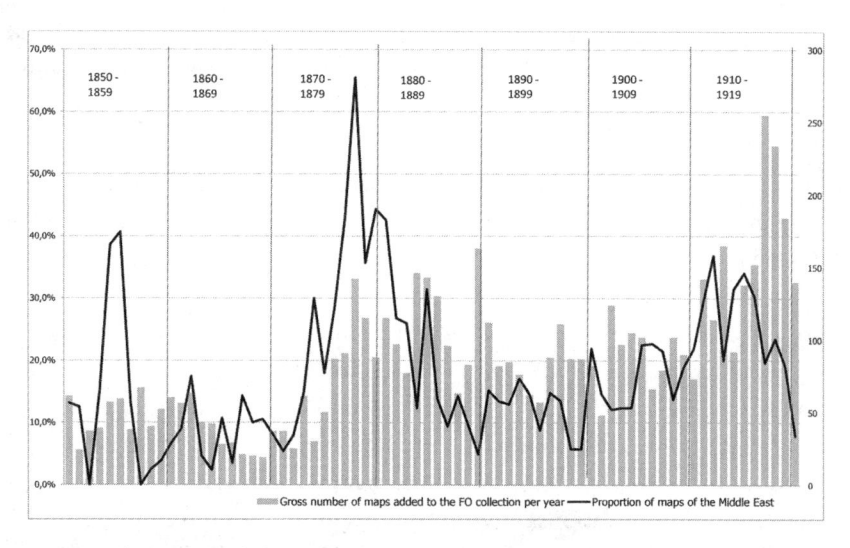

FIGURE I.2. The Foreign Office's map library, 1850–1919. © Daniel Foliard.

constant attention oscillating between 10% and 30% of the annual demand for new maps.

The last issue to be addressed is that of periodization. Chronological division is inevitably the vehicle for presumptions, orthodoxies, and prevailing narratives. It may even eclipse with its necessarily colonial assumptions the perceptions of time proper to the populations that came under Western domination.[46] I nonetheless have chosen not to dispense with this necessary historical tool of analysis. To freeze in time the 19th-century British experience overseas would be a dead end. As noted by John Pickles, maps are unstable text. It is essential to consider the historicizing conditions which produced them.[47] Periodization, despite its limitations, is the only way to break down the genealogy of the so-called Middle East "into components and phases."[48] Consequently, the following chapters do not stand alone. They pursue analytical threads and evolutions through diachrony and itinerancy. They are therefore mutually dependent on each other. Our first period (chaps. 1–3) begins with the outbreak of the Crimean War (1853–56). From that point onward, the Eastern Question was to become one of the first old chestnuts of the industrialized press system. This initial stage ends with the opening of the Suez Canal in 1869, a canal that was soon to become vital to the British Empire. In the second period (chaps. 4 and 5), British interest in the region intensified from the 1870s onward. In 1876 a French-British Dual Control was established over Egyptian finances. Disraeli signed the Cyprus convention two years later, giving Britain its first territory in the eastern Mediterranean. In 1882, Gladstone, staunch critic of Beaconsfield's vision of the East, eventually intervened militarily in Egypt, transforming it into a de facto protectorate. On the Ottoman Empire's eastern flank and in Persia, Russia's advances legitimated the Anglo-Indian interventionist policies.[49] In the Persian Gulf, protectorates were established in Aden (1873) and in Bahrain (1880). The third phase (chaps. 6–8) stretches from the mid-1890s to 1914. A first shift is to be dated from 1892. The ratification of the Exclusive Agreements brought to an end Whitehall's opposition to a forward policy in the Persian Gulf. A new subempire emerged both there and in Mesopotamia. British interference in the region also materialized in the Nile basin. Kitchener's victory in Omdurman in 1898 and Curzon's viceroyalty in India, which started the same year, were further signs of a greater British involvement in the area between the Mediterranean and India. German attempts to develop a closer relationship with Istanbul only served to consolidate this trend. This gradual transformation of British Eastern policy fostered greater public awareness of the region's international position. A number of crises, such as the Ottoman anti-Armenian repression (1895–97) and the Balkan Wars, enabled European audiences to place the East on the map. The Middle East was invented and became one of the better-covered areas in the news, along with India.[50] The last chapter focuses on the First World War and its immediate aftermath, through to the 1921 Cairo conference.

PART ONE

FROM SEBASTOPOL TO SUEZ (1854–1869)

CHAPTER ONE

THE MID-VICTORIAN PERSPECTIVE: A FRAGMENTED EAST

In May 1854, Thomas Best Jervis, a recently retired major in the Bombay Engineers (East India Company), published a map of the Crimean Peninsula in ten sheets (fig. 1.1).[1] For years he had stubbornly tried to convince the British government of the crucial importance of topographical intelligence. The Crimean War (1853–56) legitimated his pleas. Lord Newcastle, the secretary of state for war, belatedly realized that cartography was a useful tool for a modern army. What is surprising, in retrospect, is that the landing of the troops and the first weeks of the conflict were planned and managed without reliable maps. While the high command bought a few travelogues, their propensity for exoticism and their intrinsic lack of clarity made them poor instruments for commanding officers. Jervis's proposal was eventually accepted. He compiled the new 1854 map in a few weeks based on a Russian chart from 1817 by Major General Semën Alexandrovich Mukhin. He could also rely on an Austrian military map and on measurements made in 1833 by General Fëdor Fëdorovich Schubert, a high-ranking infantry officer who was commissioned by the czar to undertake the first triangulation of the Crimea. Because Britain had no actual topographical department, Jervis had to use non-British sources.

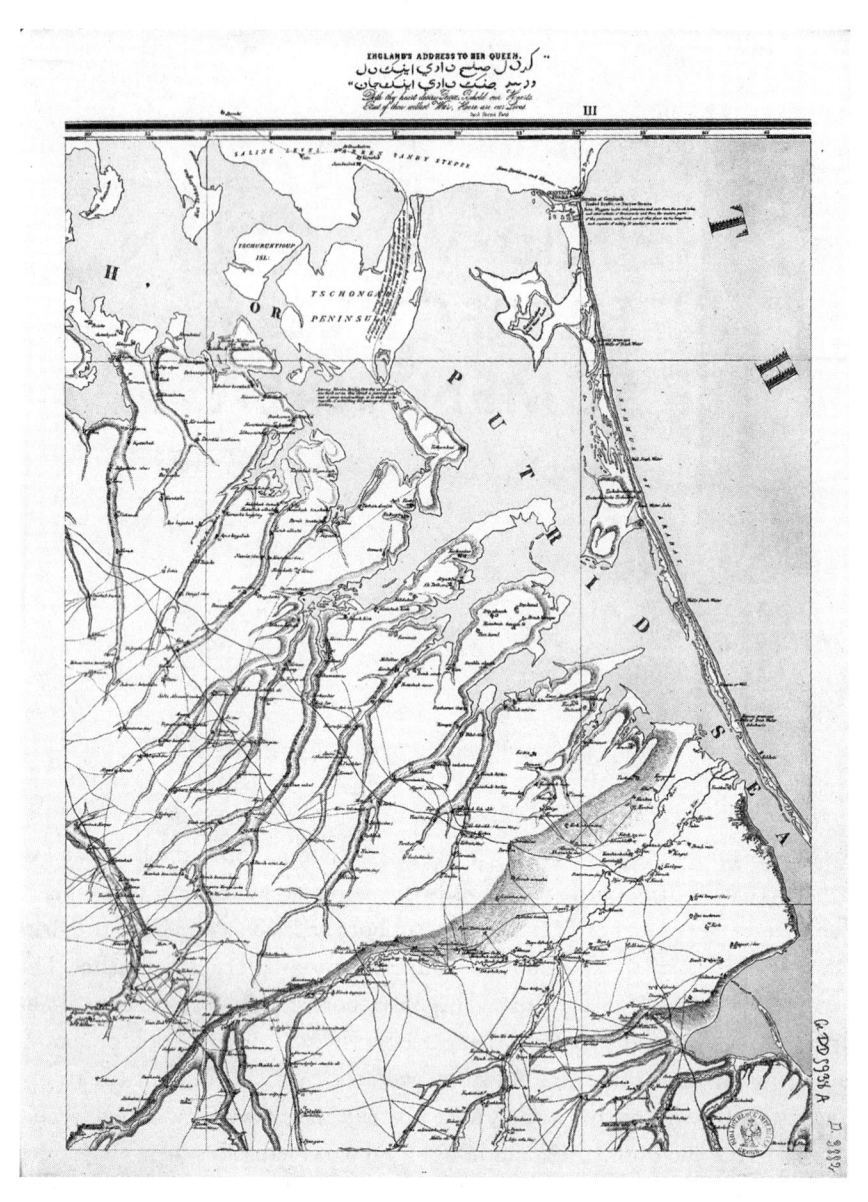

FIGURE 1.1. Thomas B. Jervis, *Map of the Krima Peninsula, sheet n° 3*, ca. 1:165 000 (London, 1855). Size of the original: 62 × 44 cm. Photograph courtesy of the Bibliothèque Nationale de France (BNF) (Cartes et Plans GE DD-5938 (A)).

The fact that he had to convert all the Russian measurements that referred to the Ferro Meridian into Greenwich-compatible coordinates is a testimony to his dependence on foreign data.[2] He was fortunate enough that Schubert was a trained surveyor. Jervis used Schubert's horizontal datum (a reference point to measure other positions), located at the bell tower of Saint Vladimir Cathedral in Sevastopol. Its coordinates were measured with an error of less than 70 feet (1″ at this latitude).[3]

Once the War Office (WO) agreed to finance the map, it took the retired officer ten days to print two hundred copies of the document. The skillful Jervis was also an accomplished lithographer. This probably explains the high visual quality of the map and his use of colors.[4] Jervis wanted his work to serve both the strategist and the tactician. However, the map was poor in terms of its feature accuracy, since Jervis chose to depict relief with hachures. He did not have the sort of hypsometric data needed for correct contour lines. The map key details the different types of roads. Those roads which "troops, artillery, cavalry and baggage waggons" could travel on were, logically, highlighted. Information was provided on the water quality of the wells.[5] Other informative insets were less crucial to soldiers. For example, Jervis deemed useful to indicate where the "celebrated Muscat grape is grown." Jervis's map was also a travelogue. It was caught between textual explanations and the graphic visualizations of the area. This was a feature of early 19th-century map culture. Cartography was, with travel narratives, one of the many modes of representing geography.

Statistical data and a small, colored geological map complemented the document. The map was a prototype, a fact underlined by Jervis when he delivered his work: "I believe it is not generally known that the present is the first war in which the British forces have been supplied with the most needful help to success, correct and suitable land-maps."[6] Jervis successfully convinced decision makers in London that maps were military instruments. A Topographical and Statistical Department under the authority of the WO was created in 1855.[7] A total of 1,476 maps were given to Jervis by the French government to serve as the basis for an archive.

What happened to the map on the battlefield? It is difficult to document its practical uses. The sheets were mounted on linen in order to survive their handling by officers in the Crimea. However, the fact that the map was divided into ten sheets showed little regard for its actual manipulation. A fragmented Crimea could prove of little interest to the officer on the field. A letter by General William John Codrington to Lord Panmure, Lord Newcastle's successor as secretary of state, nevertheless refers to the map being sent to the British high command, which took the document and other cartographic information into account with sometimes catastrophic consequences.[8] Ultimately, old-fashioned ad hoc intelligence, a typical feature of British warfare in the mid-19th century, was the British command's most reliable source of data. It seems nonetheless that lower-ranking officers did have the map while on the ground. W. H. Russell (1820–1907), who covered the Crimean war for the *Times*, had heard about the document and its relative inadequacy. The third and fourth sheets show a narrow strip of land named the Arabat Spit. The cartographer located various wells and specified the quality of the water to be found there. The wells did not exist, but, fortunately for British troops, they could quench their thirst anyway, according to Russell: "A curious instance of the ignorance of our chart-makers was discovered on referring to the sites of wells marked on the maps. There were no wells on the Spit at all, and that for the simple reason that they were not required. The water of the Sea of Azoff close to the Spit is quite fresh."[9]

Jervis's map and its uses raise several issues. The first has to do with the data he compiled. Given the lack of any intelligence department in London before the Crimean War, the cartographer relied on external sources, a fact that reveals one of the features of this nascent military intelligence: despite Jervis's efforts, Britain was surprisingly ill informed.[10] Another aspect of the emerging military intelligence is demonstrated by the painstaking process through which Jervis eventually convinced the WO of the relevance of his claims: the major geostrategic power of the mid-19th century could proceed to wage wars without anyone, except for a retired officer, seeing fit to gather a few maps before landing thousands of troops in front of the Russian army. It is true that the Crimea was an unexpected battlefield. France, Britain's ally in this conflict, was no better prepared. However, the absence of an effective intelligence department in London is revealing of the limited means in the service of mid-19th-century British influence. The fact that a Bombay officer remedied the situation also brings to light the distinct cartographic culture of the Anglo-Indian world. Jervis's impressive range of surveying and drawing skills is an illustration of this specificity. Such disparities testify to the intrinsic polyphony of the British cartographic vision of the world. Jervis's map of the Crimea questions the traditional narrative of the cartographic discovery of the Orient, which is often described as an incremental and linear process in 19th-century accounts. Scientific positivism pervaded the rhetoric of geographers and cartographers, but it is misleading.[11] Surveys and explorations in the East, far from being cumulative and systematized, were the products of a nexus of heterogeneous and conflicting factors. The WO had its own agenda.

The Admiralty charts reveal yet another view of the region. India had its own independent cartographic perspective on western Asia. Not only was there a diversity of official outlooks; there was also a range of semiofficial societies, religion-driven explorers, and adventure-obsessed individuals who were producing their own geographical accounts and maps of the Orient. Jervis's map was just one reflection of a multitude of Easts mirroring the plurality of mid-Victorian representations.

The present chapter tests these propositions. It first shows the diversity of the agendas underlying the survey of the Middle East–to-be in the mid-19th century. It then examines the random and haphazard processes that lay behind the mapmaking. The position of fieldwork in the dynamic processes that shaped geographical imaginations is of central importance to the understanding of the cultural production of the area. The second section looks into the factors and the technical restraints behind survey expeditions. It describes how the surveyors captured the topography and nomenclature of a place, how spatial knowledge was first collected and recorded before its translation into the printed authority of cartography. This study of exploratory practices emphasizes the limitations and technological determinism that shaped spatial representations. The third section tackles the question of the interactions between the local population and British agents. It pursues one of the main lines of reasoning of this book: the dialectical elaboration of cartographic and geographical knowledge. The chapter goes on to consider one of the first systematic surveying campaigns in the region, that is, the Palestine Exploration Fund project, with a view to observing the interactions of the aforementioned processes through a case study. The last sections describe the final stages of the construction of knowledge, from mapping to mapmaking as such, from the field notes to the cartographic document. It also analyzes to what extent, and how, cartographic documents were used by officials after they were drafted and printed.

1.1 THE MOTIVES BEHIND MAPPING

Mapmaking and mapping are two interdependent elements of a single process of arranging the world, but they have different constraints and rationale.[12] Cabinet cartographers may compile a new map without setting foot outside their office. Surveys required financial support, instruments, and arduous work in the field. Before I study how the East was visualized through maps themselves, this section provides an overview of mid-Victorian surveying in the

East. It is intended to unravel the underlying factors that guided the mapping of the area.

THE VIEW FROM THE METROPOLE: AN IMPERIAL GAZE?

One of the most obvious uses of maps and surveys in the colonial age was to expand geostrategic influence and domination. In that context, British explorations of Palestine, Mesopotamia, or Sinai are often understood as processes of control. The region that was eventually named the "Middle East" in the early 20th century lay between India and Europe. It controlled the overland route between the East and the West. One would thus expect to discern a coordinated approach in the mapping of the East by governmental departments in the 1850s and 1860s. No such all-encompassing outlook existed in the mid-Victorian era. Each government office interested in the region developed its own approach to the East. Three areas were of special interest from the WO's point of view: the straits and the Crimea, the Isthmus of Suez, and the borders between Russia and the Ottoman Empire. Each area was mapped on an ad hoc basis, with no indication of an attempt at a prospective and systematic collection of knowledge. Surveys and compilations were in general undertaken after major crises.

Under Jervis's direction, the Topographical and Statistical Department tried to fill in the gaps by compiling, purchasing, or copying maps of the Ottoman Empire. New surveys were conducted in the region of Erzurum by J. Geils. Heinrich Kiepert's *Karte von Armenia, Kurdistan und Azerbeidschan* was bought. A copy of C. F. Skyring's 1841 map of Beirut was drawn.[13] However, this type of comprehensive and prospective approach remained an exception. Under Henry James, the Topographical and Statistical Department returned to a state of somnolence, before the French-Prussian war of 1870 again revealed the archaic state of British intelligence. Prussian advances were attributable partly to the organization of its intelligence and to the quality of its maps.

Even a crucial location like the Isthmus of Suez, despite its growing strategic importance as Ferdinand de Lesseps's enterprise was moving forward in the 1860s, gave rise to little cartographic interest on the part of the WO.[14] The Suez project had initially been considered a utopia by some British officials (Lord Palmerston was among them). The WO eventually realized that it was lacking in up-to-date knowledge of the isthmus and the neighboring Sinai. Charles Warren (1840–1927) and Charles Wilson (1836–1905) surveyed the area in a semiofficial context. Still, Palestine and the Levant in general were strategic sideshows for the WO. The Eastern map of the military was fragmentary and distorted.

The Admiralty's perspective was quite different from that of the WO, but no less piecemeal. The Mediterranean remained partly uncharted until the 1810s.[15] Francis Beaufort, a renowned hydrographer, was sent on a charting mission in the eastern Mediterranean in 1811. Captain William Henry Smyth (1788–1865) succeeded him, followed by Lieutenant Thomas Graves (1802–56) in 1824. Graves came back to the Mediterranean in 1832 after a mission to Tierra del Fuego and charted the region for a decade. He created a modern corpus of sea charts but overlooked Syria. The Levant was eventually surveyed by C. H. Dillon after the British intervention in Beirut in 1840.[16] After this burst of interest in cartography, the Syrian coast was left aside again until A. L. Mansell charted the coast between Jaffa and Rosh Haniqra and mapped a few coastal cities.[17] Egypt was better known for two main reasons. First, the opening of the Suez Canal required more effective charting of the Egyptian shores. Port Said, Alexandria, and Suez itself were vital stations on the sea route to India. Second, Anglo-Egyptian trade grew considerably in the 1860s.[18] The American Civil War compromised cotton exports, causing the British textile industry to look for new suppliers. Egypt was one of them. Merchant shipping relied on the hydrographic expertise of the navy. Both evolutions called for better hydrographic knowledge of the Mediterranean. A series of charts were published from the late 1850s to the 1870s.[19] Was there a strategic dimension to the new surveys? For the navy, mapping the sea was one. Nothing really distinguished the Levant from Tanzania. The purpose of the new charts was mostly to supply practicable information in response to a new demand.

The East was also a latecomer to the Admiralty's catalogs of charts.[20] Syrian shores were overlooked, a likely consequence of the prevalence of French influence in the area. The Levant was partly charted in the 1840s but without any systematic approach. In the mid-19th century, the eastern Mediterranean between the Sinai and Iskenderun was less documented than the Pacific islands.[21] Even the president of the Royal Geographical Society (RGS) underlined this gap in his 1858 address when he hoped that the time had "arrived when the shores of Palestine and Syria [would] no longer be permitted to form the opprobrium of our maps."[22] Moreover, surveys and cartographic documents were generally of consequence to military action, as was the case after the Beirut intervention.

The Foreign Office (FO) produced its own flow of cartographic and geographical data. Diplomatic reports often included sketches and maps. However, cartographic representations were mere appendixes to the main body of written communication. Little had changed in a century: the "old diplomacy" survived the industrial age.[23] The inclusion of this type of information in correspondence remained unsystematic. While some agents could be more active geographers than others, it was largely a matter of chance. For instance, James Brant, who was consul in Erzurum from 1836 to 1840 and consul in Damascus at the time he retired in 1860, was a member of the RGS and published various articles and maps.[24] Austen Henry Layard, who spent part of his career in the East, made prolific use of cartographic documents.[25] However, his geographical activism was not the consequence of any centralized process. More often than not, British agents set up surveys of their own design. For example, C. Murray, minister at the Court of Tehran in the 1850s, sent two diplomats to travel the region between the Persian capital and the Caspian Sea. One of the main accomplishments of this enterprise was to ascertain the altitude of Mount Demavend: a heroic ascent, with little input, however, in terms of field intelligence.[26] The variety of the information collected, which included economic and ethnographic data, did not always make up for the diplomats' lack of expertise. The quality of their sketches and accounts varied considerably according to their zeal and training. Moreover, the network of British consulates in the East was of uneven density. Few diplomats were in a position to give reliable intelligence regarding the hinterland.

There were instances of planned surveys in the East. Demand for coal and raw materials rapidly increased with the Crimean War in the early 1850s. The FO organized an expedition to map potential mines in the region of Nicomedia.[27] Henry Poole, a member of the British Geological Survey, embarked for Turkey in June 1855. He surveyed Palestine, where he made archaeological findings in Qumran and took multiple altimetric measurements.[28] However, this type of programmed exploration was the exception. Uncoordinated initiatives were more typical. The compilation of new material and the purchase of new documents was most of the time an ex post facto decision. The department improved its library as a reaction to emerging crises. Even modest forms of anticipation were still wanting. The FO's map of the world and more specifically its cartography of the Orient were fragmented.

There was one last opportunity for Whitehall to gather more information. Britain developed an expertise in border delimitation in the 19th century.[29] The FO became a very active arbitrator in the region as a consequence of London's involvement in Eastern affairs.[30] Another source of topographical and geographical information was thus established. The border between Persia and the Ottoman Empire was one of the first to be examined from that perspective, from the midcentury onward. Great Britain favored a settlement between the two empires in order to avoid a war that might weaken two states bordering Russia. Britain acted as a mediator in the negotiations that led to the second Treaty of Erzurum (1847). It was agreed upon that the border had to be measured and mapped by a quadripartite commission involving Persian, Ottoman, Russian, and British negotiators. The surveys lasted from January 1849 to September 1852. The British surveyors were headed by William Fenwick Williams (1800–83). British expertise was far from neutral. Surveys were used as an opportunity to collect as much data as possible. A geologist, William Kenneth Loftus (1820–58), accompanied the commission and reported on his geological discoveries.[31] Fenwick Williams and H. A. Churchill (1828–86) went beyond the border region to accumulate ethnological, geological, and archaeological data. Second, a network of British agents in the East—such as Canning, the British ambassador to the Porte, Henry Rawlinson, or Layard—supported Ottoman claims and used every document and map in their possession to make their case.[32] Russia, on the other hand, was backing Persia. A map known as the "Carte identique" was eventually issued by the Russian and British commissioners in 1869 to try to settle the discussion, with little effect at first.[33] The Turco-Persian dispute established a precedent anyway. This boundary-making expertise placed British surveys and maps at the heart of a slow process of legal codification and linearization of the area's borders. For British policy makers, state delimitation by European standards first aimed at stabilizing Ottoman-Persian relationships. It helped shape developments in the area and potentially fostered British interests. It also participated in a larger effort to bring interstate relations into the emerging framework of international law.[34]

Overall, there was only a very limited form of coordination between the three offices that were particularly involved in mapping the area. Their imperial gaze was uncertain: what showed up were diffracted images. The gaps in their map of the East were numerous. It took times of crisis to remove them, and when the maps were eventually drawn and lithographed they were often already obsolete. What does this say about the making of the Orient? Whitehall's disparate worldview was a consequence of what the British Empire was in the mid-19th century: a liberal, unfinished, and underfinanced structure. Fostering trade was more relevant to British imperialism in the East than systematically controlling and mapping the area. The opening up of new markets by the 1838 Convention of Commerce between the Ottomans and Britain, or by the 1857 Paris Treaty with Persia, was far more relevant to the imperial project than coordinated strategic projections. The piecemeal official cartography reflects this. Ultimately, what the British government wanted in the East was a series of "well kept inns."[35] Inexpensive exploration and opportunistic establishment of positions in the area were haphazard but effective enough, in the eyes of mid-Victorian policy makers, to open the area to the bettering influence of British free trade.

THE CASE OF INDIA IN THE ORIENT

The Indian government was at that time the only institution with a more active approach to the surveying and mapping of western Asia than the previously mentioned government departments. Matthew Edney underscored the specificities of the Raj's cartographic ideal.[36] Here again, though, there is no evidence of any systematic planning despite the strategic importance to Anglo-Indian supremacy of an area lying at the crossroads of Eurasia. Individual initiatives and opportunities guided the surveyors' and hydrographers' efforts.

The Indian navy was a prolific producer of charts of the Arabian and Red Sea coastal regions. Hydrographic surveys of the Persian Gulf were carried out from 1820. The operations extended over more than a decade and resulted in fourteen charts and a memoir being published.[37] The Red Sea was imperfectly documented. The need to find a coaling station in the area encouraged the Indian government to fund surveys from 1829 onward. The survey brig *Palinurus* was sent to the Gulf of Aqaba under the command of Robert Moresby. James Felix Jones was the cartographer of the party. The Arabian coast was explored later by Stafford B. Haines in the 1830s. Charles Cruttenden, a surgeon attached to the *Palinurus*, traveled to the legendary city of Sana in 1836.[38] The Suez anchorage was charted by Captain William Charles Barker in the late 1840s. Memoirs, maps, and plans from the Red Sea survey were progressively published.[39] The growing body of hydrographic knowledge helped S. B. Haines convince the Bombay government of the strategic position of Aden, not only as a coaling depot but also as a commanding post on the Red Sea. Aden was occupied by the East India Company in 1839. Haines seized the territory by force, in the absence of any real official sanction. The annexation of Aden was agreed upon ex post facto, and Haines became its first political resident.[40] He immediately encouraged the members of the Arabian Coast Survey to proceed with further surveys.

The exploration and charting of the Gulf region and of the Rea Sea answered the strategic need of the Indian subempire for better control of the sea routes to the subcontinent. The plague of piracy was an obstacle to the promising development of trade. There were also more philanthropic reasons behind the growing Anglo-Indian interest in the region in this period. The British campaign against the slave trade, a major feature of the Arabian Peninsula's economy, was one of the motives behind this interventionism. International competition also had a part to play in the intensification of the Anglo-Indian gaze. In the mid-1850s, French attempts at charting the island of Perim, a location which commanded the strait between the Red Sea and the Gulf of Aden, contributed to the Indian navy's focus on the region.[41] British India saw the Gulf as a strategic hub for the area. The elaboration of a nexus of alliances and protections between India and the most influential tribes inhabiting the shores of the Gulf required better geographical and cartographic tools.

The construction of telegraphic lines between Britain and India created more opportunities for the collection of information. Patrick Stewart, a telegraph expert appointed by the India Office (IO), surveyed the territory between Tehran and Bushire from 1860 to 1862. An Indo-European Telegraph Department under Stewart's authority was created by the Bombay government in 1862. The idea was to connect India with London through Baluchistan, Persia, Mesopotamia, and finally Constantinople. When Stewart died in 1865, Frederic Goldsmid replaced him. Goldsmid had worked under Stewart's command when a submarine cable was laid in the Gulf in 1864. The new director general of the Indo-European telegraph immediately negotiated a convention with Persia. He directly supervised the construction of the line until 1870. At that point, Goldsmid had acquired unparalleled erudition on Persian geography.[42] As a result of his efforts, the longitudes of several reference stations in Persia were fixed with unprecedented accuracy thanks to the telegraph when the line was connected to India in 1863.[43] Chronometers could then be set or compared with Greenwich Time.

The Anglo-Indian gaze also turned to Mesopotamia, a vital segment of the overland route between India and Europe. In 1835, the East India Company financed one of the most impressive survey expeditions ever organized in the region. Francis Chesney (1789–1872), then lieutenant colonel, was to survey the Euphrates and assess its navigability. Parliament granted him the very large sum of £20,000.[44] Two dismantled steamers were transported in pieces from northern Syria to the Euphrates. The boiler, mounted on a wagon, was painstakingly carried across this hilly region. The expedition reached Bushire on the Gulf in 1836. Steam navigation on the Euphrates was proved feasible, even if one of the steamers was lost in the process.[45] Henry Blosse Lynch (1807–73) continued the work of this first expedition and surveyed the Tigris River from 1837 to 1842 aboard two steam vessels, the *Nitocris* and the *Nimrod*.[46] He went on to create with his brother one of the most successful British trading companies in the region. The Lynch family, thanks to their topographical expertise, progressively became very influential actors in British policy from Armenia to the Gulf.[47] The Survey of India's interest in Mesopotamia remained consistent afterward. James Felix Jones assumed leadership of the Mesopotamian survey in 1843. William B. Selby, Lieutenant James B. Bewsher, and William Collingwood took over in 1855.[48]

While no coordinated scheme presided over these surveys, British India was much more supportive of mapping enterprises in the East than London. The Raj was extending its reach in an area that controlled the route to India. The East India Company established a network of bilateral agreements with local sheikhs in the first half of the 19th century. By the early 1850s, the Trucial Coast system was established in the Gulf. Trade routes in Persia and Mesopotamia were secured. The favorable outcome of the Anglo-Persian War (1856–57), an unprecedented mapping opportunity, opened the way to an increase in trade.[49] The personnel involved in this informal expansion gained expertise which fostered British India's influence in the region. Many of them were surveyors. Through mapping and maps, Bombay labeled and categorized the transitional East. In doing so, it started to impose its own specific shade of British imperialism. The Gulf and other territories of the Orient started to look like the princely and proxy states the Raj had created on the subcontinent. Their exploration and visualization in maps rationally inscribed them in a specific Anglo-Indian subimperial sphere characterized by the adaptability of its indirect rule. Hydrography and topography were instrumental to the struggle against slavery in the Red Sea and the Gulf. Maps placed the area within the confines of an expanding British legal order.[50]

OUTSIDE THE GOVERNMENT: CULTURES OF EXPLORATION AND THE LEARNED SOCIETIES

Various learned societies promoted further exploration in the East. They provided their members with reports of surveys and original maps. The Hakluyt Society, founded in 1847, published translations of momentous travel narratives and related maps, such as the *Travels of Ludovico de Varthema*. The Geological Society of London had numerous maps of Asia in its library.[51] The Royal Asiatic Society sometimes published geographical accounts and maps, such as the Mesopotamian surveys of the early 1850s.[52] But the most influential of these associations was the Royal Geographical Society (RGS).[53] Its co-opted members were often part of the elite. Only recommendations by longtime members could get someone into the club. Many of its members were high-ranking officers and civil servants, famous explorers, well-respected missionaries, and cartographers.[54] The society quickly became the center of a political and scientific network.

Under Roderick Murchison (1792–1871), who repeatedly presided over the RGS from the 1840s to the 1860s, the society's council rarely met without an expert of the Ottoman provinces or a Persian traveler sitting in. Murchison's own interest in the East was undeniable. He was involved in the elaboration of British policy in Syria in the aftermath of the 1860–61 massacres.[55] The RGS funded a variety of Oriental surveys, such as G. A. Wallin's, W. G. Palgrave's, and Lewis Pelly's explorations of Arabia.[56] Was the RGS the informal coordinator of British mapping efforts in the mid-Victorian age at a time when the government's endeavors were characterized by their unsystematic nature? In truth, the society's role was pivotal but not yet central. RGS members exchanged a wealth of information. Commercial mapmakers such as J. G. Bartholomew used it to obtain direct access to updated information with which to compile their own maps. However, despite the accomplishments emphasized by the various presidents of the RGS in their annual addresses, no orderly approach was yet on the horizon. For obvious financial and technical reasons, the RGS could not fill in single-handedly the blanks left on the map by governmental and military organizations.

The geocartographic outlook developed from within the RGS was shaped by the Victorian "cultures of exploration" described by Felix Driver.[57] RGS-sponsored individual adventures left little room for anything other than rough traverse surveys. This is especially true of the Orient, where exoticism shaped Western readers' expectations. The explorer was to move boldly on an axis rather than painstakingly explore a surface. Blanks thus left on the maps of the Orient were more seductive than the comprehensiveness of trigonometric surveys. Since mid-Victorian maps of Arabia, Persia, and Mesopotamia were rarely more than appendixes to an article or a travelogue, their incompleteness strengthened the sense of adventure that pervaded narratives of exploration. It was not uncommon for narrators of journeys in the East to address their readership, enjoining them "to look at the map" or even "draw a line" across it.[58] Maps resonated with the text, not to control a territory, but to foster the reader's imagination in his immobile travel.

Adventures in the East had become extremely fashionable in the mid-19th century. The "Eastern traveler" had even become a type, immortalized by Thackeray's Bedwin Sands, "an elegant dandy [. . .] who had published his quarto, and passed some months under the tents in the desert, [. . .] a personage of no small importance."[59] Benjamin Disraeli himself had his own Byronic moment in the East in 1830–31, which partly inspired his *Tancred*.[60] The Eastern traveler was a versatile and resourceful man. He had to be a scholar, an expert in Oriental languages, but not of the armchair type. His knowledge was also gained in the field. A man guided by his military virility, his taste for adventure, and his will to power, he was also, thanks to his Oxford education, unreservedly respectable and phlegmatic. In his eyes, his presence could civilize what he believed to be untamed Arabs. He could furiously embrace nationalist causes, like his illustrious predecessor Byron, the romantic hero of the Greek War of Independence. His Eastern travels could make him an expert in the eyes of London. In the mid-19th century, most explorers in the Ottoman Empire, Egypt, or Persia had to fit the profile. Too dull an account would have deprived the surveyor of any opportunity to transform his travels into a tangible reward. He had to be Byronic to a certain extent.[61] This romantic exoticism made maps secondary to a travel story. They were enmeshed with narratives of adventures that above all revered the peripatetic dimension of the Eastern experience.[62]

The RGS contributed to the cultivation of this Byronic perspective on the region. Richard Burton (1821–90) was one of its most famous champions. The society supported his expedition into Arabia in 1852. Burton disguised himself as a Pashtu pilgrim and improved his Arabic in Cairo to undertake the hajj. He reached Mecca in 1853. His topographical contribution to the understanding of Arabia was hardly notable, which might be explained by the limitations imposed by such journeys. Burton recalls how he concealed his notes and sextant in his garb to hide them from the potentially fatal curiosity of the Bedouins, who were quick to take them for a sorcerer's artifacts, according to Burton. He was himself well aware of the limited accuracy of his scientific attempts at surveying.[63] Instruments, such as the barometer, could break during the journey. Chronometers and compasses could be stolen. Rather than maps and measurements, it was the vicissitudes of his journey which enlivened his epic tale.[64] Burton, like many of his coexplorers, could not expect funding from the learned societies and respectable sales for an upcoming travel book without giving his narrative a Byronic twist. The scientificity of maps and mapping was merely an adjuvant to this representation.

Other would-be adventurers followed in those steps in the mid-Victorian age. In turn, they collected data without any official or institutional sanction. These numerous prototourists sometimes published accounts and maps. Of the many lesser Eastern travelers, John MacGregor was probably not the least eccentric. He was a canoe devotee who traveled the world in a skiff of his own design. He published a series of books entitled *The Rob Roy* in praise of one of his most famous ancestors. He toured the Levant from Suez to Acre in his canoe. MacGregor, equipped with a compass purchased in London, was well versed in geography and wanted to add his own maps to the travel book he published in 1869.[65] As shown in his publications, the dilettante's perspective could significantly differ from that of the appointed surveyor. MacGregor's maps and geographical account left more room for ethnographic information than did the official documents. Interestingly, such individuals were not held in low esteem by cartographers, who often used their works in their own compilations.

These Eastern travelers helped shape a conception of the Orient that was still imbued with romanticism and exoticism. The RGS, being the only body capable of conferring some sort of coherence to British mapping enterprises, did not yet favor more comprehensive approaches.

A MORE MILTONIAN ORIENT: MISSIONARIES AND GEOGRAPHICAL INFORMATION

The missionary enterprise prompted some of its agents to collect and produce their own information. The largest institutions regularly issued maps and geographical accounts for their specific purposes. The Church Missionary Society (CMS), for instance, published atlases based on a wide spectrum of material that could differ in nature from what more official institutions would consider useful. As stated in one of these publications, the "population of the world may be viewed in reference to Christian Missions, politically, ethnologically, or religiously." This conception involved a specific form of cartography, with more regard for what would be called later "human geography." The East itself was an area of particular interest to major missionary societies. The competition with Rome, which maintained strong ties with the various Christian churches in Syria, Armenia, and Mesopotamia, fostered Protestant involvement in the region. Evangelical concerns for the fate of the East contributed to this dynamic.[66] One of the first envoys of the CMS in the Levant, William Jowett, explored the region as early as the 1820s.[67]

Among the several geographically productive missionaries in the region, George Percy Badger was certainly considered one of the best experts on the East.[68] He was born in 1815 in Malta. He had family ties with Mesopotamia: his sister, Matilda Badger, married Christian Rassam, a Chaldean Christian who became vice-consul in Mosul from 1832 to 1872. Badger's fluent Arabic made him a prized translator for the CMS in the late 1830s. He was eventually ordained an Anglican priest in 1842 after missionary training in Islington. He then journeyed to the East, where he was to fulfill religious and diplomatic functions for three decades between Mesopotamia, Aden, and Persia. A number of officials sought his expertise. He had a keen interest in cartography and wrote dozens of works. He rarely missed an opportunity to include cartographic documents in his publications.[69]

His linguistic skills and firsthand experience of Arabistan allowed him to publish maps that were sometimes better informed than official documents. Badger explored the Gwadur area (Baluchistan) in the prospective phase of the Indo-European telegraphic line. He was also one of the first British travelers to provide an ethnographic portrayal of the Yazidis.[70] Badger was able to determine place-names very accurately. His map of Oman was in this respect more reliable than the chart of James Raymond Wellsted (1805–42).[71] He was an accomplished cartographer, as shown by his plans of Egyptian fortifications.[72] His missionary activity was only one of the facets of his Eastern experiences. Badger was a well-respected member of both the RGS and the Hakluyt Society. On account of his specific status, his outlook on the territories and populations he observed was naturally different from the perspective of the hydrographer or the royal engineer. He stood at a crossroads where religious incentives and more secular geographical ideals could pave the way to an improved human geography of the East. Contrary to France, where missionary accounts were often held in low esteem by academic circles, British geography was not overly compartmentalized. Badger and many of his colleagues were elements of the disorderly networks that drew the mid-19th-century map of the East in Great Britain.

Professional surveyors were the exception. Most of the time, the Eastern traveler combined numerous roles. He could be a missionary and a cartographer, a surgeon and a geographer. Hazy motives are to be found behind British maps of the Orient: geopolitical worries, religious imperatives, outright curiosity, careerist ambitions, investment opportunities, or militant sea charting. All these factors determined the nature of the search for knowledge in the field. However, despite the varying agendas of the men on the ground, the very nature of the geographical and cartographic representations of the East largely depended on the same technical restraints.

1.2 MOTIFS: LINES AND OUTLINES

Not only was the shape of the East as viewed from a British standpoint determined by an array of abstract reasons and representations; it also had to comply with the form of knowledge which could actually be collected on the field.

Two perspectives were instrumental in drawing the region: one from the land, the other from the sea.

THE LANDSMAN'S VIEW: THE REIGN OF THE TRAVERSE SURVEY

The cartographic ideal of a panoptic mastery of overseas territories under British influence remained largely rhetorical for most of the 19th century. The codification of an unknown or partially known land into a map generally started with the intrinsically sketchy route survey or the more accurate closed traverse survey (fig. 1.2).[73] A large proportion of the new topographical and geographical information gathering in the Middle East was based on this type of measurement in the mid-19th century. The compiler who wanted to draw small-scale maps of Eastern territories was confronted with an intricate network of routes. Few could rely on the fixity of triangulation. The route survey or link traverse was the inescapable and imperfectly accurate primitive form of Western geographical knowledge.

Its geometrical shape was that of a line composed of segments. The explorers would first determine the coordinates of their starting point. Finding its latitude by celestial observation of the North Star with a sextant was relatively straightforward. However, determining its longitude could be a thornier issue. A chronometer could be used to measure the time difference between a time sight made at one location, for example local apparent noon, and a reference longitude or meridian. Establishing longitude could become a very inaccurate operation because of alterations in the chronometer's rate. The mechanisms of portable chronometers in the mid-19th century were not flawless. Reference time could be erroneous and longitude calculations defective. Once the starting point was fixed, route surveyors would then walk from this first point to their next stop and determine its coordinates. In order to measure the distance between the two camps, their most viable option was pacing. Perambulators were costly, heavy, and imperfect. Eastern travelers were lucky enough to be able to count on camels and their extremely regular pace, a feature already noted by Eratosthenes. In their absence, counting one's steps was the only solution. A first segment was thus calculated. Each day of exploration would add another series of measurements, so that a succession of stations and segments would be created. The computation of linear and angular measurements along these lines would then allow the surveyors to establish a route map. Altimetry estimations were even more approximate. Mercurial barometers were relatively accurate but unsuitable for adventurers, who were confronted with the issue

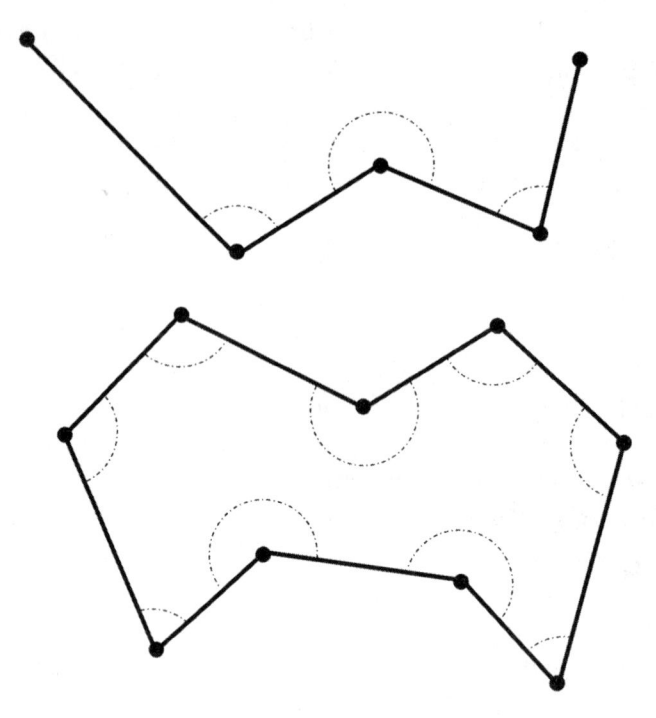

FIGURE 1.2. Open traverse and close traverse. © Daniel Foliard.

of portability. The glass tubes, an essential component of the barometers, could easily break. Boiling-point thermometers remained the favored solution—if not always a reliable one. Traverse surveyors were encouraged by the specialized literature in the field to draw some sketches of the relief that was visible along their segmented exploration and to gather as much information as they could on the climate, the natural features of the land, and its inhabitants. The field book was for that reason another necessary instrument—the last of a long list that made covert surveying a very hard task. Fieldwork in traverse surveys was a very lonely—and inaccurate—task. The close traverse, which implied travelers making a round trip from point A to point A, made it easier to adjust measurements, but its degree of accuracy was still low compared with triangulation.

After their adventures in remote places, explorers could then start their office work or delegate this task to a specialist. Their field books did not contain many graphic representations, except for a few sketches. In order for the field books to be transformed into a cartographic document, the segments and stations had to be plotted and protracted on paper. Additional data such as relief, place-names, rivers, and settlements would then be added to this skeleton. A route map initially looked like a corridor, progressively lightened by the traveler but surrounded by darkness. External data could be appended to fill in the blanks on the map. Sometimes explorers would compensate for their lack of direct information by adding data gathered from local inhabitants.

Lewis Pelly's route map of Arabia (1865) illustrates this (fig. 1.3). Pelly became chief political resident in the Persian Gulf and explored inland regions to Riyadh in central Arabia. It was a rare instance of an officially approved mission to meet with the Nejd lords. Pelly seized the opportunity to survey the area. Pelly was an ardent promoter of trade with Arabia and the Persian Gulf. His surveys often substantiated his views on the matter.[74] His route is shown on the document. The cartographer of the RGS at that time, London-based Edward Weller, used the political resident's field book and the sketches of Lieutenant Edwin Dawes, one of Pelly's appointed "geographers." The map did not stand alone. Like most cartographic representations in the 19th century, it accompanied a lengthy memorandum based on Pelly's field notes.[75] Pelly determined latitude by "stellar observation" from within their small tent, "pitched at night with its entrance towards the North so that their proceedings would go unseen."[76] Measurements made in Riyadh under the cover of darkness and with basic instruments succeeded in fixing the coordinates of the Wahhabi capital with only a small margin of error of less than two miles. He had taken a very reliable chronometer with him. In contrast with this astonishing accuracy, relief features, roads, and villages that were far away from Pelly's route are inaccurately drawn on the map. He noted that he was careful "to collect information as to stages along routes which I did not actually traverse; and I venture to hope that those points [. . .] may have been laid down with sufficient accuracy."[77] Local information, clearly identified as potentially inadequate by Pelly himself, added confusion to an already patchy image.

Apart from the inescapable limitations of mid-19th-century instruments, the landsman's view suffered from a lack of coherence. For decades the standardization of surveying techniques was merely a project. Though the RGS published its *Hints to Travellers* from 1854 onward to advise explorers on field observation and instruments, the trustworthiness of the many individual reports remained questionable.[78] Even more official surveying bodies lacked any coordinated approach. The Royal Engineers had developed surveying expertise. Chief Draftsman Samuel B. Howlett, for instance, had a career-long commitment to the promotion of surveying-technique improvement. In the late 1830s, he explained how to operate a basic set of instruments such as the prismatic compass or the plane table in order to survey and map extensive ter-

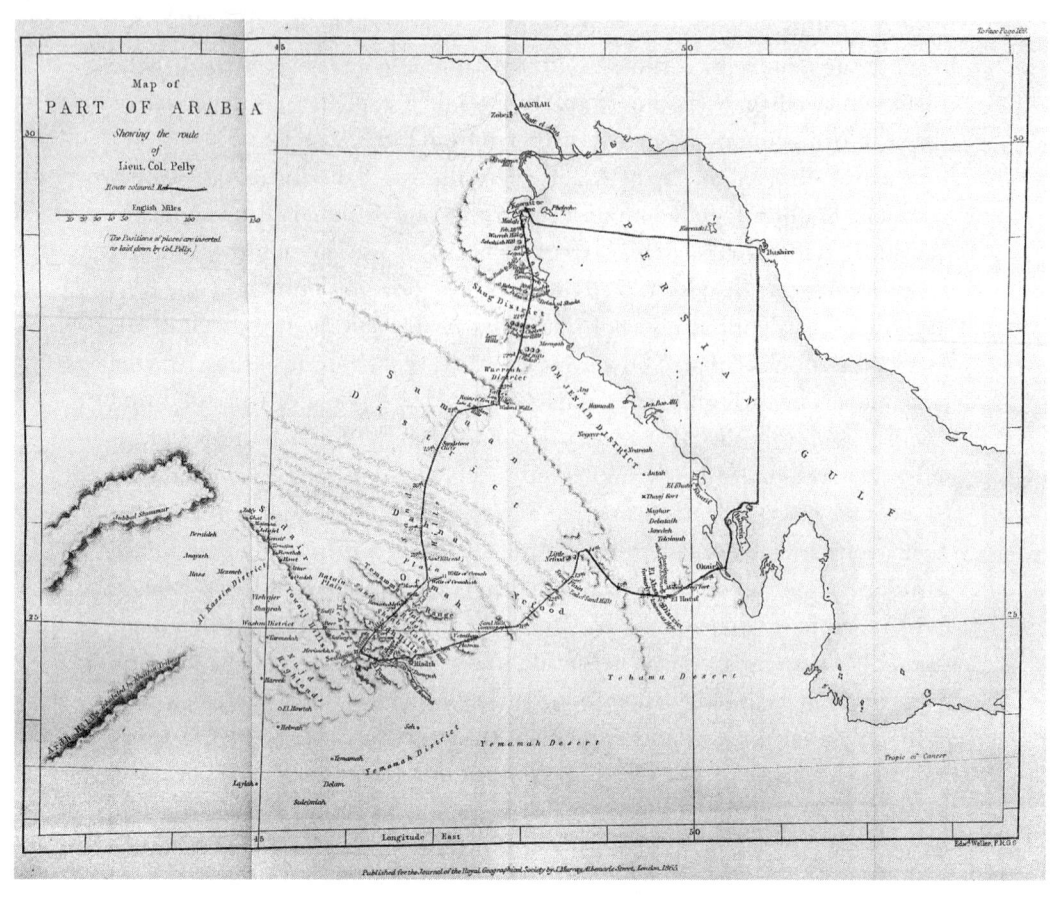

FIGURE 1.3. Lewis Pelly, *Map of a part of Arabia*, 1 inch = 80 miles, in Lewis Pelly, "A Visit to the Wahabee Capital, Central Arabia," *Journal of the Royal Geographical Society of London* 35 (1865): 169. Size of the original: 20 × 23.5 cm. Photograph courtesy of the Bibliothèque Interuniversitaire des Langues Orientales (BIULO PER.5134).

ritories.[79] This type of knowledge was disseminated through specialized publications, the *Papers on subjects connected with the duties of the Corps of Royal Engineers* for instance. However, the very incomplete professionalization of the Royal Engineers in the topographical field was an ongoing issue until the 1860s.

Haphazard surveying sometimes became controversial. Not only could field books be incomplete, measurements approximate, and local information defective, but explorers themselves could also lie. The explorations of William Gifford Palgrave (1826–88) testify to the latter. His expedition in central Arabia lasted some two years, from 1862 to 1863. Palgrave served in a Bombay regiment before his conversion to Catholicism. He became a Jesuit in 1848. He contacted Napoleon III and started to advocate a mission toward the Wahhabis in the Nejd. Palgrave underlined the potential benefits of this enterprise for France. Central Arabia was still unexplored, and an alliance with the leading tribes in the area could be advantageous. A French-Wahhabi alliance could help the Second Empire on the Red Sea.[80] He published the account of his travels in two volumes entitled *Narrative of a Year's Journey through Central and Eastern Arabia (1862–1863)*. Kiepert compiled several maps to illustrate them in 1865.[81] Being one of the few Westerners since Niebhur to have been able to explore this far into the peninsula, Palgrave was, at first, applauded in Britain. The *British Quarterly Review* praised his accomplishments: "He has a complete mastery of his subject, [. . .] his account of desert travelling is the best, the fullest, the most graphic that has ever been written."[82] However, some of his data were visibly inaccurate. The Jesuit explorer avoided too many questions by saying that he had lost his notes.[83] Lewis Pelly criticized Palgrave's work as early as 1865. His own survey aimed at proving the doubtful nature of his predecessor's measurements. In turn, Richard Burton wrote a few derogatory remarks in his preface to the third edition of his *Pilgrimage to Al-Madinah and Meccah*. He went as far as to use the French idiom "vieille coquette, nouvelle dévote" ("old coquette, new bigot") to describe Palgrave's career.[84] Both Wilfrid Scawen Blunt and Charles Doughty questioned the Jesuit's achievements as well: it seemed obvious to them that Palgrave had lied.[85]

The problems of the mid-Victorian geography and topography of the Orient, while perhaps unexpected for the casual observer familiar with heroic tales of discovery, was not a surprise to 19th-century British agents. Such flaws were common for British surveyors. One should not be misled by the claims of domination and progress found in writings of the period: most of their authors knew that the use of contemporeanous British maps could get them lost. Historians rely on written words that rarely mention the experience and know-how of the people on the spot. Agents on the ground generally knew how to assess the lack of reliability of the information at hand. In practice, the need to compensate for inaccurate maps most certainly developed their multifarious skills, which could be as effective in understanding a territory as any panoptic knowledge. Individual abilities to understand and read a map are as crucial as the inherent quality of the cartographic document itself. Most 19th-century explorers with some training knew how to locate themselves roughly on the ground and, to a certain extent, could probably do a lot better with a bad map and their skills than helpless modern travelers equipped with their GPS.

THE SEAMAN'S VIEW

Traverse surveys, reports of exploration, and ancient maps were not the only sources of information for cartographers compiling new maps of the East. They could rely on a second type of graphic representation that differed significantly from the landsman's view: the hydrographic chart.

As stated earlier, the main producer of such cartographic documents was the navy. A specially dedicated department, the Hydrographic Office, was created as soon as 1795. A first catalog of the charts was published in 1825. Hydrographic charting had its specific requirements and logic. Competition with France had increased the accuracy of the surveys and favored the development of new equipment in the early 19th century. The works of Charles-François Beautemps-Beaupré, one of the main instigators of the modernization of French hydrography at that time, were translated into English.[86] One of the reference works for British hydrography was edited in 1849 by John Herschel (1792–1871), an astronomer.[87] It describes hydrographers' instruments, surveying techniques, and what type of information to collect. Logically enough, the book focuses first on bearings, meteorological observations, soundings, astronomical measurements, and the various dimensions of the hydrographer's craft. The last sections, however, such as J. C. Prichard's remarks on ethnology (section XIII), were exhortations to adopt more of a landsman's view. They promoted the collection of knowledge that was not purely hydrographic and involved investigations on land. It is nonetheless a tautology to state that hydrographers generally look at the land from the sea. As a result, the hinterland was generally left unexplored and thus uncharted. The navy charts mostly showed contours.

Stafford Bettesworth Haines's Red Sea surveys on the *Palinurus* in the 1830s were, for instance, mostly made from the ship's deck. Village outlines, relief, or vegetation were in general only partially documented (fig. 1.4). The depth of field from such a viewpoint was intrinsically limited. Hydrographers gleaned as much data as they could without going further than a few miles inland. Jeremy Black underlined the fact that the perception of the world derived from hydrographic surveys was ineluctably from and for the sea.[88] The hinterland was left as a blank. The hydrographic charting of the East reinforced the image of an Orient open to exploration. The map of the Orient was made of lines and of a few points—stations such as Aden, whose existence on the map was enough to give a sense of control—rather than surfaces.

The representation of partially known territories such as the East was largely determined by technical considerations, regardless of the apparent rationality of maps and surveys. It was also enmeshed in other practices, such as travel writing, art, and literature. Traverse surveys and charts could give the Orient its contours, but the image of its interior was still characterized by a gradual blurring of the vision, which left much to Victorian imaginations replete with biblical and classical resonances.[89]

1.3 THE LOCAL ELEMENT

Maps and surveys were never unilateral constructions. The traditional view of the British Empire generally focuses on the red-colored surfaces on late 19th-century maps of the world, supposedly hung on the walls of many a school. Those maps reflected a very metropolitan outlook. The many forms of local input are generally erased from or overlooked in the empire's records. Condescending attitudes and archival effects thus contribute to obscuring how dependent explorers and traverse surveyors actually were on local support when gathering knowledge. Their isolation made interaction imperative. The heroic advance of Western science in an objectified East is a creation of the narratives of exploration of the time. Few were the travelers on the ground who could deny the existence of local forces that could work to their advantage or disadvantage, depending on a constantly changing nexus of factors.

Mid-19th-century travelers were worried that Easterners might not want Europeans "writing down their land."[90] British explorers often anticipated resistance. They seized any opportunity to survey under some form of protection. For example, Henry Leeke from the Indian navy took advantage of the occupation of Bushire during the Anglo-Persian war to survey the area. He and his officers worked very quickly under difficult climatic conditions for fear of missing this window of opportunity.[91] Surveyors often had to seek local help. Charles Golding Constable, son of John Constable, charted Oman in 1862. The chiefdom supported his endeavor, and local tribes did not interfere with his work. In a classic tale of Westerners educating curious Easterners, Constable describes hundreds of people amassing around him and taking turns to look through his telescope.[92]

British agents in the East did not have the same approach to local information. Years of accumulated experience helped the Anglo-Indian officers to negotiate their presence better than metropolitan explorers could. James Felix Jones, an Indian navy officer trained by Robert Moresby, exemplifies this Indian specificity. He mastered all the different steps of the cartographic process, from survey data collection to drawing. In addition to his cartographic

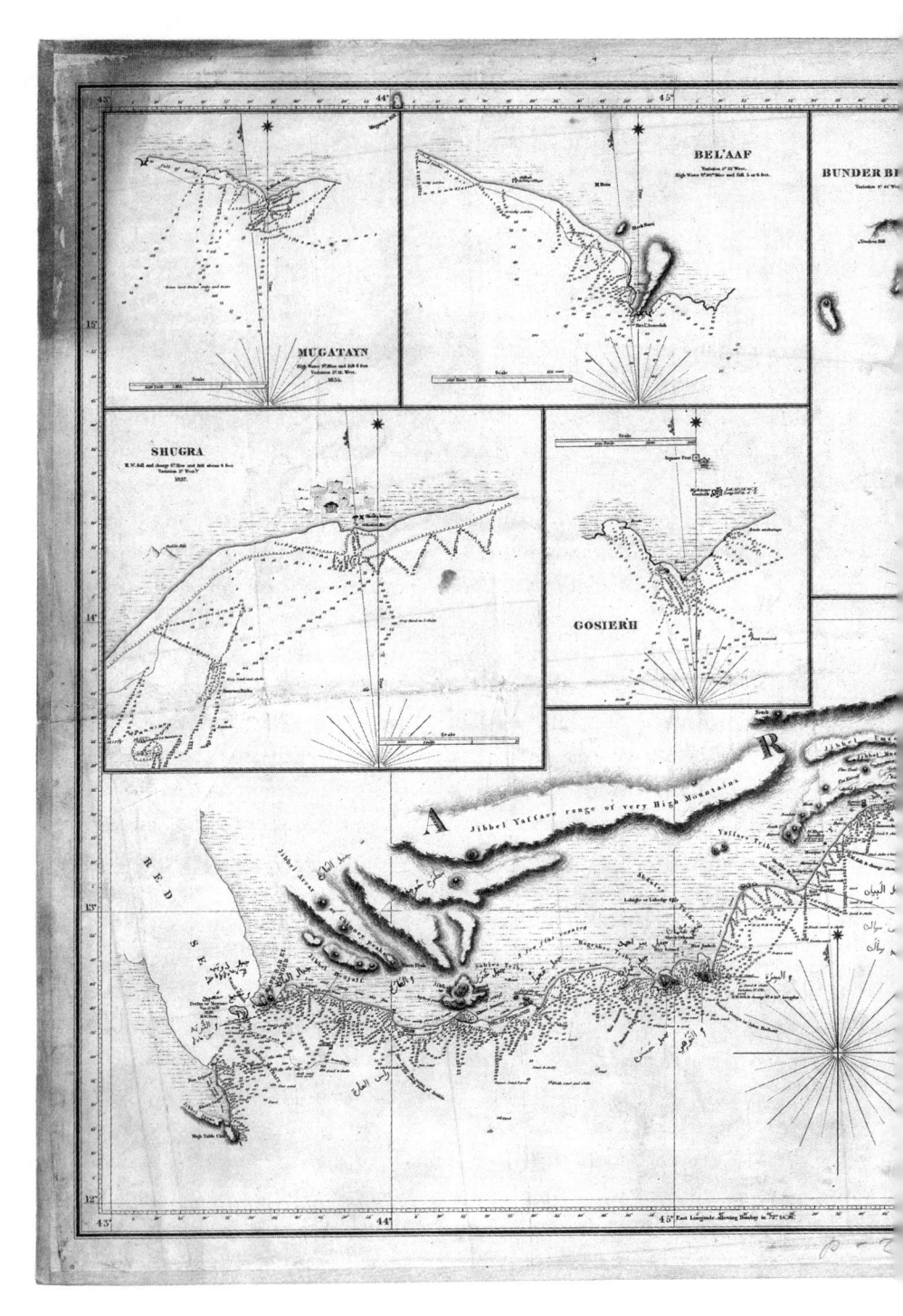

FIGURE 1.4. Samuel B. Haines, *Chart of the East Coast of Arabia*, scale not given (London, 1839). Size of the original: 95 × 61 cm. Photograph courtesy of the BNF (Cartes et Plans GE C-3053).

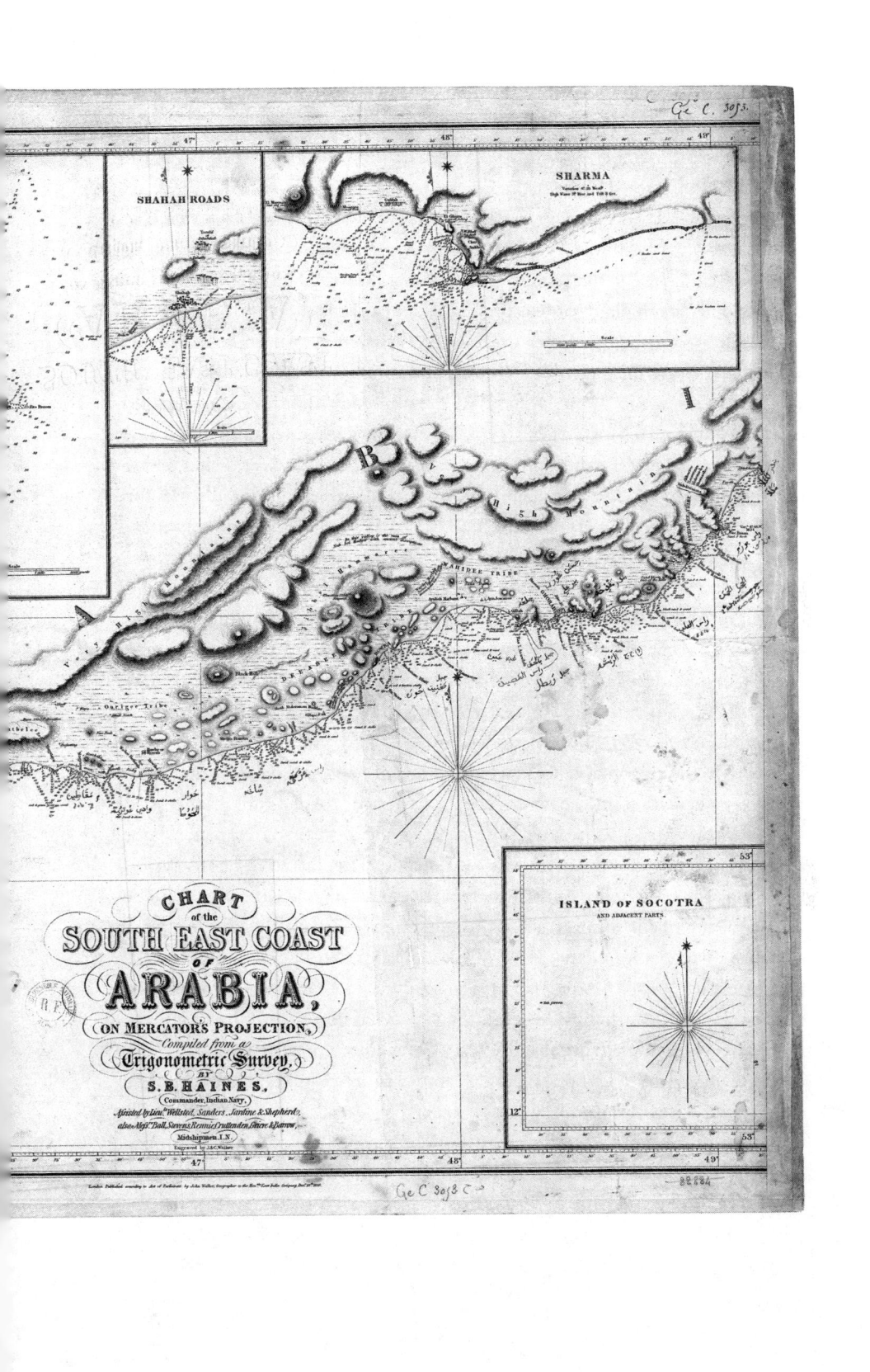

CHART of the SOUTH EAST COAST of ARABIA,

ON MERCATOR'S PROJECTION,

Compiled from a

Trigonometric Survey,

BY

S. B. HAINES,

Commander, Indian Navy,

Assisted by Lieuts. Welsted, Sanders, Jardine & Shepherd, also Messrs. Ball, Stevens, Rennie, Cruttenden, Snow, & Barrow, Midshipmen, I.N.

Engraved by J. & C. Walker.

SHAHAH ROADS

SHARMA

ISLAND OF SOCOTRA
AND ADJACENT PARTS

expertise, he had firsthand knowledge of Mesopotamia and Persia. He was instrumental in opening the first steam lines on the Tigris River from 1842 to 1846 and assisted Henry Rawlinson in demarcating the Turco-Persian border in 1844.[93] He accompanied Austen Henry Layard to map Nineveh in 1852. He designed the first modern map of Baghdad in 1853. A first trigonometric survey of the area allowed him to publish a map of the ancient city in 1855.[94] Jones is a good example of the Anglo-Indian adaptability to local circumstances. The flat and barren plains of Mesopotamia could prove difficult to survey. Explorers needed to be on higher ground to take their measurements. The best solution to establish a point of reference was to use minarets as stations. Jones was able to convince Mosul's imams of the innocuousness of his topographical proceedings. In Mosul rumors spread that he wanted to build an artillery battery or even transform the mosque into a church. However, Jones had secured the imam's protection thanks to his fluent Arabic. He was allowed into the Great Mosque and established a datum at the location of the minaret.[95] Jones worked with few instruments and had little time to make repeated measurements. He fixed the longitude of the minaret with a chronometer as well as with astronomical observations. He then established a baseline and more than ten stations. He also used a chain to measure the circumference of the Nebi Yunus mound, which the villagers apparently found nothing short of hilarious. His inconspicuousness greatly contrasted with the propensity of metropolitans to create diplomatic incidents. Layard, whom he knew, became infamous in the region because he attacked one of Mosul's imams with his stick while on a boat trip on the river. The affair almost turned into a large-scale riot. Contrary to Layard, when obstacles were too great to overcome, James Felix Jones knew not to swim against the tide. The narrative of his surveys in Baghdad is revealing from that perspective. Ottoman authorities were becoming wary of his surveys. They forbade him to explore the city's intricate network of roads. Jones asked one of his assistants, William Collingwood, to take measurements.[96] Collingwood wrote down his bearings and the number of paces he took on the inside of his shirt to avoid detection.[97] Jones collated them afterward and eventually drew a very accurate map of Baghdad.[98] The Anglo-Indian school favored this adaptability more than did its metropolitan counterpart.

Western surveyors were identified as such. Most of the time, direct support from local inhabitants came at a price. In the Orient, Western visitors could be used to oppose the rule of Eastern sovereigns, namely, the rule of the Ottoman Empire. Charles van de Velde (1818–98), a Dutch missionary, provided a revealing anecdote that sheds light on the complicated relations between the surveyor and the surveyed. Van de Welde explored Syria and Palestine where Ottoman rule was not always popular. He was "quite astonished to find people so willing." People supposedly said to him: "You are coming to survey the country, because your nation will ere long take it away from the Sultan." For the missionary, the inhabitants of Palestine clearly perceived "that a survey of the country helps towards forwarding a future conquest" and observed that they co-operated "towards the completing of the survey."[99] The missionary thought this was one of the manifestations of what he believed to be Israel's forthcoming restoration. Local sheikhs were not millenarians: European encroachments in the Levant could serve them well against Turkish rule.

Depending on local circumstances, mapping had, on occasion, to be covert. Open surveying could become extremely dangerous in central Arabia for instance. Lewis Pelly, fearing the wrath of the Wahhabis in central Arabia, asked his geographer, Lieutenant Dawes, to burn all his sketches and to leave no obvious trace of his exploratory concerns.[100] The Westerners' lives were not the only ones at stake. A sheikh who was too collaborative could also be in danger. Distrust of any form of foreign enquiry characterized some parts of Arabistan. Whoever had little regard for prudence could be harshly reminded of that reality. W. B. Selby met with such an experience while exploring the marshes of southern Mesopotamia.[101] Ma'dān Arabs inhabited the area. They were wary of attempts to establish outside control over their lands. When Selby explored the region, it was a terra incognita for the Ottomans, who had only theoretical authority over these territories. Selby and his assistant, Collingwood, were under constant surveillance and could do little to survey the marshes correctly. According to them, they almost lost their lives because of the ruthlessness of the sheikh, named Faasil. However, his tribe's distrust of the surveyors was even more intense than Faasil's own suspicion. He apparently signed his own death warrant by allowing Selby and Collingwood to go untouched. He was killed by his own tribe because he had let them spy "on the fat of the land."[102]

The local point of view was always a factor that determined the quality of the survey and the map. Recent assumptions underlining the purportedly unilateral construction of knowledge in the imperial 19th century tend to miss this point. The explored usually knew how to manipulate or even stop the often solitary explorer. The mapping of the Orient, almost inevitably, was an encounter. Much like their 18th-century counterparts in South America, studied by Neil Safier, Victorian surveyors had to conciliate different types of spatial knowledge to produce scientific data.[103] The collection and classification schemes of European spatial representations then inexorably suppressed the hybrid origin of the mapmaker's data.

1.4 AN EXCEPTION? BRITISH SURVEYS IN PALESTINE

Of the many institutions involved in surveying and mapping the Orient, the Ordnance Survey might seem the most unlikely. It had developed as the main mapping organization for the British Isles and was placed under the authority of the WO in 1855. Its surveys had been strictly limited to the United Kingdom until that point. Despite this, the institution participated in one of the most systematic mapping projects of the mid-Victorian age in the East. Henry James, its director between 1854 and 1875, was keen to enlarge the scope of its surveys and, in a break with its normal practice, sponsored a survey of Palestine. The first expedition explored Jerusalem in 1864–65. It was organized by the Jerusalem Water Relief Fund, a merger of two previous entities, the Christian Asylums Committee and the Syrian Improvement Committee. One of the leading advocates of the exploration of the Holy Land, John Irwine Whitty, had explored Jerusalem by himself in the early 1860s. He promoted a modernization of the water distribution system in the Holy City. Having convinced the Prince of Wales of the importance of his project in 1862, he published a report in 1863 to demonstrate the feasibility of his proposed scheme.[104] The publication caused quite a sensation and the Jerusalem Water Relief Fund was created in 1864 in its footsteps, with a £500 grant by Lady Burdett-Courts to survey and map Jerusalem. The surveyor, Charles Wilson, a member of the RGS and a royal engineer, was paid for by the WO. The operation was placed de facto under the authority of Henry James. Wilson was able both to survey Jerusalem and make altimetric measurements from there to the Dead Sea. His work was published quite rapidly.[105]

Charles Warren continued the surveys from 1867 onward under the aegis of the newly established Palestine Exploration Fund (PEF), which was created in 1865 on the initiative of Walter Morrisson, George Crove (an erudite Evangelical scholar), and other prominent Victorian authorities. The PEF was founded partly in reaction to the growing interest of other European nations in the field of biblical archaeology. The identification of the holy sites, the object of a competition between Protestants and Catholics, called for better instruments and further expertise. During the survey, Warren explored underground Jerusalem. A survey of Sinai was undertaken between 1868 and 1869. Charles Wilson, H. S. Palmer, and photographer James MacDonald gathered geological, archaeological, geographical, and topographical data that would supposedly help develop a better historical critique of the book of Exodus.[106] The institutional structure behind the explorations was highly ambiguous. The Ordnance Survey and its Royal Engineers were working for nongovernmental associations. Such British maps of Palestine in the 1860s were thus the product of dissonant agendas.

Archbishop William Thomson's speech is frequently taken as an illustration of the PEF's predominant stance.[107] His 1865 inaugural address seems to resonate with imperialist presumptions: "This country of Palestine belongs to you and to me, it is essentially ours. It was given to the Father of Israel in the words: 'Walk through the land in the length of it, and in the breadth of it, for I will give it unto thee.' We mean to walk through Palestine in the length and in the breadth of it, because that land has been given to us. It is the land from which comes news of our Redemption. It is the land towards which we turn as the fountain of all our hopes; it is the land to which we may look with as true a patriotism as we do to this dear old England, which we love so much."[108] However, this often-quoted extract is sometimes wrongly interpreted as heralding further British involvement in Palestine. Was the only mid-Victorian attempt to systematically survey and map a part of the Orient an imperialist undertaking? The heterogeneous membership of the fund was in itself an obstacle to a coherent and unified orientation. Its members came from a variety of social, religious, and political backgrounds. Lord Shaftesbury's millenarian Evangelicalism, Layard's secularist stance, and Edward Pusey's Tractarianism did not partake of the same vision. The fund even had Catholics among its members. Besides, notwithstanding the ascendancy gained over the long term by the critical and archaeological approach advocated by James Fergusson (1808–86), against more fundamentalist readings of the uncovered biblical remains, the conflicting assumptions of its founders enduringly characterized the PEF.[109]

William Thomson's words are easily misunderstood if taken out of their context. "Walking through the land" did not mean actually traveling through Palestine. Thomson, certainly inspired by an Augustinian understanding of the nature of Jerusalem and the Holy Land, did not see the latter as a concrete territory. He considered "that in the last and most mysterious revelation of God to man the very realms of blessedness and glory are designated by a name and specified by allusions (Rev. xxi. 22) which warrant our recognizing in the Holy City on earth, the 'Jerusalem that now is,' a type of that 'Heavenly City which God hath prepared for the faithful' (Heb. xi. 16), a similitude of the Jerusalem that is above, a shadow of the incorruptible inheritance of the servants and children of God."[110] The fact that the first meeting of the PEF took place in the Jerusalem Chamber in Westminster Abbey is yet another sign of the underlying theological assumptions that, for some of its members, presided over the foundation of the association. The legend held that Henry IV died in front of the chamber's fireplace in 1413. He had planned to go to Jerusalem, where he was to die, according to a soothsayer.[111] When he asked where he was, this domestic version of the Holy City was the answer to his question. The educated participants to the PEF meeting could hardly have been ignorant of the symbolic dimension of such a location. The Jerusalem they were looking for was of a celestial nature. The Holy Land was not a territory waiting to be conquered and colonized; it was a palimpsest, an accumulation of remains and ruins for the faithful to travel and reflect upon. And there was no need physically to go there. Modern tools, such as triangulation, could be used to bring the vestiges home. This stance was clearly articulated by A. P. Stanley, dean of Westminster, guide to the Prince of Wales during the latter's 1862 tour in the East, and founding member of the PEF, who professed in a sermon that it was "possible to be just, and holy, and good, without coming to Palestine," because it was not "the earthly, but the heavenly Jerusalem" which actually mattered to the believer.[112] Exploring the Holy Land was a means to authenticate and substantiate the biblical record in that regard, not necessarily a colonial claim. It had more to do with William Blake's "mental fight" to build Jerusalem in "England's green and pleasant Land."[113]

There were of course other motives behind the project. Stanley, for instance, became convinced of the powers of cartography when he first saw a map of London drawn for the Metropolitan Sanitary Commission to improve the sewage system in the capital.[114] The prospect of modernizing Jerusalem's antiquated water distribution system was a reflection of metropolitan sanitary anxieties.[115] The Sinai survey was a welcome addition to the body of knowledge on Moses's journeys, but it was also a way for the WO to gather relevant geographical information on an area that partly controlled the Suez Canal under construction. However, one should not overestimate the impact of one motive in particular. Despite this renewed interest in the Holy Land, actual Palestine remained a strategic sideshow for London. The PEF's maps were above all the expression of a Victorian biblicism that permeated the British perspective on the East.[116]

1.5 THE SURVEY AND THE MAP

Cartographic documents and the data that support them are separate things.[117] Sometimes, a skillful individual, like James Felix Jones or T. B. Jervis, could be simultaneously a geographer, a topographer, and a cartographer. The three operations of gathering data, designing a map, and engraving it responded to different logics. The translation of the traveler's written notes and sketches

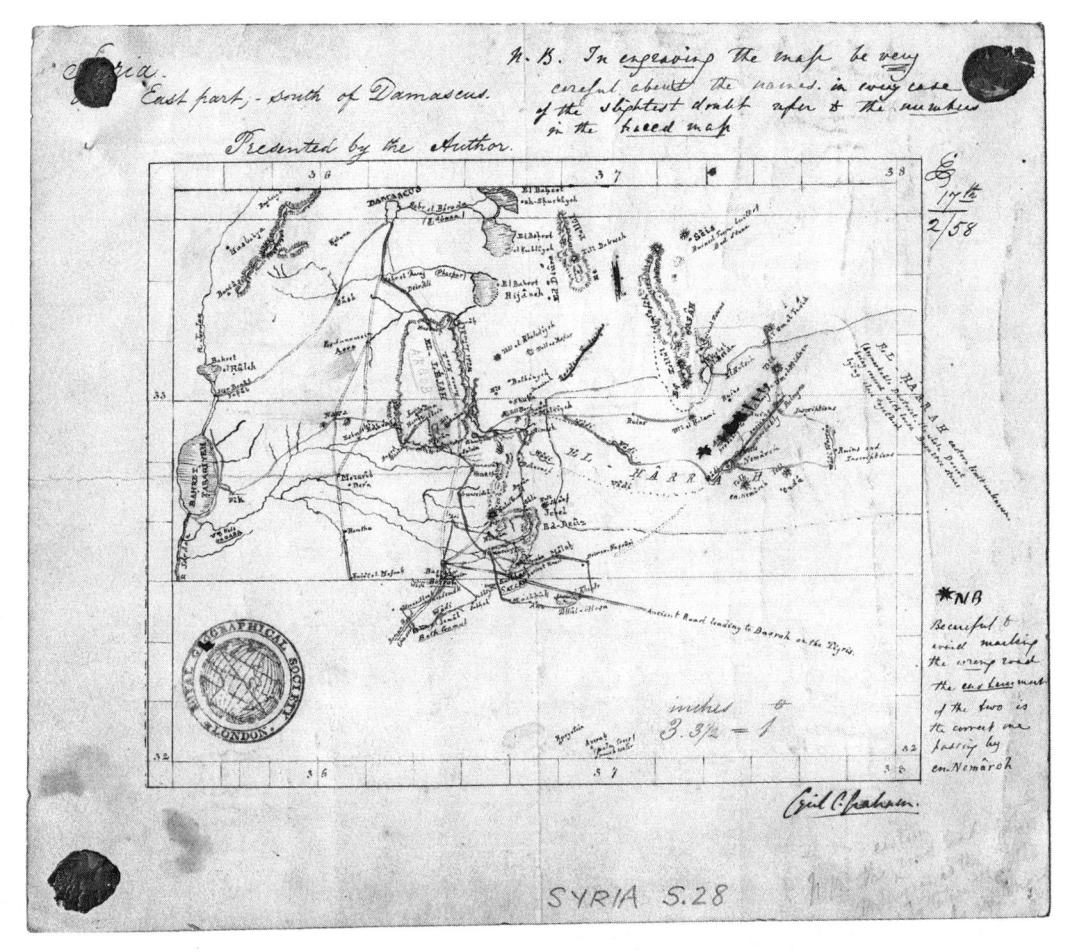

FIGURE 1.5. Cyril C. Graham, *Syria. East part; South of Damascus*, 1:1 300 000 (London, 1857). Size of the original: 14 × 20 cm. Photograph courtesy of the RGS (Map Room mr Syria S.28).

into a map was the source of many difficulties and approximations. There is little to be found in the archives about this issue. Drafts and notes were not always kept. Fortunately, a rare opportunity to compare the work of the agent in the field with its transformation into an engraved map is provided by the RGS collection of manuscripts, in the form of an 1857 sketch by Cyril Graham showing part of Syria (fig. 1.5). Graham published an article in the *Journal of the Royal Geographical Society* (*JRGS*) after returning from an exploration of the Hauran.[118] A map of the area was published in the same issue of the *JRGS* (fig. 1.6). A comparison of what the traveler designed and what was actually printed is particularly instructive. It shows how paratextual uses of maps were entangled in intricate relations between explorers and publishers.[119]

THE MID-VICTORIAN PERSPECTIVE · 45

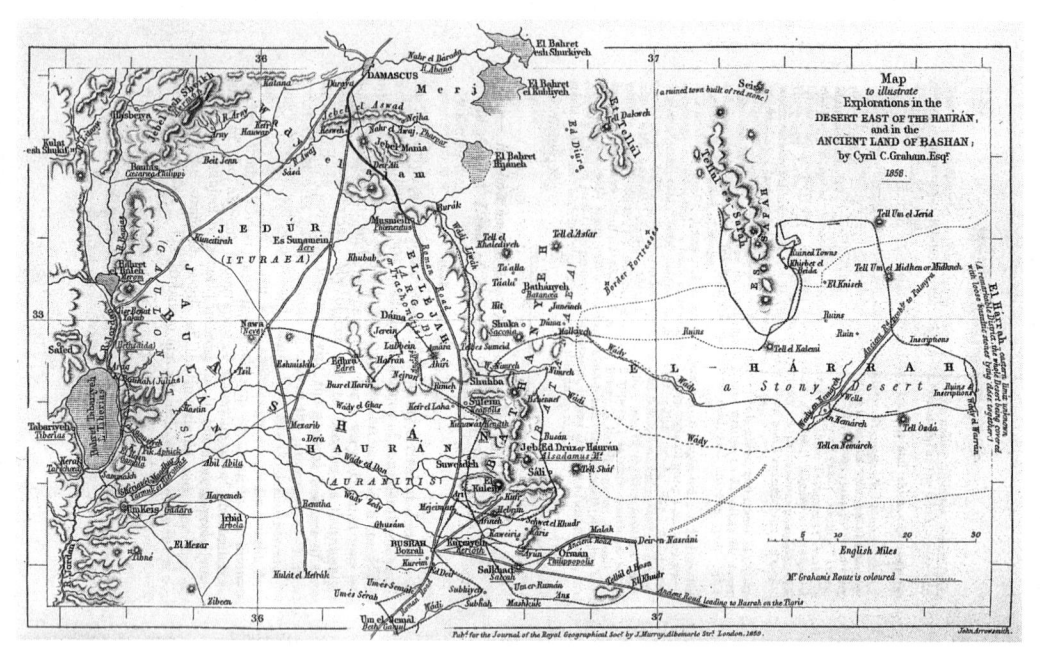

FIGURE 1.6. Cyril C. Graham, *Map to illustrate explorations in the desert east of the Hauran*, 1 inch = 19 miles, in Cyril C. Graham, "Explorations in the Desert East of the Haurán and in the Ancient Land of Bashan," *Journal of the Royal Geographical Society of London* 28 (1858): 226–63. Size of the original 19.3 × 11.1 cm. Photograph courtesy of the Bibliothèque Interuniversitaire des Langues Orientales (BIULO PER.5134).

Some indications written by Graham, such as a road described in his sketch as leading "probably to Palmyra," could not be transcribed on the published cartographic document for lack of authority. On the other hand, the same cartographer abhorred a vacuum: he added numerous elements that were not on the traveler's submission to begin with. Relief in particular mobilized his skills and his imagination. Graham had hachured a few *tells* but knew that to entrust his badly drawn map to an armchair technician might prove problematic. He thus wrote a note in the margin asking the cartographer to be "very careful about names." This did not suffice. The course of the Jordan tributaries, which had already been re-created by Graham's untamed pen, was reinvented by the engraver. The El Kubliyeh and Esh Shurkiyeh lakes were relocated to make them fit into the chosen projection, which differed from the one Graham used. Finally, the cartographer designed his own version of the Hauran massif when attempting to clarify the overwhelming number of place-names the explorer had burdened him with. Given the pivotal role of the RGS in British and European cartography, the accumulated approximations of the map were often compiled in smaller-scale documents. The resulting inaccuracies could then survive for years or even decades because of the very slow updating of topographical information in areas so remote.

Independent travelers were perhaps less inclined to overlook the defects of the maps they generally transported in their luggage. They had no stakes in the promotion of the official agents' missions. John MacGregor, the canoeing explorer, had pinned his hopes on Western maps. These hopes were soon dashed. His frail boat left no room for the traditional dragoman. He could only count on portable Western knowledge to get his bearings, so he embarked on his adventure with the most esteemed works of expert cartographers. He had Van de Velde's, Carl Ritter's, and August Petermann's maps, as well as Josias Leslie Porter's guide. This panoptic apparatus left him at a loss for the right directions: "It is evident enough that none of them have been made by personal survey from each side. Porter declines to imagine where he has not inspected, and rightly merges the lake in the desert without any southern outline, though Hijaneh has a very distinct shoreline all round it. Van de Velde's map is distinct, but rather inaccurate. Petermann's is worse, for the whole is imagined and while this is distinctly done, the result is certainly not accurate. Ritter's, however, is the worst of all, for it 'lumps' the three lakes into one, and marks all sorts of bays and capes as if they had been accurately surveyed. This pretentious accuracy is equally fallacious in his delineation of the Abana and Pharpar, the Jordan, the Lake of Hooleh, and the Sea of Galilee."[120] MacGregor was not the only one to feel disoriented by the European maps. Warburton, while exploring Syria, experienced the same difficulties: "My maps were with my luggage, and I had only a slight sketch from Arrowsmith's very inaccurate map to guide me over the waste. In following its guidance, I repeatedly lost my way."[121]

This assertion confirms the impression given by another travelogue written by the Reverend H. B. Tristram (1822–1906), a prolific biblical scholar. In one of his most popular books on Palestine, *The Land of Israel*, he points out the flaws of the available maps of the region: maps of Palestine were full of blanks. Blanks on maps actually warranted stricter scientific imperatives. By the 1750s, Jean-Baptiste Bourguignon d'Anville's innovative use of blank space did not denote "a want of intelligence" anymore.[122] It excluded legends, monsters, and inaccurate data from the map. D'Anville mapmaking practices were very influential on subsequent British mapping. But regardless of the Enlightenment cartographic ideals, pilgrims could not rely on the still blurry cartographic representation of the biblical lands: "With map before us—and our map of the shore was, in size, worthy of the ordnance scale, having been enlarged for us from Van de Velde on a scale of an inch to the mile—we reclined round the watch-fire, discussing geography, and listening to Abou Dahuk's stories of his former experiences [. . .]. We saw that the tracing to the south of us was as inaccurate as that to the north of our position had proved, and that headlands, and bays, and sandspits were all unrecorded."[123] European maps could get you lost. They certainly did not replace good dragomans or accurate directions from local inhabitants.

British cartographic practices and techniques had not yet proved to be unquestionably autonomous and effective. The map of the East was full of blanks to be roughly filled by the engravers' or readers' imaginations. Not only were entire parts of Arabia, Persia, or Turkey blind spots for the British gaze, but decision makers could also have difficulties in locating existing documentation.

1.6 USING MAPS

Maps had been tools of government long before the industrial age.[124] However, the imperatives of Britain's growing international involvement in the second half of the 19th century turned them into clear instruments of global influence. John Gilbert's 1855 depiction of a cabinet meeting, *The Coalition Ministry*, seems to imply exactly this.[125] Palmerston, the architect of mid-Victorian British foreign policy, is pointing at a map that is centered on the Black Sea. His shadow covers Asia Minor. Lord Aberdeen, the unpopular prime minister blamed for the military setbacks suffered on the battlefields of the Crimea, is looking the other way. The map is not at the center of the drawing but is the instrumental artifact which legitimizes Palmerston's expertise. The modernity of the map substantiates his argument in front of his colleagues' limited cartographic awareness. In fact, in the mid-1850s some members of the cabinet doubted Palmerston's genuine interest in cartography as a decision-making adjunct. Clarendon, the foreign secretary, wrote to John Russell in 1854 that "Palmerston is perpetually making new maps of Europe, but that is mere child's play."[126] Was Gilbert's drawing propagandizing a new model of foreign policy making? Were maps increasingly becoming part of the governing process? Palmerstonian propaganda could be deceptive. A look at how cartographic and geographical information was really organized in the various offices interested in the East yields a different picture.

In 1856, Jervis tried to consolidate the department he was in charge of, when he wrote to Lord Panmure to request the recruitment of more staff and to urge for the improvement of military intelligence standards to Continental levels.[127] He also compiled and lithographed additional maps, with a specific focus on the Russian and Ottoman Empires. However, his death in 1857 signaled a return to the previous inertia. Henry James, his successor, did not give the department the same impetus. A catalog of maps was published under his aegis, but this should not hide the curtailing of the cartographic activity of the department in the 1860s.[128] The situation was similar in the FO, where there was no internal department to design new maps. However, the new buildings inaugurated in 1868 incorporated a specially dedicated map room. The FO personnel could rely on other departments, although part of its cartographic information derived from foreign sources. The example of the Ottoman Empire is particularly striking. Approximately half the maps of Turkey in the FO were foreign productions.[129]

The last governmental office potentially involved in Eastern affairs, the IO, was created following the Indian Mutiny of 1857. The 1858 India Act modified India's governmental structure: the East India Company lost its autonomy and the IO replaced it to supervise the administration of British India from London. The relative role of maps and geographical intelligence in the IO itself remained secondary for years. The official outlook was not of a visual nature. Notes, dispatches, and reports were the preferred form of communication among the officeholders rather than maps. Some, nevertheless, promoted a different approach. One of the staunch advocates of a better use of cartographic tools was Clements Markham. He was first employed in several of the many departments of the IO, which was a small government in itself, where he developed a firm belief in the power of maps. He was transferred in 1862 to the Public Works Department, where he realized the state of the documentation was far from ideal. He started to call for the creation of a geographical department, which was to organize and supplement the disorganized and little-cared-for collection of thousands of maps and geographical documents already in the possession of the IO. In 1867, the creation of the Geographical Department was approved. Markham restored and supplemented the map room with the help of various assistants, among them Trelawney Saunders (1821–1910), a cartographer who had worked for Edward Weller, a renowned mapmaker, in the early 1850s.[130] Saunders eventually became the official cartographer of the IO. He compiled a great variety of maps for the IO, the PEF, and the Indian government.[131] He also mounted most maps on linen in order to facilitate their consultation and archiving. The Geographical Department provided information to the Indian secretary, to other government offices, and even to the press. This first attempt was nonetheless inconclusive in some ways, notwithstanding

Markham's accomplishments. He was never appointed head of the Geographical Department. His work with maps was never fully acknowledged, despite his continuous requests for better wages. He eventually retired in 1877 without his department's having been comprehensively consolidated.

British administration in India had its own perspective on cartography. The map culture in Bombay and in other presidencies was far more advanced than in the metropole. However, Anglo-Indian cartographic archives were hardly better managed than the map rooms in Whitehall. The Indian navy's surveys lacked coordination. Printing and cataloging were organized randomly. Documents were often lost.[132] The Mesopotamian survey's reports were apparently destroyed. The presidencies did not always work in perfect harmony with the Bombay headquarters. Copies of the charts printed in Madras might well never be sent to Bombay and could thus fall into oblivion. Notwithstanding the Great Trigonometric Survey's attempt at a panoptic vision of the Raj, data on the emerging Indian subempire was still collected in an ad hoc fashion. There was nonetheless a headline for "Western Asia" in the *Catalogue of Maps of the British Possessions in India and Other Parts of Asia* that was published in 1870. It comprised, among other items, a map showing the routes and telegraphic lines between Constantinople and Felix Jones's maps of the Assyrian vestiges, one map of the Persian Gulf, and three maps of Arabia.[133] They hardly added up to a coherent perspective.

Given its previously described pivotal role, the RGS was the only potential coordinating institution in London. The RGS was partly funded by both the Treasury and Parliament (through a £500 annual grant in the mid-19th century). According to Henry Rawlinson's testimony before the Royal Commission on Scientific Instruction, an arrangement between the government and the RGS allowed the staff of the various relevant government offices in London to consult maps and books in the library of the RGS. This agreement did not provide for a coordinated use of the RGS's resources. Despite the rising influence of some of its members who held official positions, the RGS was not yet the imperial hub it became in the early 20th century.

Overall, mid-Victorian cartographic intelligence and its uses were unsystematized and surprisingly haphazard. The disparateness of official and semiofficial map collections mirrored the piecemeal mapping of the transitional East, despite its key position on the route to India. Decision makers still heavily relied on an old information order based on a muddled aggregate of travelogues, romanced exploratory accounts, pictorial maps, and an embryonic range of seemingly more expert instruments of knowledge. The promoters of a new model intelligence were still preaching in the desert.

CONCLUSION

William Cowper's verse "I am monarch of all I survey" became a common phrase to describe the position of the British explorer overlooking his geographical discoveries.[134] To the casual reader, Cowper's words could legitimately illustrate what the nature of 19th-century Western knowledge was, whether in the Middle East or in other parts of the world: an all-powerful gaze that surveyed and categorized at its will. Mid-Victorian maps of the area are a test case in that regard. British power in the region did not yet project itself in clear-cut imperial terms in the 1850s and 1860s. The addition of uncoordinated surveys and mapping projects never amounted to an unambiguous expression of power. The first trigonometric enterprises, in Mesopotamia or Palestine, were still unable to correct the myopia that characterized the British official outlook on the region. While some, in the RGS or in ministerial departments, promoted a more coherent approach, the Western knowledge of the Orient was an aggregate of fragmentary information collected under very different agendas.

Early surveyors and explorers were also missionaries, merchants, and soldiers on leave. They were not the agents of a systematized spatial rationalization. Some of their surveys were driven by broader beliefs in the virtues of trade. Mid-Victorian liberalism and the will to open markets underpinned the Lynch brothers' various enterprises in the area, for instance, and Lewis Pelly's explorations. Agents, working in a nebula of communication networks, were looking out for trade opportunities, new routes, and occasions to open up the way to Britain's civilizing creed through commerce. The East was slowly pulled by the gravitational force of the expanding British world system. This was far from a benign process, as shown for instance by the Anglo-Persian war, where maps proved very valuable to Britain, both to defeat the shah and to plan liberal reforms. But other motives were deeply entangled in producing the East. Biblical scholarship and romantic exoticism played their parts. They depicted a completely different Orient, one that was in fact protected against modernity's encroachment. Blanks on the maps were not justifications of the *terra nullius* legal fiction in that regard. They were the promise of untouched ruins and authentic ways of life instead of figuring empty territories for the taking. Other enterprises testified to the nascent expansion of the Raj system of indirect rule in the transitional East. Indian maps and mapping served, justified, and actuated early attempts at protecting the High Road to India. This variety of motives cannot be collapsed into a single expression of power. It reflects the ambiguities and mutability of British influence in the area.

CHAPTER TWO

LABELING THE EAST

Where did Africa end? Where did Europe begin? Where was the East? If we look at some examples of mid-19th-century British attempts to delineate the frontiers between Asia and Africa, we find that there were a variety of solutions available to the cartographer. In his 1870 *Public School Atlas* George Butler, M.A., principal of Liverpool College, chose a traditional line from the Mediterranean to Suez. Sinai was thus still in Asia (fig. 2.1). A decade earlier, in his *The Imperial Atlas of Modern Geography*, W. G. Blackie applied the same delimitation along the Isthmus of Suez, a straight line between the city of Suez on the Red Sea and Port Said (fig. 2.2). The desert of the Exodus was not yet in Africa. Sinai was partly relocated outside Asia in Alexander Keith Johnston's 1869 *Half-Crown Atlas of General Geography* (fig. 2.3). The stones used to apply color were difficult to manipulate. One of the layers was slightly offset, giving the impression that the coasts of Arabia were not part of the Asiatic Ottoman Empire. Johnston's reference *Royal Atlas Of Modern Geography* separated Africa and Asia by drawing a segment from El Arish to Taba in 1861, or from El Arish to Suez in 1879 (figs. 2.4 and 2.5). When geopolitical circumstances changed, Continental definitions could coincidentally evolve, and the commercial map was thus turned into an involuntary territorial claim. In the wake of the Scramble for Africa, which saw a definition of European spheres of influence on the continent, Stanford printed a map designed with a new perspective on where the Ottoman Empire ended (fig. 2.6). The eastern shore of the Gulf of Aqaba was no longer under Turkish sovereignty.

These ever-changing boundaries are illustrations of what Martin Lewis and Kären Wigen demonstrated in *The Myth of Continents*: geographical delineations are the result of a variety of historical legacies, physical realities, and imagined ethnicities.[1] Place-names are reflections of these evolving constructions. They can be purged, erased, and transformed. From this perspective, the invention of the Middle East was the final stage of a slow geographical and cartographic elaboration of the transitional East in the 19th century. This chapter analyzes some of most relevant elements of the nomenclature and delineations of this volatile area before early 20th-century discourses rearranged it. It therefore applies a different focal length to the subject of this book. Most of the following sections consider geographical constructs of the intermediate East that predated the invention of the "Middle East" from a small-scale perspective, in contrast to the large-scale maps studied in the previous chapter. In doing so, it examines toponymy and the collection of place-names from a critical perspective.[2] It demonstrates how names and limits reflect some of the clusters of representations that participated in the process. This chapter thus provides a nuancing to the conventional correlation between cartographic knowledge and imperial and modernist projects. Recording, preserving, and figuring the past of the Orient were integral to early British surveying. Mapping as an instrument of modernization was often peripheral to the mid-Victorian cartographic perspective on the East.

2.1 EUROPE AND ASIA

The globe was divided into two hemispheres defined by the meridian in most 19th-century European atlases. The Western Hemisphere was centered between the Americas, whereas the Eastern Hemisphere showed Eurasia. Maps using these divisions virtually isolated Europe on the Far West of the East. The British Isles faced a gigantic continental mass. The Ottoman Empire, Persia, and Arabia belonged to a subcategory of the Eastern Hemisphere that was loosely named "East" or "Orient." This nomenclature was characterized by a web of cultural legacies, some of which reached back to Greek antiquity and to the Roman Empire. Carl Ritter (1779–1859), one of the most influential

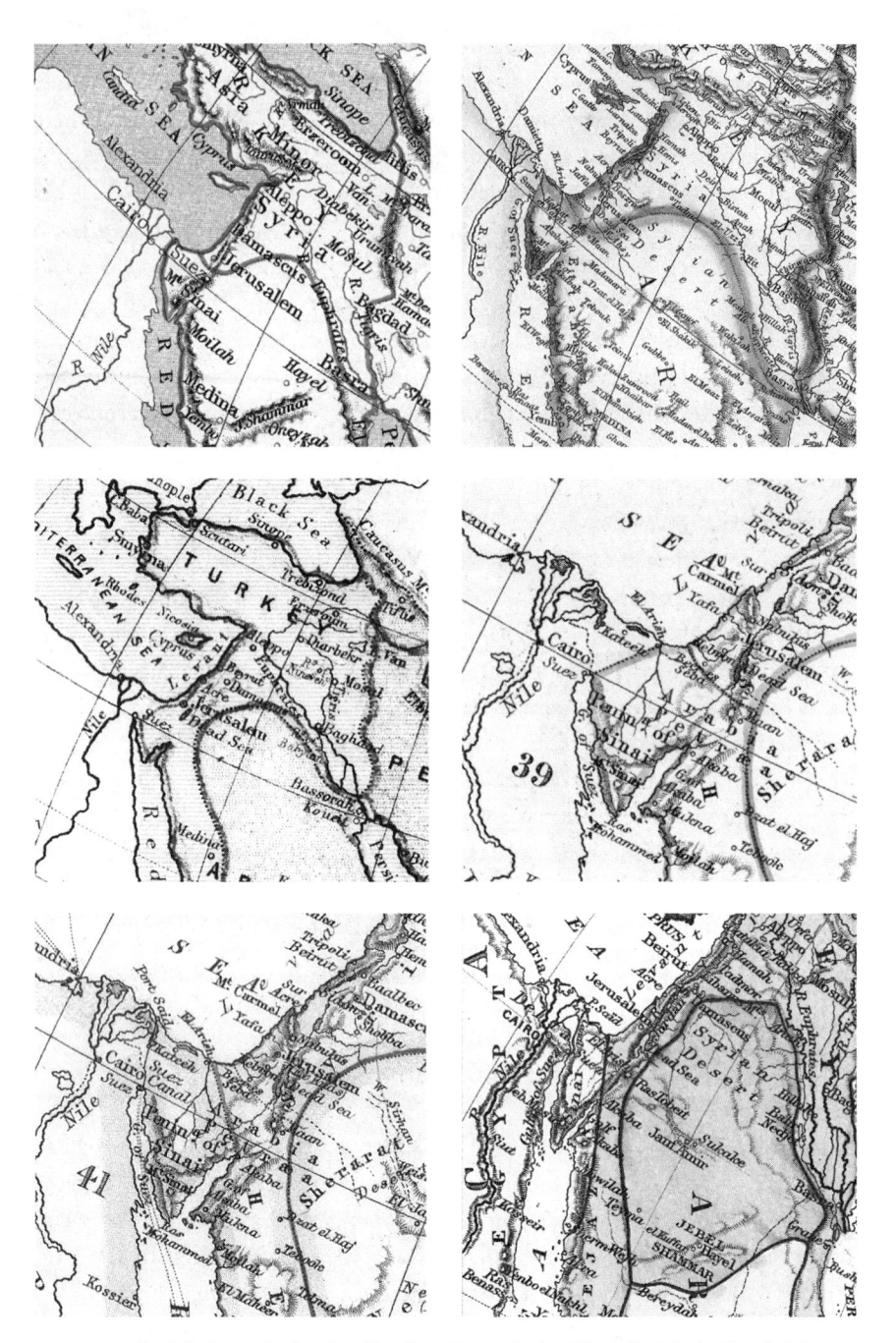

FIGURE 2.1. (*top left*) George Butler, *Asia* (detail), in George Butler, *The Public Schools Atlas of Modern Geography* (London: Longmans, 1872), map no. 3. Photograph courtesy of the BNF (Cartes et Plans GE FF-9101).

geographical authorities on the Continent, stated in the English translation of his *Geographical Studies*: "It is evident from that contrast between Orient and Occident, or, to use terms which are not so limited by usage to the Eastern Hemisphere, between the East and West, that the relative positions of the great masses of land on the globe have exercised a deep and abiding influence [. . .]. Ancient and modern, past, present, and future, are in the Orient, the cradle of the world and of the world's culture and history; but progress is in the Occident."[3] From Eliot Warburton to John Ruskin, this cultural division between Orient and Occident was obvious to the Victorian observer.[4]

This binary opposition was not reflected by any clear-cut delimitation. Mid-19th-century British cartographers seldom referred to these categories in their titles, or their nomenclatures. They were used in travel writing and geographical accounts, but rarely on maps as such. The uncertainty of their contours did not facilitate their graphic representation. The emergence of the expression "Near East," a translation of the German *der Vordere Orient*, in the 19th century, did not resolve the issue.[5] In its first occurrences, the "Near East" referred only either to the Balkans, or, albeit less frequently, to the entirety of the Ottoman Empire. The expression was a counterpoint to the concept of "Far East," as exemplified by an article for *Fraser's Magazine* published in 1856: "The Far East—in contradistinction to the Near East—for the integrity of which we went to war with Russia—contains a population of six hundred millions of people, or perhaps more."[6]

The insertion of the Balkans and Greece into the Eastern sphere implicitly suggested that the Orient could be partly European. The division between

FIGURE 2.2. (*top right*) William G. Blackie, *Asia* (detail), in William G. Blackie, *The Imperial Atlas of Modern Geography* (London: Blackie and Son, 1860), pl. XXXVIII. Photograph courtesy of the BNF (Cartes et Plans GE DD-2285).

FIGURE 2.3. (*middle left*) Alexander K. Johnston, *Asia* (detail), in Alexander K. Johnston, *The Half-Crown Atlas of General Geography* (Edinburgh: W. & A. K. Johnston, 1869). Photograph courtesy of the BNF (Cartes et Plans GE FF-7958).

FIGURE 2.4. (*middle right*) Alexander K. Johnston, *Asia* (detail), Alexander K. Johnston, *Royal Atlas Of Modern Geography* (Edinburgh: W. & A. K. Johnston, 1861), map no. 28. Photograph courtesy of the BNF (Cartes et Plans GE DD-1855).

FIGURE 2.5. (*bottom left*) Alexander K. Johnston, *Asia* (detail), in Alexander K. Johnston, *Royal Atlas Of Modern Geography* (Edinburgh: W. & A. K. Johnston, 1879) map no. 28. Photograph courtesy of the David Rumsey Map Collection (Image no. 0377032).

FIGURE 2.6. (*bottom right*) Edward Stanford, *Asia* (detail), in Edward Stanford, *London Atlas of Universal Geography* (London: Stanford, 1882) map no. 21. Photograph courtesy of the BNF (Cartes et Plans GE DD-12).

East and West was not superimposed on the separation between Asia and Europe. Territories and populations west of the Bosporus could be categorized as "Oriental." An 1856 article for the *Quarterly Review* described the Bulgarian man in a stereotypical way: "[He] is a Slavonised Oriental, of a more gentle temperament apparently than the Tatar; and his tranquil agricultural habits contrast strongly with the roving pastoral habits of the Servians and the Bosnians."[7] The very term "Eastern Question" generically referred to Balkan vicissitudes, which semantically assigned that area to a non-Western sphere. The expression was of course used to describe the various crises associated with the decline of Ottoman sovereignty in other parts, such as Syria or Armenia.[8] However, Russo-Turkish confrontations, as well as convulsions in the Balkans, were focal points.[9] The growing concern in Britain and more generally within the Western powers, over southeastern Europe increasingly drew it closer to the West. Europe moved eastward, as Ottoman authority underwent a long and sometimes violent decay. This process was similar to that which had redefined the position of Eastern Europe in the 18th century from a Western standpoint.[10]

The first challenge that the mid-19th-century cartographer faced when mapping the East was the geographical scope of the Ottoman Empire. It overlapped Africa, Europe, and Asia. Its very capital, Istanbul, was a synecdoche of the sultanate. It stood at a crossroads between continents. Which centering of the map could best render it in the atlases? The question raised by the multicontinental nature of the empire was habitually solved by a division of this political entity. There were two Turkeys on the maps: Turkey in Europe and Turkey in Asia. This dichotomy was an almost systematic feature in 19th-century cartography. Most mid-Victorian atlas makers chose this option.[11] Comprehensive cartographic representations of the Ottoman Empire in its entirety were somewhat less common.

Then, as now, definitions of Europe and Asia were based on a substrate of political, religious, and economic assumptions. The delimitation between the two entities was volatile. The most commonly used delineations were described in the *Gazetteer of the World* in 1856: "On the NW, Asia is separated from Europe only by an imaginary line, the course of which is variously traced by different geographers. The boundary usually assigned to Asia in this quarter is the Arabian gulf, and the isthmus of Suez, which separate it from Africa; the Mediterranean and Aegean seas; the Hellespont, and the Propontis; the Bosphorus of Thrace; the Euxine or Black sea, the straits of Kaffa, the sea of Azoff, the Don, the Wolga to its confluence with the Samara [. . .]. Others draw the boundary line from the straits of Kaffa, up the Kuban and down the

Terek River to the Caspian sea; along the NW shore of that sea to the mouth of the Jaik; up that stream to the Ural mountains, and thence to the Icy Sea."[12]

Most geographical works granted the Ottoman capital city a European component, notwithstanding the fluctuations of imagined delineations. European writers rarely used Istanbul, the native name, which was already well established in the sultanate's official documentation. When used in British of French travel writing, the toponym merely refers to the area adjoining Topkapi palace, in opposition to the more European Pera district.[13] George Fisk's 1884 travelogue is an illustration of this narrow meaning: "The objects of positive interest in Constantinople are, after all, really but few; and such as they are, one looks on them only as remnants of a passing pageant. We made many excursions to Stambool, crossing from Pera, in one of the light graceful caiques, just at the mouth of the golden horn, opposite the Seraglio point."[14] Benjamin Disraeli used the place-name Stambool in his *Tancred* to designate the entire city, but it was merely an exoticizing choice of names. "Istambol" or "Istanbul" were uncommon before the late 19th century. "Constantinople" was an enduring appellation. Its position in the European nomenclature was a reminder of Byzantium, Rome, and Christianity. Significantly, when the Turkish name was considered worth printing on the maps, it was often put between brackets, as shown on the map of Turkey in Asia from the *London Atlas of Universal Geography*, first published by John Arrowsmith from 1832 to 1861.[15] Other European place-names resisted Turkization. The Turkish transliteration of the name Smyrna, Izmir, though obvious to most subjects of the Sultan, was not adopted by British writings until the early 20th century. The resistance of an occasionally obsolete nomenclature made it necessary to publish an English-Turkish dictionary for the use of sailors, who actually needed accurate directions.[16] The reluctance to use Turkish names was also visible in the way major monuments in the Ottoman Empire were called in European literature. St. Sophia in Istanbul was converted into a mosque when the last Byzantine emperor was defeated by Mehmet II in 1453. The Christian name of the monument was persistently used in 19th-century European literature as a means to symbolically reclaim the place.[17] An article from the *British Educator* summarized the prevalent view: "The Turkish geographical names, being often but corruptions of the Greek, are amongst the least striking of those of the oriental nations. The Armenian are nobler and more indigenous."[18] Indigenous place-names were rarely favored when Christian toponyms already existed.

Ottoman Asia, or Turkish Asia, was often subdivided into four regions: Asia Minor, Turkish Armenia and Kurdistan, Al-Jazira or Mesopotamia, and finally Syria.[19] Asia Minor's persistence as a geographical category is another example of the underlying cultural references which presided over the naming of the Orient. The expression was widely used in atlases, parliamentary debates, and newspapers. Its popularity demonstrates how pivotal classical culture and traditional humanities could be in the cultural elaboration of the Orient. It conveyed the souvenirs of the Roman Empire rather than the geopolitical realities of the 19th century. A few mapmakers tried to put forward an updated nomenclature. They felt unsatisfied with the widespread lack of interest for Ottoman place-names in commercial maps in the mid-19th century. Sir Francis Beaufort, a leading figure in British cartography and geography, insisted upon the limitations imposed by such inertia: "The name of Karamania is commonly applied by Europeans [. . .] but, however convenient such a general appellation may be as a geographical distinction, it is neither used by the present inhabitants, nor is it recognized by the seat of government."[20]

The volatility of the Orient as a geographical category was not specific to the 19th century. Mid-Victorian maps echoed centuries of shifting definitions. However, the 19th and early 20th centuries undoubtedly departed from this perspective. The erosion of Ottoman authority both in Greece and in the Balkans and the outbreaks of mass violence which punctuated the agony of the "sick man of Europe" favored a redefinition of the limits between East and West. Stafford Canning, who was the British ambassador to the Porte from 1842 to 1858, summarized this in a striking paragraph: "For the origin and very roots of the question we must refer to the character of the Turks as a race [. . .]. To their Tartar blood they are indebted for the despotic temper which facilitates the exercise of their power, but tends to shorten its duration [. . .]. Their fanaticism impelled them to conquest; their despotism enabled them to hold the conquered in subjection; but the effect of these two principles was to keep them in a state of isolation as to countries not yet brought under their yoke, and of utter antagonism with a large majority of their fellow subjects."[21] Such a Turcophobic rant, from someone as central to the diplomatic relations between the Great Britain and the Ottoman Empire as Canning was, is in itself proof of the diplomat's underlying belief, fostered by many Victorians: the Balkans and Greece were destined to find a way back into Europe.

Indirect signs show that the geographical conception of the Orient was undergoing mutations from a British perspective in the mid-19th century. In most Grand Tour travel books before the 1850s, Athens, the Cyclades, and even Montenegro were indisputably Eastern. The first *Handbooks for Travellers* published by John Murray followed the same path. Greece and the Ottoman Empire were covered by the same travel guide in 1845.[22] Less than ten years later, Greece, Albania, and Macedonia were distinguished from the sultanate.[23] Murray's editorial policy was another symptom of evolving delimitations. Maps and geographical works participated in the evolution of the representations of the Ottoman Empire. They often anticipated redefinitions to come. They carved up the Osmanli Empire years before its fall.

2.2 ASIA AND AFRICA

To many 19th-century Europeans, the Isthmus of Suez seemed an obvious separation between Asia and Africa. It was a convincing natural feature. However, classical legacies were an obstacle to this simplification. Edward Gibbon, inspired by the works of great writers of antiquity such as Pliny or Sallust, did not hesitate to make Egypt Asian: "By its situation that celebrated kingdom is included within the immense peninsula of Africa; but it is accessible only on the side of Asia, whose revolutions, in almost every period of history, Egypt has humbly obeyed."[24] Even if Ptolemy or Strabo fixed the isthmus as the limit between the two continents, many British scholars believed Egypt's fortunes to be closely linked with Asia, whether geographically or politically. Biblical narratives supported this assumption. The book of Genesis states that God gave "this land, from the Wadi of Egypt to the great river, the Euphrates" to Abraham's descendants.[25] The Egypt of the plagues was part of the biblical ecumene. Some, such as George Rawlinson (1812–1902), believed that Egypt's inhabitants "were certainly of Asiatic origin; and the whole of the valley of the Nile has been peopled by the primeval immigration of a Caucasian race."[26] Ethnography complemented scripture and located Egypt in close relation to Asia.

If Egypt was moving in and out of Asia, Sinai's standing proved to be even more complex. Exodus made it a hyphen between Egypt and Canaan, a feature Victorian culture could hardly overlook.[27] Most mid-19th-century maps of the area were illustrations of the scriptures rather than appendixes to modern geographies. A myriad of Sinai-centered maps were thus printed to illustrate scriptural geographies in order to trace Moses's tribulations. This projection transformed the Sinai into an articulation between the two continents, as on the map of the peninsula published by A. P. Stanley in his *Sinai and Palestine, in connection with their history*.[28] The line had in any case to be drawn somewhere. The mouth of the Wadi El Arish, "the Brook of Egypt," was the traditional reference for European biblicists for the location of Canaan's southern end. Where was the limit to go to from there? As seen earlier, it could go toward Aqaba on the Red Sea or to Suez, either in a straight line or not. Depending on the preferred solution, Sinai was African or Asian. The 1877 index to the publications of the RGS even located it in North Africa.[29]

From Whitehall's standpoint, confusion could be helpful. Too categorical a division could have been harmful to British interests. The borders of the Egyptian *pashalik* were never clearly established in the 19th century, leaving the Sinai in a largely undefined status.[30] The London Convention (July 15, 1840) was the first attempt at delimitation.[31] The border was to be laid down between El Arish and Ras Roway on the Red Sea. A firman issued by the sultan confirmed the decision a few months later. Significantly, the map accompanying the official documents was subsequently lost. This accident allowed confusion to reign—a situation beneficial to everyone except the Ottomans.[32] The sultan then granted the Egyptian khedive sovereignty over all the Sinai, Aqaba, and part of the Hedjaz *vilayet* (an Ottoman province) in 1866–67. The border had suddenly moved eastward. This type of uncertainty was at times a very deliberate feature of the British mapping of the East. It left room for negotiation and interpretation. It postponed negotiations and removed obstacles. In the absence of definitive historical, religious or physical borders between Asia, Africa and Europe, several subcontinental divisions were put forward in the 19th century to create what Lewis and Wigen have called a "transitional East."

2.3 NAMES FOR AN INTERMEDIATE EAST: HITHER ASIA, SOUTHWEST ASIA, THE NEAR EAST

East and West were far from fixed categories. A more secular approach to the East, underlining the continuities between the two geographical entities, co-existed with the scriptural viewpoint. The growing concern for the geography of races, or the modernization of transportation, redefined the Western gaze on the Orient. The most popular of these constructs was a subdivision of Asia named Asia Anterior or Hither Asia. Hegel, whose *Vorlesungen über die Philosophie der Weltgeschichte* were translated into English by John Sibree in 1857, did much to revitalize this concept. Farther Asia was characterized as stationary, whereas Hither Asia and the Persian world participated in the revelation of the *Weltgeist*.[33] This appellation was used indiscriminately by Continental geographers.[34] The concept was also adopted by British scholars. An 1864 article entitled "The Races of the Old World" described the political unification of Hither Asia under the reign of the "Aryan" Cyrus in these terms: "A network of administration brought all south-western Asia, from the Indus to the Hellespont, and from the Caspian to the Cataracts of the Nile, into one dominion."[35] Others, such as James Rennell, identified Hither Asia as "being the tract generally between India and Europe."[36]

Persia and the Ottoman Empire were thought to be racially connected with Europe. British maps illustrated this imagined connection on several occasions. An 1856 document showing the global distribution of races, published by Walter McLeod, a fellow of the RGS and headmaster of the Military School, shows "Indo-European Nations."[37] It was certainly inspired by an "ethnographic map of the world" published by A. K. Johnston in 1852 showing an Indo-European continuity from "Iceland to the Ganges."[38] Others, such as David Page in 1864, believed this subdivision of Asia to be a "European-Asiatic transition province" from a zoological perspective.[39] Ethnographic and religious categorizations also underpinned notions such as "the Arabian world," which was popularized by Gibbon, and "the Muslim world," used by Edward Lane in his *Account of the Manners and Customs of the Modern Egyptians*.[40] The conflation of the pluralistic Orient under a single label was therefore well underway before the early 20th century.

The opening of the Suez Canal in the late 1860s constituted yet another factor of geographical redefinition. It involved the use of more utilitarian projections. The Red Sea was no longer a cul-de-sac. The overland route to India was doubled by a sea route. Pre-1869 maps had already been centered on the Red Sea or the Gulf, thus giving some substance to the transitional East connecting the Mediterranean to India. It was the Anglo-Indian input which, more than any other factor, had helped metropolitan cartographers apply new projections.[41] The contract signed in 1837 between the British government and the Peninsular and Oriental Steam Navigation Company established a mail route between England and Calcutta via Egypt. The first permanent service for travelers between Calcutta and Southampton was opened five years later. In 1862, the British India Steam Navigation Company created a service between Bombay and Basra. The Suez Canal gave the region bordering the Red Sea an additional appeal for the mapmaker. New telegraphic lines complemented the process. The route to India progressively became a legitimate cartographic topos.[42] New projections began to be applied from the 1870s onward, substantiating James Wyld's first attempts to demonstrate the emergence of a new communication nexus in the East. F. Goldsmid's 1874 map "Telegraphs of India" was centered on northern Arabia.[43] Years later, J. Bartholomew's *British Colonial Pocket Atlas* displayed another example of this new genre of map showing the modernized route to India.[44] Once again, the transitional East stood at the crossroads of British formal and informal empire in Eurasia. This East was certainly less enchanted than the biblical Orient of the pictorial maps. Its definition also required a new vocabulary. Other toponyms such as Southwestern Asia made their way into the literature of the late 19th and early 20th centuries.[45] The journal of the RGS published various articles drawing on this subdivision. It was described as the territory comprising "all Asia lying this side of the broad continental isthmus which divides the Black Sea from the Persian Gulf."[46] The possibility of an Orient stretching from Cyprus to the fringes of Persia was materializing.

2.4 ENTANGLED TOPONYMS: THE CASE OF PALESTINE

> Palestine Proper had Lebanon on the north, the Mediterranean on the west, the deserts of Arabia on the south, and the river Jordan on the east. The Israelites had, however, a considerable tract of country on the eastern side of the Jordan, extending from Mount Hermon on the north, to the river Arnon on the south. Taking in the whole of this territory, it extends from 30° 40' to 33° 36' N. latitude, and between 33° 45' and 36° 20' E. longitude.[47]

Despite Walter McLeod's definitive delimitation of Palestine in 1858, the region's labeling and definition were no more consensual in the 19th century than in the 20th century. The area is an illustration of how place-names reveal the underlying geographical imaginations presiding over the conception of a territory.[48] The multiple layers of place-names amassed over millennia made it an exceptional geographical construct in the 19th century.

In Britain, because of a widespread biblical perspective, most writers privileged the use of geographical denominations that were inspired by the scriptures, such as "the Holy Land," "the Land of Judah," "the Land of Israel," or even "Canaan." However, less religiously imbued choices of words were also at hand. The Admiralty, for instance, retained "Syria" in its publications. It referred to a vast overlapping area caught between the Taurus and Sinai. It is not surprising, therefore, to find two Palestines in most British atlases: the ancient and the modern. This duality induced a fairly systematic superimposition of the biblical Holy Land on the contemporaneous realities of this territory. The ancient often masked the modern.

The 19th-century maps of ancient Palestine served as a means for religious edification. They often were a combination of text, topography, and images for educational purposes. Gilbert's 1849 *Map of Palestine*, and Edmund Evans's cheaper *Pictorial Map of the Holy Land*, are striking examples of these combinations.[49] Vignettes showing episodes of the Old and New Testaments lavishly illustrated the cartographic document. *Sola scriptura* geography, a purely biblical and historical approach to these lands, inherited for centuries, survived into the industrial age. Most Holy Land geographies from the first half of the 19th century were written by armchair travelers with little regard for scientific accuracy.[50] The advocates of higher criticism slowly transformed the scene in the mid-19th century. David Friedrich Strauss's *Life of Jesus*, a pivotal work in the historical interpretation of the Gospels by a controversial German scholar, was translated into English by George Eliot in 1846.[51] Renewed debates on the verisimilitude of the scriptures entailed a different balance between positivism and theology.[52] Contending Catholic and Protestant geopieties of the Holy Land eventually fostered modernized approaches to the geography of Palestine.[53]

Place-names were instrumental in these enterprises of intellectual repossession of the sacred land. In 1889, a volume indexed hundreds of ancient place-names with their contemporary identifications.[54] This minute toponymy was an instrument used for the confirmation of biblical records. Edward Palmer (1840–82), who joined the Sinai survey in 1869, insisted on the fact that "Arab tradition is 'fossilized' in their nomenclature, and often furnishes undying testimony to the truth of Scripture."[55] The PEF collected as many Arabic toponyms as possible, proceeding primarily from an etymological standpoint. British biblicists believed "traditional Latin topography of holy places untrustworthy, as well as oral traditions of the local Christian population. Local Arabic traditions were their preferred sources."[56] The transliteration of Arabic place-names was a time-consuming process for the surveyors. They distrusted the Bedouins. Charles Warren explained in an article how he systematically cross-checked the information provided by the local inhabitants. Both he and his dragoman collected the place-names from carefully selected interlocutors. Their lists were compared every night. A specialist in Arabic dialectology, Dr. Sandreczki of the Church Missionary Society in Jerusalem, reviewed the data.[57] Warren underlined that the informers could be contradicting one another. Names considerably varied. Two reasons account for these uncertainties. First, the local inhabitants had their own understanding of their territory. It followed a logic that differed from Western approaches to the land. A river could have many names, depending on where you were. Second, local inhabitants kept part of their knowledge to themselves for fear of Ottoman intrusions. They also distrusted British surveyors, because they suspected Westerners wanted to measure "the Peninsula with a view to purchasing it for the Christians, and ejecting the true believers of their patrimony."[58]

It was through these filters that modern Palestine made its way into the mass of British scriptural geographies. Its evolutions were thus overlooked, even when it was no longer the deserted space described in early travelogues. In the second half of the century it was in fact a rapidly changing reality. A stronger Osmanli rule pacified the land from the 1860s. Economic and demographic development ensued.[59] Palestine was partly an administrative reality under the Ottomans. What the Europeans called Syria or Palestine was divided into administrative entities called *sanjaks*. The Holy Land was composed of three districts: Jerusalem, Nablus, and Acre. Almost no reference to them is to be found in European maps. British cartography took little notice of these evolutions. Biblical concerns appeared to be insurmountable obstacles to a fully updated visualization of the area: terrestrial Palestine was buried under the rubble of the Holy Land.

CONCLUSION

Mid-Victorian British maps of the East were caught between the sacred and the secular. The biblical perspective shaped the Victorian geographical imagination. Scripture coalesced Palestine/Syria, Egypt (Exodus), Arabia (Kings, Isaiah), Mesopotamia (Jonah), and Anatolia and Greece (St. Paul) into one historically coherent area. Echoes from the past were not only sacred; for many British agents, especially in India, Alexander the Great was also a model. His empire, which extended from Egypt to India, was one of the early references many officers had in mind.[60] Edward Weller's map *Kingdoms of the Successors of Alexander* sold with the *Weekly Dispatch* in the 1860s illustrates how historical landmarks could frame the area.[61] There again, history and education played a part in shaping embryonic delineations of the Middle East. Gibbon's outlook on the development of the Roman Empire in the eastern Mediterranean was part of the intellectual baggage of the educated Victorian as well. It substantiated the historical and geographical homogeneity that could be assigned to the transitional East by 19th-century map readers. Modernized approaches were yet to prevail. The novelty of the Suez route had not erased the traces of Moses's Sinai from the Victorian standpoint. 19th-century realities were incidentally reflected on the maps which functioned as the "theatres of memory" Raphael Samuel so brilliantly analyzed.[62] Mid-Victorian maps of the Orient were very often about placing the past. However, the secularization of the area's geographical depiction was in its early stages. The East was still the Orient, full of reflections of a biblical and bygone past. Names on the maps as well as territorial delineations often located a scriptural space, a romantic and pictorial area. Place-names, both in Palestine and in the rest of the biblical Orient, participated in the rhetorical construction of a parabolic space, not yet a colonial space. It was rarely shown as a contemporary political reality, brought closer to the West by the opening of the Suez Canal and commercial treaties. Westernization had no yet redefined the position of the area in British geographies. As much as we would expect maps to be instruments of a wider culture of persuasion that furthered the virtues of modernity for the East, toponymy and delineations labeled an imagined space enmeshed within a network of historical references. Names on the map reflected the mapmakers' anticipation of readership interest in what the East was, not what it could be. Significantly, autochthonous place naming was systematically distrusted despite its potential usefulness. The suppression of the Bedouins' knowledge by the PEF's surveyors and the professed distrust for Ottoman nomenclature exemplify how "out of place" local inhabitants and their experiences could be comprehended to be in this regard. For most, it was apparently not yet necessary to use maps to actually travel and locate oneself. Nomenclature located a ubiquitous and spiritual Orient. This layer of representations nonetheless coexisted with emerging redefinitions of the East. Embryonic constructs tried to answer the need for clearer distinctions within Asia. Ethnography, strategy, and trade participated in the elaboration of new continental divisions. As India's crucial situation in the imperial system became more and more obvious after the Mutiny, what lay between the Raj and Europe could not be a vague land anymore.

European place labeling gave rise to the existence of places over a long time frame. The homogenization of nomenclature through the elaborate mechanisms of transliteration and through the application of standardizing schemes had more to do with the history of the area than just the appearance of names in letterpress on maps. Naming Syria, Arabia, Palestine, or Mesopotamia indirectly participated in the shaping and invention of the territories themselves. Johan Büssow has demonstrated, for instance, that the Ottoman government reacted to the growing European religious and intellectual interest in the Holy Land. High-ranking members of the sultanate had several discussions in the second half of the 19th century on the implications of the creation of a united Palestinian province. The proposed reconstitution of an area that was historically characterized by its chaotic administrative structure and its lack of coherence as a territory from an Ottoman standpoint was not foreign to the resurrection of the Holy Land as imagined by Western scholars.[63] The use of the phrase "Ard Falastan" on the map of Asia Minor ("Bar Al-Sham") published in the *Cedid Atlas* (1803) confirms the imprint of European topographies on Ottoman conceptions of the region. It was translated from British references, such as

William Faden's *General Atlas*.[64] Nineteenth-century uses of the word "Filastin" in Ottoman cartography were thus partly reflecting European uses. The same reasoning applies to the naming of places and populations in Mesopotamia, and specifically to the ethnographic classifications of Eastern Christians. Labeling played its part in the actuation of an ethnographic diversity that was not necessarily crucial to local experiences of place, with far-reaching consequences on how various populations perceived themselves in the long term.

CHAPTER THREE

MAPS FOR THE MASSES?

On the 12th of August 1854, the many readers of the *Illustrated London News* (*ILN*) could unfold a large map of the Ottoman Empire (fig. 3.1). The pictorial weekly newspaper, founded in 1842, had surpassed the sales of the London *Times* a few months before thanks to its unrivaled coverage of the Crimean War. Its circulation peaked at 150,000 copies during the conflict. Not only did it offer its readers engravings showing the valley of Death, where the charge of the light brigade tragically ended, and others battlefields of the Crimea; it also displayed numerous maps of the East, most of them specifically designed for the weekly. This *Map of the Ottoman Empire, Kingdom of Greece and the Russian Provinces on the Black Sea* was the work of an engraver and mapmaker, John Dower (1825–1901). He published his own *School Atlas* while also working for renowned establishments such as Weller and Petermann.[1] The cartographic document was mostly based on earlier work for his atlas. The centering of the map is a reminder of the projections used by Kiepert and Justus Perthes on slightly earlier maps of the Ottoman Empire.[2] The mapmaker had to adapt to circumstances in order to make the confrontation between the sultanate and the Russian Empire visible. The integrity of the Ottoman Empire, often inevi-

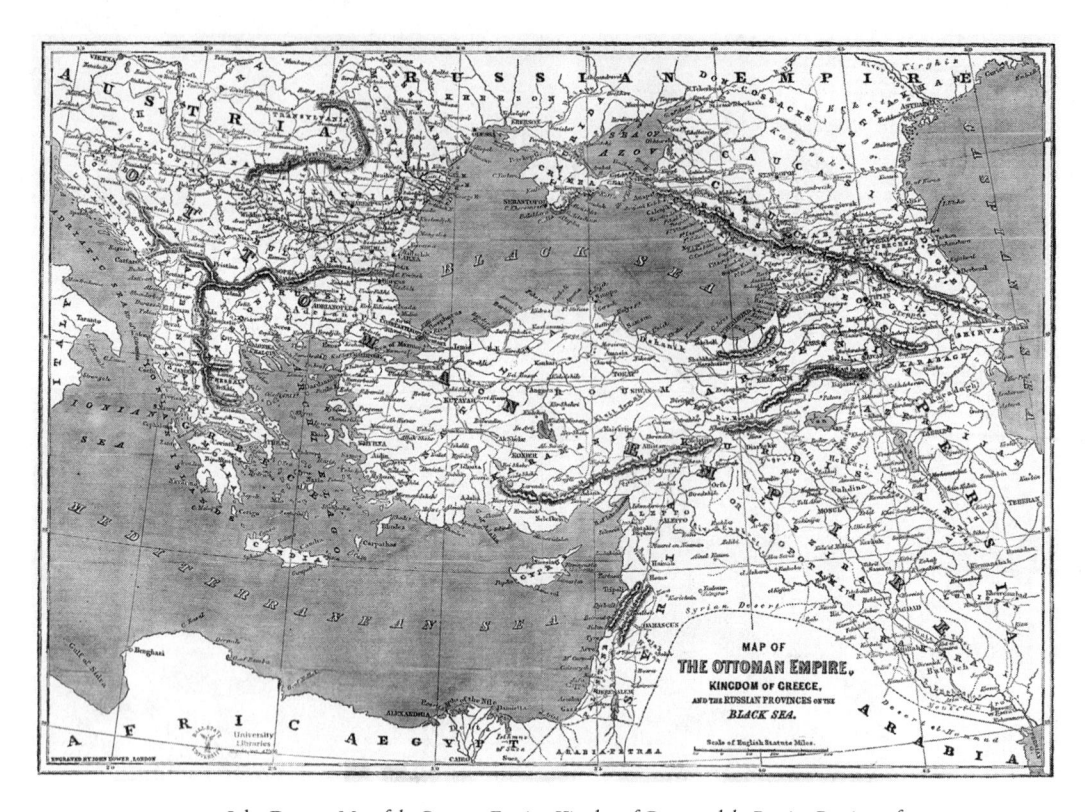

FIGURE 3.1. John Dower, *Map of the Ottoman Empire, Kingdom of Greece and the Russian Provinces of the Black Sea*, ca. 1:5,900,000, in *The Extra Supplement to the Illustrated London News*, August 12, 1854. Reprinted in *The Extra Supplement to the Illustrated London News*, April 21, 1877. Size of the original: 36 × 52 cm. Copyright 2013, Ball State University. All rights reserved (G7430 1877.W5).

tably divided into Turkey in Asia and Turkey in Europe by the very same atlases which Dower worked for, was preserved. The very label was a hyphen on the map between the two continents, running from Croatia to the Persian Gulf. A closer look at the document shows its hesitations. The map was not purely political. Occasional ethnographic indications were to be found. The feared "Don Cossacks" were threatening the Caucasus, "independent tribes" lived in Kurdistan, and "Montefik Arabs" ruled the marshes at the mouth of the Tigris-Euphrates: a welcome touch of exoticism to glamorize the map for the middle-class audience which could afford the sixpence copy of the newspaper.

The document and the many subsequent maps printed by the *ILN* fostered the geographical literacy of its readership. The war opened the East to a generation.[3] The events in the Crimea nurtured a widespread Turcophilia in Britain. Dower's map was not an isolated instrument in the popularization of both geography and ethnography. Marx himself contemplated the enthusiasm of the liberal press for the Ottoman Empire with some perplexity: "[It] goes on in dithyrambic strain; so far as the *Daily News* can be dithyrambic, with an apotheosis for Turkey, the Turks and everything Turkish."[4] A market was building up, as exemplified by the success of Christopher Oscanyan's and Serovpe Aznavour's Oriental and Turkish Museum in London. The two entrepreneurs displayed an Oriental version of Madame Tussauds's wax museum in St. George's Gallery, Piccadilly. The number of visitors of this ethnographic exhibition, which had both wax figures and live dancers, soared thanks to the war.[5] The geography of the Crimean War was never overlooked. An audience eager for novelties could embark on Charles Marshall's *Grand Moving Diorama of a Tour from Blackwall to Balaclava* in Leicester Square or travel East thanks to Albert Smith's *Diorama of Constantinople* in the Egyptian Hall.[6] Another testimony to the growing interest of the Victorian public in a leisurely education on the Turkish world could be found in the variations on the genre of the "map of the seat of war." *Stanford's Bird's-Eye View of the Seat of War in the Crimea* went through several editions during the war.[7] James Wyld published his own map of the area, which drew extensively on Thomas Jervis Best's work.[8] The Crimean War and the early 1850s marked a shift in the popularization of the map. The East was only one of the many objects to be displayed by a press imbued with midcentury educational ideals. It did, however, hold a particular position, because it continually provided Fleet Street with sensational news, crises, tales, and opportunities to unveil the region through maps and exotic illustrations. The increasing circulation of maps exposed a larger audience to geographical knowledge.

This chapter examines the reception of these cartographies and geographies of the Orient by the wider public in the 1850s and 1860s. The first section considers commercial mapmaking and the contrasted history of its gradual modernization and inherent conservatism. It demonstrates that British mapmakers' pretensions to modernity were often contradicted by cost-related restraints and lack of updated data. The following sections provide a detailed examination of how maps became integrated into increasingly widespread social practices. They explore the less scholarly forms of maps with a view to demonstrating how the exhibition of the world's outlines became one of the facets of 19th-century leisure. Another focal point of this chapter is the issue of geographical literacy and the educational uses of maps of the East. We shall discover that that the Orient held a specific position in British curricula despite the very uneven opportunities offered to mid-Victorian pupils.

3.1 AN EVOLVING TRADE

There was a geography of British mapmaking in the 19th century. Knowledge has its loci. London map printing clustered in the Charing Cross area, where the establishment of Thomas Jefferys, a well-known cartographer in his days, already had its premises in the mid-18th century. The most famous establishments were a stone's throw away from each other. Charles Smith, "Map Seller Extraordinary to HRH the Prince of Wales," James Wyld's establishment, Trelawney William Saunders's shop, which was taken over in 1853 by Edward Stanford, A. Petermann's Geographical Establishment, all stood there in the 1850s. The English capital was not the only cartographic center. Edinburgh was the other heart of British mapmaking, where two dynasties, the Bartholomew and Johnston families, developed profitable cartographic establishment in the 19th century.

Such business clusters are an ancient feature of city life. In the mapmaking business, they favored interconnections which were to prove vital for the first stages of the elaboration of a new cartographic document. The Royal Geographical Society held its meetings in Regent Street, minutes away from the Charing Cross area. Most of these mapmakers were members of the society and represented vital elements of a network of learned societies. These institutions provided them with updated information and unpublished material. Some of them had direct connections with the explorers on the ground. A 1860 subscription list to fund John Petherick's expedition from Khartoum up the White Nile shows the names of A. K. Johnston and Aaron Arrowsmith, for example.[9] Nor was Whitehall very far either: these cartographers worked directly for the government offices which externalized part of their production. Some of them were official sales agents for governmental publications. Stanford, for instance, sold Admiralty charts and IO maps. The commercial cartographer could make the best of these concentrated information inputs. They were also in touch with German and French colleagues. Edward Stanford and John Bartholomew Jr. (1831–93) both collaborated with August Petermann, an influential German cartographer who had worked in Edinburgh at W. and A. K. Johnston, before spending time in London between 1847 and 1854. The largest establishments were thus able to create reference libraries, with a variety of foreign material, for their own compilations. Bartholomew's collection of maps and plans for Turkey comprised Kiepert's *Turkischen Reich in Asien* (1853) and *Map of the Caucasus and Armenia* (1854), Reimer's *Karte von Klein-Asian* (1854), and Justus Perthes's *Karte von Armenien, Kurdistan und Aberbeidschan* (1858).[10]

Not only were these eminent mapmakers pivotal in a growing network that was producing geographical information in the mid-Victorian period; their craft also saw a remarkable transformation for several reasons, including changes in both supply and demand. New print technologies allowed for an industrialization of mapmaking. In 1864, one well-established company was described thus: "The most extensive establishment in this variety of the publishing trade is that of Messrs. W. and A. K. Johnston, which gives employment to about 160 persons, and keeps in use 48 presses [. . .]. The departments carried on in this large concern comprise engraving on steel, copper, and zinc; copper-plate, lithographic and letter-press printing; map coloring and mounting, globe-making; constructing plans, &c."[11] New printing processes changed the mapmaker's business in the mid-19th century.[12] The traditional intaglio on copper plates was still used.[13] It had the advantage of allowing easy corrections and detailed etching. This solution was adopted for the most complicated and costly maps. However, hampered by the technical limitation resulting from the fact that after a few thousands potential impressions the soft copper lost its sharpness, the process could not compare with lithography and its almost endless number of copies. The planographic method used in lithography was much more productive than intaglio printing. The engraver drew directly on the stone, which was then etched with acid. The generalization of steam-powered automated presses improved the printers' output. Chromolithography revolutionized color printing as well.[14] Transfer lithography, which was introduced from the mid-19th century, could save the time-consuming process of reverse drawing on the stones. It simplified the engraving phase, which was time consuming and a source of errors. Henry James, T. J. Best's successor as the head of the Topographical Department and director of the Ordnance Survey, claimed to have invented a new process using much lighter zinc plates.[15] This process, called photo-zincography, was designed to make accurate copies of any type of documents. The image, a carbon print of a photographic negative, was transferred onto a zinc plate. It was first designed as a cheap process to accelerate the map production of the Ordnance Survey. The alliance between printing and photography progressively caused engraving to be phased out. The more cost-effective photomechanical processes became widespread among the British commercial mapmakers from the 1880s.[16]

The cartographic document, once a costly vector of social prestige, became cheaper and more accessible. A 1864 article from the *Bookseller* celebrated this development: "Mr. Alexander Keith Johnston has devoted himself for more than a quarter of a century to the advancement of our knowledge of the earth we inhabit [. . .]. In all forms and prices, from the small quarto map, selling at sixpence, up to the splendid folio atlas at twelve guineas, he has, during that period, been constantly before the public as a constructor of maps, and has undoubtedly done more to popularize geographical knowledge in all departments than any man now living."[17] The author may have been somewhat overenthusiastic. A sixpence map—a little more than one pound in 2014—was nothing like a popular commodity. It was not, however, an aristocratic artifact anymore.

Was the East in particular, and the world in general, brought to the general public in both a more popularized and a more systematic way? Such narratives of technological progress and scientific accuracy were also a dimension of the mapmaker's business. They often advertised their productions as "compiled from the more recent authorities." None but the lexical field of modernity was used in the promotion of their prints. However, contradictory forces sometimes froze the mid-19th commercial map into an obsolete document. In order to protect their profit margins and sell less expensive maps, many well-established mapmakers resorted to the recycling of old copper plates. The compilation and engraving of a completely new cartographic document were the costliest stage of mapmaking. A preexisting copper plate could be altered and corrected at much less expense than that required for the elaboration of a completely new document. The most outdated elements of the engraving could be erased by hammering and repolishing. The worn-out plate could also be copied.

Some plates consequently enjoyed a considerable life expectancy. The protracted story of the origins of a mid-1850s map of the Ottoman Empire by Wyld is a good example of these reuses. Wyld's father, James Wyld the Elder (1790–1836), bought William Faden's establishment in 1823. His predecessor's stock of plates proved to be valuable. Wyld made sure it continued to yield a profit for decades. In 1824, he printed a hand-colored map titled "European Dominions of the Ottomans or Turkey in Europe," which was a corrected version of a document published thirty years earlier by Faden.[18] The identical illustrated cartouche can be seen on the lower right corner. James Wyld the Younger eventually published a new edition of the map in 1842, with very few alterations. He printed a last edition of the map in the mid-1850s, despite its obvious obsolescence (see fig. 3.2). A careful look at the document shows that Wyld the Elder and Wyld the Younger made few changes to the map, except for the lettering. They polished part of the vignette to engrave their names. The puffs of smokes from the cherub's pipe were visibly deleted in the process. The last version of the map shows the alterations of the plate. Details were getting fainter due to the excessive number of prints. One of the main additions to this last edition was a line on the map showing the new Smyrna-Aidin railway.

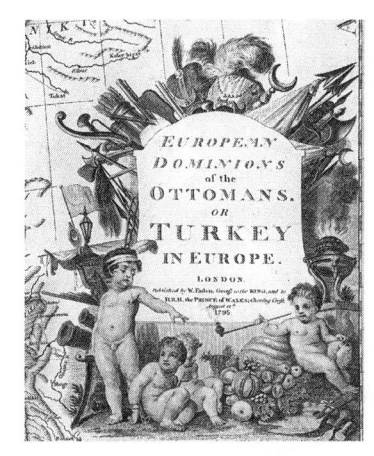

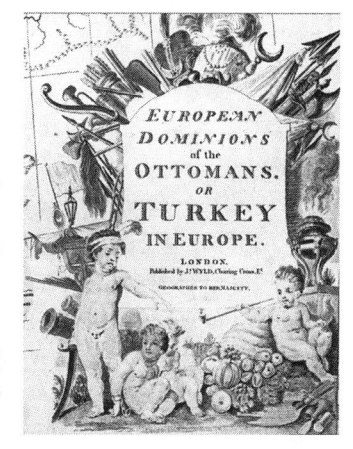

FIGURE 3.2. *Left*, William Faden, *European Dominions of the Ottomans or Turkey in Europe* (detail), 1:2,900,000 (London, 1795). Photograph courtesy of the David Rumsey Map Collection (Image no. 2104040). *Center*, James Wyld the Elder, *European Dominions of the Ottomans or Turkey in Europe* (detail), 1:2,900,000 (London, 1824). Photograph courtesy of Wayfarer Books. *Right*, James Wyld the Younger, *European Dominions of the Ottomans or Turkey in Europe*, 1:2,900,000 (London, ca.1854). Photograph courtesy of the BNF (Cartes et Plans GE DD-1855).

The concession for the new junction was granted in 1856. The sultan had decided to implement reforms, the *Tanzimat*, to modernize the Ottoman Empire. His intended audience in this case did not care for a completely up-to-date map. The little more than sixty-year old document was thus slightly updated.

Wyld's *Ottoman Empire* illustrates how slow and nonlinear the production of cartographic information was. His map was a palimpsest echoing 17th- and 18th-century visualizations of the East. Small-scale maps of the area were often compilations of decades of topographic works. Many of the maps I refer to in this work were reused in following works for a generation or more for lack of any useful updates. That is why the study of 19th-century maps is essential to the understanding of the deep history of the "Middle East" taken as a geographical object. Cartography from that regard was not different from other forms of representation of the area. Travel narratives or early photography endlessly recycled the same tropes and patterns. As a consequence, 18th-century ornate mapping survived well into the 19th century despite the mapmakers' claims to accuracy in the industrial age. From that regard, Wyld's map not only testifies to the haphazard evolution of cartographic techniques, but also reflects the nostalgia and inertia that shaped part of the geographical figuration of the East. This is one of the many examples of enduring echoes of earlier cartography in the supposedly more accurate commercial maps of the mid-19th century. For instance, outdated maps of the Holy Land circulated well into the 1880s despite the unparalleled accuracy of the ordnance survey of Jerusalem and the Sinai in the 1860s. Such documents substantiated an antiquated Orient rather than a well-surveyed East.

Another way to save money on the compilation or the engraving stage of map production was to buy existing plates from foreign producers and change the letterpress. The less decent mapmaker could even infringe copyright and produce a copy of a competitor's map. This time-saving ploy was not uncommon. In 1853, A. K. Johnston brought a case against A. Fullarton in Edinburgh. The latter was charged with piracy, his *Companion Atlas* maps being obvious copies from A. K. Johnston's *National Atlas*. The jury found in favor of the claimant, and Fullarton was held liable.[19] Mergers provided a more legal solution than piracy for the acquisition of a stock of already existing plates. John Bartholomew Jr. considered for a time the option of buying George Philip's establishment.[20] For all these reasons, the commercial mapmaker was not a transparent intermediary between information collection and map creation. He translated data, while taking into account the need for a profit. The claimed accuracy and modernity of his maps could amount to nothing more than publicity.

Did this mean that a renowned mapmaker such as Wyld was not able to print a fully updated map? All depended on how he understood his clientele's expectations. Obsolete, multiauthored cartographic documents coexisted with the newest compilations. Behind the semblance of unbiased truth, contradictory forces determined the complicated production of tradable geographical and cartographic facts. Maps were not standardized. They echoed the variability of the expectations and tastes of multiple target audiences. The commercial mapmaker produced as many Easts, Americas, or Chinas as necessary to meet what was often a heterogeneous market. Wyld also published up-to-date documents on the region insofar as fresh information was available. His 1853 chromolithographed *Map of the Ottoman Empire, the Black Sea and the Frontiers of Russia and Persia* was compiled from very recent works such as Edouard Taitbout de Marigny's *Pilote de la mer Noire et de la mer d'Azov* as well as "the consular reports and the information furnished by private individuals."[21]

If we adopt an overview of the demand, it is obvious that map consumption evolved from the 1850s onward. The 1850 Public Libraries Act, and the end of taxes on knowledge, were among the many signs of changing attitudes to the question of public access to information and education. British mapmakers seized these opportunities to expand their market. Stanford, for example, created a collection of maps specifically designed for the newly built public libraries.[22] Its monumental *Library Map of Asia* was thus published in 1862, specifically designed to be displayed on the walls of reading rooms around Britain.²³ Philanthropic enterprises played another central role in developing the geographical literacy of the general public. From 1829, the Society for the Diffusion of Useful Knowledge produced more than two hundred plates and sold more than 3 million maps until the series were sold in 1856 to E. Stanford, who recycled them in various atlases.²⁴ The Society for the Promotion of Christian Knowledge published its own atlases and maps, mostly supplied by Stanford. For the promoters of "useful knowledge," maps were an ideal artifact: "One thing is indispensable to the study of geography, and that is good maps [...]. In the copiousness and completeness of the information it communicates, in precision, conciseness, perspicuity, in the hold it has upon the memory, in vividness of imagery and power of expression, in convenience of reference, in portability, in the happy combination of so many and such useful qualities a map has no rival."²⁵

Both societies sold their maps at very competitive prices. The Society for the Diffusion of Useful Knowledge's *Companion Atlas*, published by Charles Knight, was sold in twelve parts. Each part, containing two or three maps and part of the index, cost one shilling. Were their educational endeavors successful? It is doubtful whether the maps reached their targeted audience. The SDUK's publications were primarily bought by the urban middle class, or the working-class aristocracy. Dividing lines between the learned and the less educated public resisted the generous assaults of the philanthropists.²⁶ However, their educational activism did contribute to a widening of the geographical interests of the general public. The East in general and the Holy Land in particular were the specific focuses of these enterprises. Evangelical philanthropy underscored the mapmaking achievements of the Society for the Promotion of Christian Knowledge.

Religious concerns and momentous international crises placed the East on the map. Its relative position in this booming mid-Victorian geographical interest was substantial. The rise of a middle culture from the 1850s onward helped to redefine the status of the cartographic object itself.

3.2 A PROCESS OF COMMODIFICATION: MAPS OF THE EAST AND LEISURE

> SCENE I.—*The Front of the Haymarket Theatre.*
> *The Box Book KEEPER standing at Box Office door.*
>
> [...]
> **BOOK KEEPER**: What have you got Under your arm?
> **PROPERTY MAN**: Of books a precious lot.

AUTHOR: What are they, pray? and who may they be for?
PROPERTY MAN: All Murray's Handbooks for the Governor. Germany, North
   and South,
France, Holland, Spain, Switzerland, up the Rhine, and back again—
Italy, Russia, Egypt, Turkey, Greece—
Some of 'em 12 or 14 bob a-piece.
And now I've got to go for loads of maps,
To Wylde and Arrowsmith's, and all them chaps.
I don't know what the Governor's about,
But if he can't an Easter piece bring out
Without this bother about foreign climes,
I'd rather get up fifty pantomimes.[27]

Mr. Buckstone's Voyage round the Globe was an extravaganza by James Robinson Planché (1796–1880), shown at the Haymarket Theatre in London from April 12, 1854. It starred John Baldwin Buckstone (1802–79), a famous comic actor of the Victorian age. An eager audience was promised "Constantinople and the Golden Horn; Mr. Buckstone visits the theatre of war, but without the slightest intention of managing it; Gallipoli and landing of the French forces; Mr. Buckstone retreats without coming to an engagement, and crossing the Dardanelles, continues his journey by the 'Overland Route' to Asia; a grand oriental spectacle; Bayaderes, Miss Lydia Thompson, Miss L. Morris and the Corps de Ballet." In the first scene, interestingly, maps and guidebooks became comical artifacts. Referential humor rarely survives the test of time for the good reason that common references are forgotten and are replaced by a new and different set of connections. It is thus a very good record of what was part of the emerging popular culture elaborated in London theaters in the mid-19th century. The trusty Planché with his dozens of plays and revues knew how to make people laugh, and the amusement, in the spring of 1854, revolved around the growing curiosity about the geography of the East. Far from anecdotal, this extravaganza demonstrates how the crisis in the East had made maps and geographical writings much more commonplace objects than before.

The cartographic representation of the world also had a specific location in the culture of display which was typified by the Great Exhibition of 1851. As the masses flocked to see the wonders of the world in the Crystal Palace, another very popular venue was inaugurated in Leicester Square by James Wyld, a mapmaker of renown and member of the Royal Geographical Society. It was a gigantic hollow globe, a georama more than sixty feet in diameter.[28] The visitor paid one shilling to enter the structure, where he could first walk through an exhibition of maps before exploring the main building, where the surface of the earth was modeled in plaster on the inside of the sphere. The visitors exited the globe into a shop where Wyld's most recent productions were on sale: Wyld's Great Globe was a commercial venture above all.[29]

The Orient was, of course, part of the display. Wyld's guide to his Great Globe described how "Mesopotamia, or Al-Gesira, as it is now called, lies between the Euphrates and the Tigris, and contained the Plain of Shinar, the great cities of Babylon, Harán, and other places mentioned in the Old Testament. On the east bank of the Tigris, and in the immediate neighbourhood of Mosul, the chief city of Al-Gesira, is the village of Nounia, occupying part of the site of the Ancient Nineveh, whence Layard has recently recovered such precious relics [. . .]. Lying between that and Persia, is the province of Kurdistan, in which the Kurds are ever at war."[30]

Wyld mentioned Layard's Assyrian discoveries, which attracted more than six hundred thousand visitors to the British Museum, where the human-headed bull from Nineveh was displayed in 1851. The discoveries had revived the enthusiasm for Oriental tales. Wyld's attraction was immensely popular, with hundreds of thousands of people visiting his educational globe.[31] The exhibition garnered rave reviews, as exemplified by Eliza Cook's remarks on the globe: "No method of teaching geography in a summary view has yet been devised more effectual than this extraordinary globe of Mr. Wyld's [. . .]. It is geography anatomized and dissected in the most beautiful manner [. . .]. Were it possible to have a similar globe erected in every large town, it could not fail to prove eminently instructive to the young as well as the old, in conveying to them, in the most striking manner, a direct and accurate knowledge of the geographical relations and conditions of the great globe on which we live."[32]

Leicester Square was not the only place in Britain where one could explore the East without leaving home. London had several dioramas of Palestine and Jerusalem.[33] In Manchester, pyrodramas in the *Zoological Gardens* reenacted the battles of Sebastopol and Malakoff.[34] A relief map of the Holy Land, modeled and embossed by Henry F. Brion, was displayed around England in the late 1850s and early 1860s.[35] The PEF organized school visits to see an exhibition by the Ordnance Survey on its expeditions in the Holy Land during the summer 1869. The geography, landscape, and outlines of the East became a commercialized entertainment. Maps in particular had the advantage of being both a vehicle of epistemic authority and a potential spectacle for the growing mid-Victorian leisure market. Wyld's Great Globe, like the relief maps exhibited in London dioramas, was categorized as an instance of the "new amusements [. . .] daily springing into existence," according to Henry Mayhew's *The*

World's Show, 1851, or, Adventures of Mr. and Mrs. Sandboys.[36] They stood at the intersection between the fairground world and scientific observation of the globe. They were a combination of educational ideals and a respectable amount of entertainment.

Cartographic toys such as Henry Smith Evans's *Crystal Palace Game* are a testimony to this transformation of the map into an amusement.[37] Evans, a fellow of the RGS, was famous for his emigration maps and guides. This board game, published by Alfred Davis in 1854, invited the player to embark on "a voyage around the world, an entertaining excursion in search of knowledge whereby geography is made easy." The East holds a decisive position. The player toured Arabia through squares 19–22. The artifact was an updated and more democratized version of previous forms of cartographic recreations such as John Wallis's *Complete voyage round the world: a new geographical pastime* (1796). Evans's game celebrates modernity, in the shape of an interconnected British colonial empire. However, it is also a reminder of more antiquated forms of cartographic expression. The vignettes, showing, for example, Arabs and camels in the desert, are reflections of early 19th- and 18th-century prints and classic Orientalist discourses. Another example of these attempts to transform geography and topography into more popular commodities was John Betts's portable globe. It functioned as an umbrella, with a steel wire frame. It could be collapsed and opened at will. Betts's apparatus sold at twelve shillings and sixpence in 1867, a rather inexpensive alternative to the costly traditional globes.

The anecdote is just one of a multitude of signs pointing to the growing popularity of geography and maps in the 1850s. Maps became elements of a growing leisure market. Cartography's hermeneutic process was in fact more and more relevant to the daily life of individuals living in an industrially developed society. The transportation revolution, telegraphic communication, photographic visualizations of the world, and a wide array of other factors brought wide audiences in touch with distant regions of the world. The inherent authority of maps, their capacity to authenticate and assert spatial facts, placed them at the center of the collapsing of space and time that characterized British lives in the 19th century. Travel guides, railroad maps, and city plans became familiar objects. This extension of the uses of spatial visualizations brought about a broad experiencing of the world through the cartographic prism. However, as the next section shows, this first popularization was of course far from complete, and its very nature was contradictory, as exemplified by the uses of the map in the press.

3.3 THE VISUALIZATION OF THE EAST IN THE PRESS

The media revolution which took place in the 1850s is a well-known phenomenon. A variety of factors combined to democratize the news. Taxes, such as the stamp duty on newspapers, were gradually abolished. The ever-improving mechanization of the printing presses was a wonder of the age. The rotary steam press had been progressively adopted from the 1830s. Charles Knight, the publisher for the SDUK, was one of its staunchest advocates. In 1856, *Lloyd's Weekly Newspaper* was the first British client of the American Robert Hoe, the inventor of a high-speed press which could print as many as twenty thousand copies per hour. The very contents of the papers had changed. The *Illustrated London News* and its competitors, such as the *Graphic*, founded in 1861, provided their readers with an unusual wealth of images. Legions of newspapers emerged and died as new audience segments turned to daily and weekly publications. Circulation of the daily press reached hundreds of thousands of readers.[38]

How did the development of a mass reading public affect the imaginations of British people regarding the East? Answering this question might prove impossible for lack of documentation. There are traces of how ordinary people read and used atlases and maps, but they are too disparate to provide a comprehensive picture.[39] What can be ascertained are general discourses of geographical knowledge. Cartographic documents were very appropriate vehicles for the educational ideal which presided over the publication of the first illustrated papers. However, Michael Heffernan has shown that this first popularization of the map by the Victorian press was yet to be fully accomplished in the third quarter of the 19th century. Inserts of maps were still intermittent before the 1880s.[40] As stated earlier, the Crimean War provided one of the first opportunities to improve readers' geographical literacy. The *Illustrated Times* produced a striking bird's-eye view of the peninsula.[41] The prototypical coverage of the conflict by major London papers by means of engraved copies of photographs and an abundance of maps even influenced the less affluent local press. In May 1854, the *Kentish Gazette* printed its own "map of the seat of war," whereas the even more adventurous *Chelmsford Chronicle* was able to present a bird's-eye view of the battlefields in January 1855.[42] Map printing in the papers peaked during the conflict. It created a precedent for the next two decades. The costly foldout map and the technically challenging insert were routinely provided to the readers when major conflicts erupted. Sales were then high enough to keep the publication of original maps profitable. The American Civil War, the

1859 Italian War of Independence, and the Franco-Prussian War were given the same coverage.

Maps of the East did not disappear completely from the press after the Crimean War. Illustrated papers printed several cartographic representations which conveyed other perspectives on the region. For instance, a map of Lower Egypt, "with the latest corrections by Captain Bedford Pim," was printed in the *ILN* in July 1859. De Lesseps's scheme had sparked much debate in the late 1850s. Bedford Pim (1826–86) had surveyed Egypt in December 1858 to assess the feasibility of the transoceanic canal. He had presented his findings to the RGS a few months later.[43] These were not in favor of the scheme. The weekly adopted his conclusions. The *ILN* map showed the projected railway lines between Cairo and Suez. An alternative route was suggested by Pim and some of the RGS members. A few years later, the *ILN* published a "Telegraph Map of the Eastern World," on July 8, 1865. The projection was centered on India. The map publicized the recent accomplishments of the engineers working on the Indo-European telegraph line.[44]

The slow adoption of maps in newspapers was acknowledged by contemporaries as a remarkable evolution in the press.[45] A 1861 article from *Chambers's Journal* acknowledged the fashion: "It is a peculiar feature of our modern newspaper system, especially that of the Sunday newspapers, that subsidiary attractions are repeatedly offered to the public, intended to implant in the minds of the purchasers the idea that the proprietors are never so happy as when they are giving something to an enlightened British public [. . .]. Another case is that of a Sunday newspaper which, three or four years ago, began the presentation of a map every week, without any augmentation of price."[46] The author refers to Edward Weller's work. Weller was a fellow of the RGS who started the publication of a serialized atlas in 1855. The increasing flow of news bolstered the demand for cheaper cartographic information. His *Weekly Dispatch* met it. Each issue included a folio map, which was a powerful incentive to buy the weekly. It sold more than sixty thousand copies each week. The regular customer could eventually acquire the entirety of *Cassell's Weekly Dispatch Atlas*. The cartographic documents were elaborated with up-to-date data. Weller used this network to compile recent information. His 1858 maps of "Syria showing all the ancient sites hitherto identified in Palestine" in three sheets included data gathered by the ordnance survey of Jerusalem.[47] He compiled part of the Indian navy charts for his 1860 map of Arabia.[48] The *Weekly Dispatch* exemplifies how the popularization of an advanced cartography could become a successful commercial enterprise in the mid-Victorian period.

The British readership in the 1850s and 1860s enjoyed unprecedented access to cartographic material. The press displayed an Orient that was still caught between the biblical legacies of a precolonial era, loosely defined forms of British mid-Victorian imperialism, bursts of free trade enthusiasm, and some sensational conflicts. These multiple perspectives constructed an image of the East that was far from uniform. How did these maps form views of the readership? Many did not care much for foreign lands. Jonathan Rose insists on the "intense localism" which framed the British working-class geographical imaginations.[49] Yet the weekly sales of the *ILN* boomed during the Crimean War, partly because war could be observed through photography and cartography for the first time. The industrialization of commercial mapmaking answered a growing need for new modes of visualization. The Eastern Question provided many opportunities to display those modes. More important, continuities still prevailed with certain early 19th- and even late 18th-century representations of the area, as illustrated by the uses of geography and maps in schools.

3.4 "A GOOD DEAL OF JEWISH HISTORY AND SYRIAN GEOGRAPHY": LEARNING THE MAP OF THE EAST[50]

Thomas Henry Huxley's words on the geography that was hammered into the brains of the 19th-century pupil might not have been an exaggeration. The biblical Orient held a particular position in the curriculum, which may have distorted the representation of the world's geography. What did the map of the East look like in the 1850s and 1860s, from the viewpoint of a pupil or a young student? A word of caution is required before we look at map uses and how they might reflect Victorian assumptions: one should be careful not to give too much cultural agency to the map. In order to be relevant, the study of cartographic discourses must be combined with context-specific analyses. Reading a map is never an ahistorical process. The main difficulty facing anyone exploring 19th-century school practices of spatial representation is that relevant documentation is lacking. Hand-drawn maps, school tests, and notebooks are rarely considered worthy of a place in the archives. This is all the more unfortunate because mid-Victorian education provided the cultural background within which Edwardian decision makers were educated. Most of the individuals involved in the creation of the colonial Middle East from 1914 to 1923 were born between the 1850s and the 1870s. George Curzon was born in 1859, David Hogarth in 1862, and David Lloyd George in 1863. Their formative years shaped, even if only to a limited extent, their interpretative frameworks of the area by the first two decades of the 20th century. Fortunately, documents that might help us provide the depth of field required to understand this cultural context were sometimes stored in local archives.

J. T. Francombe's 1859 lesson book is one of these rare illustrations of what the study of geography meant outside governmental and elitist circles (see figs. 3.3 and 3.4). The young man, age sixteen when he kept this document, was not a neophyte anymore. He was a pupil teacher at Redcliffe Boys School. He eventually became the headmaster of St. Mary Redcliffe, Temple Colston, and Pile Street schools. In 1923, he was elected lord mayor of Bristol. Francombe's education and career may well not have been representative of the common lot.[51] His cartographic accomplishments would please most 21st-century teachers. They do, however, give a rare insight into English education in the 1860s.

The diligent trainee teacher hand-drew various maps of Asia, Palestine, and Jerusalem. His map of the Holy Land was copied from Walter McLeod's popular scriptural geography, which was in its tenth edition in 1858.[52] Francombe even applied the same colors as the original document. This hand-drawn duplicate is a testament to the rote-learning tradition which characterized part of geographical education in the 19th century. The outlines of Palestine, carefully learned and transcribed, had no secrets for him. His map of "Jerusalem on the date of its destruction" exemplifies the same didactic principles. It is a copy, framed by a chronology of the Judges, the instrument of a very scrupulous memorization. He learned the contours of Asia as well. They were less certain to him than those of the Holy Land. His Europe extended well into the Ottoman Empire. Afghanistan was larger than life. He knew how to place Medina, Aden, Muscat, and Mecca quite accurately.

Francombe's exceptional lesson book raises two main questions. First, to what extent were British children educated in geography? In the absence of a centralized schooling system, the variety of institutions and experiences, as well as the socially distinct educational opportunities, makes it extremely difficult to arrive at any form of quantitative assessment of what a young boy or a young girl might have learned. Official reports and various other sources nonetheless give indications of the specific position of the East in their formal or informal schooling. I will thus try to determine the relative position of the Orient in school geographies. Second, what lessons did the pupils learn? Francombe's 1859 notes seem to demonstrate that while the Holy Land and ancient geography held a distinctive place, the student did know how to draw a map of Asia. He could even locate a remote coaling station on the Red Sea. Was a more

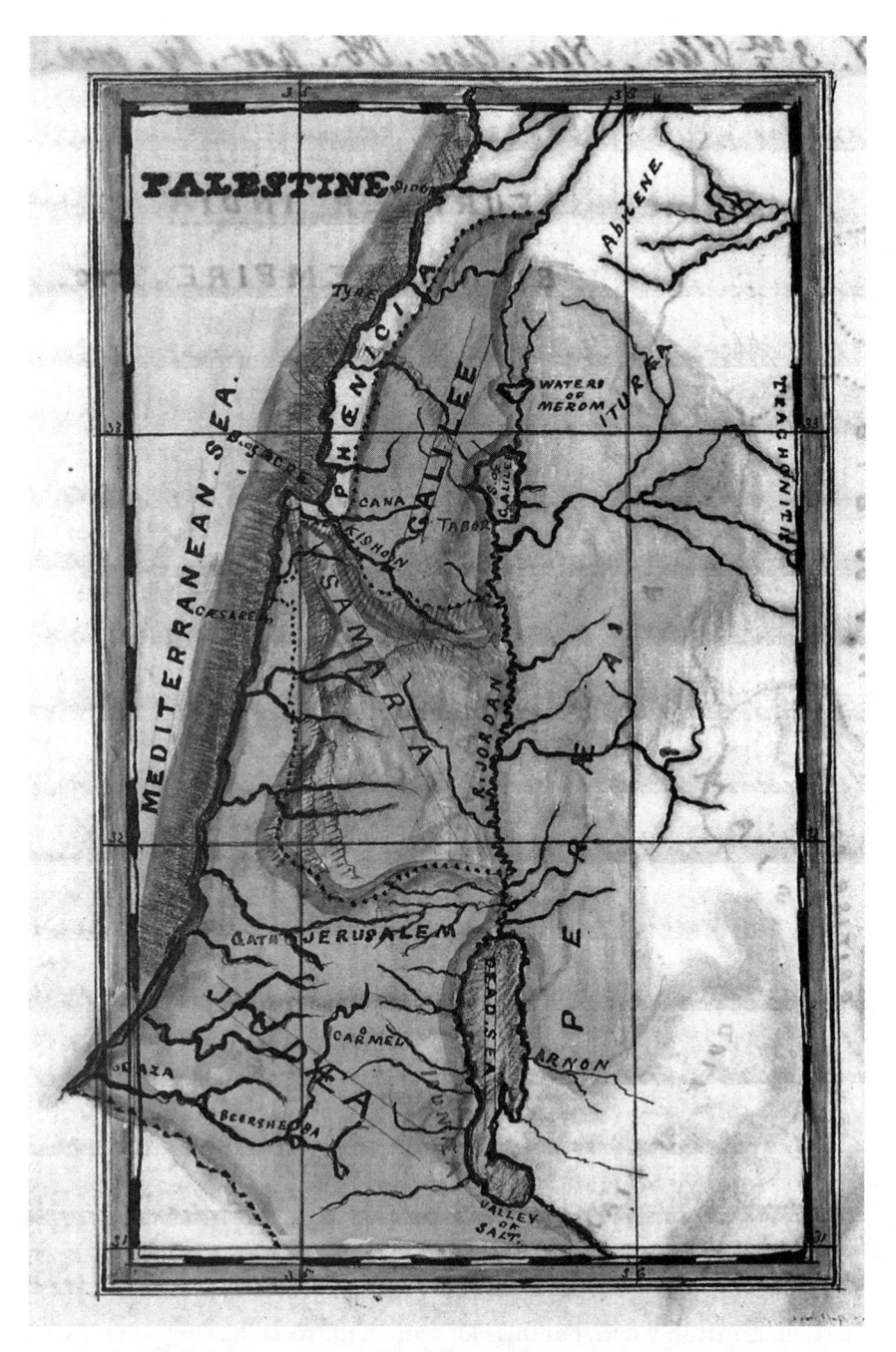

FIGURE 3.3. James T. Francombe, *Hand drawn map of the Holy Land*, scale not given, from J. T. Francombe's notebook, 1859. Photograph courtesy of the Bristol Record Office (40477/F/2).

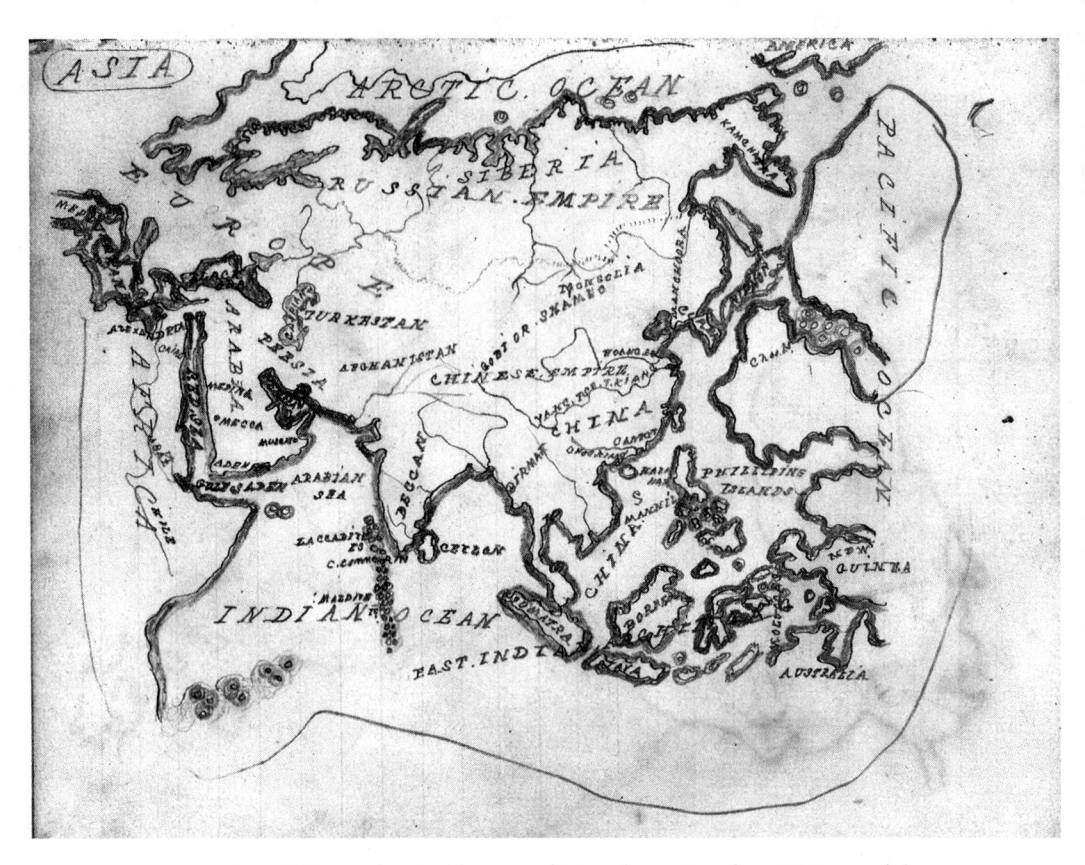

FIGURE 3.4. James T. Francombe, *Hand drawn map of Asia*, scale not given, from J. T. Francombe's notebook, 1859. Photograph courtesy of the Bristol Record Office (40477/F/2).

utilitarian and modern geography making its way into the curriculum? What East was outlined? Was it an antiquated biblical land or a secularized entity showing the empires and states on the route to India?

While mid-19th-century Britain was not geographically illiterate, the place of geography in the disparate curricula varied considerably. In the public schools, attended by upper-class children, geography was a very secondary subject, according to the 1864 Clarendon Commission, which examined Eton, Westminster, Winchester, Charterhouse, St. Paul's, Merchant Taylor's, Harrow, Rugby, and Shrewsbury.[53] Their 2,700 pupils represented a very small portion of British youth, but their location in the social hierarchy made such schools and their curricula models for the hundreds of grammar schools in the kingdom. Modern geography was consistently overlooked in these establishments. Despite growing criticisms leveled against what began to appear as a defect in the 1860s, M. E. Grant Duff (1829–1906), a fellow of the RGS, hoped, for instance, in 1864 that MPs would not consider that the commissioners reporting on public schools "have done anything superfluous in insisting upon Geography being taught in regular lessons." Modern geography remained a subsidiary subject.[54] There was a noncollegiate class in Shrewsbury in which mathematics, modern history, and modern geography represented a significant proportion of the classes, but the curriculum was designed for the less able pupils. When these subjects were finally adopted on a larger scale, the school did not think it necessary to recruit a specialized teacher. Nor did the comparison with Continental educational systems, where maps and geography had a comparatively more favorable position, function as a catalyst for change.

By contrast, there was room for much ancient geography. Aaron Arrowsmith's publications for the public school pupils clearly demonstrate that mid-Victorian upper-class education epitomized this classical vision of the world. His *Atlas of Ancient Geography for the Use of King's College School* and his *A Compendium of Ancient and Modern Geography, for the Use of Eton School* went through several editions in the second half of the 19th century.[55] These books were largely outdated by the midcentury. Charles Grenfell Nicolay (1815–97), a lecturer in geography at King's College in the 1850s and a fellow of the RGS, updated the *Compendium*, but its conservative outlook remained untouched. Chapters on Asia Minor, Syria, and Palestine focusing on Roman provinces, Greek colonies, and fallen kingdoms were complemented by a few pages and notes evoking the modern Orient. Nicolay's updates echoed James Wyld's exotic descriptions of the East: "Al Gezira, or The Island, corresponds generally with the ancient Mesopotamia [. . .]. It contains much desert country, but its desolate condition is not so much attributable to this as to the oppressions and exactions of the Turkish government, which entirely prevent anything like improvement, and rather tend to encourage the ravages of the roving Kourdi and Arabs who infest it, than to repress their rapacious and plundering cruelty."[56] Between the maps of the Holy Land and the stereotypical depictions of the Ottoman Empire, the upper-class student inevitably had a very distorted and antiquated vision of the East which, from the Mediterranean to eastern Persia, occupied no less than one-fifth of the book, almost as much as Britain itself.[57]

Geography and maps were given a similar treatment in the network of endowed grammar schools. The Anglican grip on this very heterogeneous assemblage of secondary schools, some of them dating back to the 16th century, ensured that their curriculum did not go against the grain. Francombe's notes are a glaring example of this traditional approach to the matter. Modern geography remained a subject of low status. The same was true of elementary schools. The map of the East was primarily biblical, before the 1870 Forster

Act created a more coherent system in England and Wales. Reverend W. H. Brookfield's 1850 report on the elementary schools of the southern district of England lamented the rote geographical knowledge acquired by the pupils: "Will they understand, or value, or enjoy the Scripture the less, or be less likely to recur to it in afterlife, because they have been taught where Egypt is; what lies between it and the Land of Palestine?"[58]

The education of the poor was based on the same principles. The late 18th and early 19th century saw the development of philanthropic societies which aimed at providing education for social groups hitherto excluded from the traditional schooling system. The two most influential organizations were the Non-conformist British and Foreign School Society and the Anglican National Society for Promoting Religious Education. Both networks heavily relied on monitors. Older students were recruited to train younger pupils. Teachers supervised the operations. A closer look at how they were trained by both societies confirms the impression that the biblical East held a peculiar position. In the Whitelands Training School (Chelsea), established by the Church of England's National Society in 1841, mistresses learned how to draw maps each Tuesday from 7.30 to 8.30 p.m. in the early 1850s.[59] Whitelands College owned three wall maps of Asia Minor showing the travels of St. Paul, three of the Holy Land, and only one of England. Smaller cartographic documents were also available for the trainee teachers. There were seven copies of a small map of Palestine but only one of America.[60] Male teachers were confronted with the same geographical obsessions. Section 5 of the general examination for schoolmasters in the training schools of the Church of England for 1851 asked them to draw maps of the journey of the Israelites from Egypt to Canaan, of Judea and Jerusalem.[61] At the Teacher's Training College of Sarum St. Michael for women, founded in 1841 in Salisbury, mistresses were also trained in geography. Notes taken by one of the students survived in the local archives and give some insight into what was taught by the trainees in the nearby model school. They mainly consist of lesson plans. One dealt with the river Thames, others with Italy or the mountains of England. Two lessons on the East, dated May and June 1856, were kept in the records.[62] The first one focused on "a caravan in the desert," the second on the Dead Sea. The pupils were to be warned of the "danger of being overwhelmed by sand." They also learned about the situation of the salted sea and the "bituminous stones" found on its shores. Exoticism and biblical erudition combined to produce a vernacular Orientalism that was more determined by the Old Testament than by modern imperialism.[63]

The same was true outside the National Schools and the British and Foreign Schools. In Scotland, for example, trainee teachers were confronted with similar geographical distortions. In the late 1840s, at the Edinburgh Normal School, the examination for first-year bursaries in geography started with questions on England and Europe and ineluctably concluded on sacred geography. The students were asked "the utmost limits of Palestine promised to Abraham's posterity," the "Roman divisions of Palestine in the time of our saviour," and the "chief cities of the Philistines."[64] In the 1860s, the training of primary school teachers by the Free Kirk in Glasgow included a detailed memorization of the outlines of the biblical lands. The inspector noted that "in the first year, the outlines of Old Testament history are carefully gone over, special attention being given to the geography of the Holy Land."[65]

This corroborating evidence might explain why in some schools more effort was put into teaching the geography of Palestine than into any other area of the world. In his 1847 "Report on the Day schools of the British and Foreign School Society," Joseph Fletcher, Her Majesty's inspector of schools, criticized this distortion when he wrote that "the outfit of maps in the British schools appears to have consisted of a few outline maps" had been superseded "by one good-sized map of Palestine."[66] Teaching methods were often devised to favor a passive memorization of the features of the Orient, since the biblical and the sacred were significant to the entire East. Scripture delineated a vast geographical assemblage from Thebes on the Nile to Nineveh. The official manual for British Schools provided a lesson on the map of Palestine that illustrates this method. The imaginary pupil had to answer an unstoppable flow of questions on Tyre: some on Nebuchadnezzar and Alexander the Great, and others on the "prophecy respecting the city" and "passages of Scripture."[67]

In the 1850s, rote learning the contours of the Holy Land was one of the objectives of the third class in the British Schools network. National schools applied similar views. In order for the pupils to be able to draw the outlines of Palestine, George Taylor, who trained the teachers in geography at the society's College in Battersea, advised his students never to attempt "to give a lesson on the physical geography of any country without drawing a map of that country on the black board."[68] He provided a blank map of the region for the use of the class. Pupils then copied it on their slates.[69] They could also buy blanks maps, such as the ones published in the 1861 *Map Book for Beginners*.[70] Examinations often included drawing a map of Palestine. This biblically centered geographical education was a key feature of an English, Welsh, or Scottish education in the mid-19th century.[71] The ascendancy of scripture geography over "modern geography" could have surprising consequences. Henry V. Johnson noted in the late 1840s in his report on the state of education in North Wales that "the scholars have no conception where Palestine lies with respect to any other country in the world, and cannot distinguish north from south upon the map."[72] Pupils attending the Sunday schools in Wales were believed "to be better versed in the geography of Palestine than of Wales."[73]

From the 1850s, reports, commissions, and essays tend to point more and more at the shortcomings of this education. A secularist movement promoted a new education under the auspices of William Cobden and W. E. Forster. From the *Westminster Review* to agnostic thinkers such as Thomas Henry Huxley, the traditional curriculum was coming under attack from various quarters. Modern or commercial geography slowly made its way into some schools. The dissenters from the British and Foreign School Society distanced themselves from the standpoint of some in the established church who believed "it would be very wrong to teach children any geography other than the geography of the Holy Land."[74] Even if Palestine remained, as stated earlier, a focal point in British Schools, a different outlook on the geography of the world can be found in their curriculum. The writings of James Cornwell (1812–1902) testify to this. Cornwell was appointed head teacher of the Borough Road Training College, the society's central institution, in 1846. He published schoolbooks on a variety of subjects from the 1840s onward. His 1847 *School Geography* quickly became a reference. The fact that he became a fellow of the RGS in 1860 increased his authority on geographical education. His book went through dozens of editions, reaching the seventy-second in 1913. It was used for informal schooling as well as in well-established institutions. Generations of children memorized his geography to a point where some of them "could repeat whole pages of it."[75] The book was an attempt at human geography. Cornwell updated his information and, as early as the mid-1860s, even referred to recent explorations such as Palgrave's travels in Arabia. It nevertheless indulged in well-worn stereotypes. The young reader was informed that Persians "on account of their politeness, have been called the French of the East." Of Persian women it was stated that "so little attractive are the women when out of doors, that some have said 'They look like bundles of old clothes going to be washed.'" Each study of a country concluded on series of exercises to help the pupil memorize. He was asked to "draw Arabia" and to talk about "pearl-fisheries" and the particulars of "internal communication; religion; population; language; government."

The geography, then called political or modern, in contrast with scripture geography, was more likely to be included in the curricula of the hundreds of trade schools and commercial schools, where it could replace Latin and classics. Modern geography and history were the humanities of the middling sorts. They were respectable enough, but not elitist enough to encourage unduly high hopes of social mobility. At the Whitechapel Foundation Commercial

School in London, for example, the subjects were "reading, dictation, British history, and modern geography."[76] All classes were taught modern geography, at least one hour a week.[77] What were the lessons learned? Richard Hiley's *Progressive Geography*, a schoolbook designed for the so-called commercial classes attending the trade schools, shows that his geography, like Cornwell's visions of the world, was still tainted with old prejudices, despite its modernist claims.[78] His description of the national character of the Turk takes us back to the 18th century: "The Turks are represented as grave, sedate, and passive; but, when agitated by passion, furious, raging, and ungovernable; full of dissimulation, suspicious, and vindictive beyond conception [. . .]. Laziness and apathy are their distinguishing characteristics."[79]

Assessing the efficiency of these attempts at a geographical education is extremely difficult, given the diversity of local and national conditions. The differences between the curriculum of a ragged school for the destitute and the courses taught in a public school were wide. However, there are several indications that geographical literacy made progress well before the age of high imperialism, despite the lack of a coherent educational system. The increase in British publishing of educational geography books is a first sign of an increasing demand.[80] The cheaper maps produced from the 1850s onward were also a contribution to the improvement in geographical literacy. A look at the training teachers shows signs of improvement in some areas. M. Fletcher's 1851 report on Borough Road Training School underlines the achievements of the training teachers of the British and Foreign School Society. It shows that 76% had results in geography that were between good and fair.[81] Pupils' attainments were obviously different. Local circumstances as well as social and sexual differences conditioned their achievements. There was a distinction between geography and domestic subjects for girls, for instance.[82] They were, as a rule, taught less geography than boys. An 1847 report shows that in fifty-two British schools, nearly 25% of boys had geography classes; the comparable figure for girls was a mere 20%.[83] All in all, if sacred geography was a common denominator, the scattered statistics included in the many official reports of the 1850s tend to show that only a minority of the schools systematically included modern geography in the week's schedule.[84]

In the mid-Victorian period, a British education included sacred geography, regardless of the social extraction of the pupils. The first signs of a secularization of the curricula did not call into question this cultural cement. The first competitive examinations to recruit civil servants in the late 1850s still included questions on scripture geography. The candidates to a junior position in the Department of the Committee of Council on Education were thus asked to "draw a map [. . .] of Palestine, according to its territorial divisions at the date of our Saviour's birth, indicating the position of the Jewish tribes."[85] The sacred facet of British geographical imaginations of the Orient percolated into the perceptions of what the modern East was. As we will see later, this educational background, the relative importance of the East in the curriculum, and the strong attraction that the biblical and historical Orient exerted on many pupils had lasting consequences. It partly shaped some of the most influential Edwardian decision makers' grip on the Middle East when the area underwent dramatic changes in the aftermath of the First World War.

CONCLUSION

British culture slowly evolved in its relationship with the East before the emergence of late 19th-century new imperialism. Early traces of a modernized and secularized imagination coexisted with a persistent biblical perspective. British Orientalism was not one, but many. It drew on 18th-century traditions, and only reluctantly did it incorporate new data. Cost constraints favored conservatism, and mapmakers often anticipated the expectations of an apparently nostalgic readership. The picturesque and the sacred still resisted scientific accuracy. The modernization of mapping techniques did not yet imply a profound overhaul in how the West charted the East. Mid-Victorian imaginative geographies of the area conflated the mapping of the past and the mapping of the present.[86] Religious and historical factors dominated the geographical elaboration of the East more than imperialist ones.

The 1850s and 1860s nonetheless saw a first and limited popularization of the map of the East. It was rooted in a prevailing traditional and biblical outlook on the area, but it also fostered a growing awareness of foreign lands. The Orient acquired a new cultural significance. It started to lose the elitist and aristocratic nature long preserved by its remoteness. While Grand Tours and pilgrimages still were the privilege of a few, the embryonic commodification of geography by maps brought the East closer to Britain. An increasing portion of the British population could visualize this area of the world, and many others. Was this process part of a wider Victorian tendency to measure, catalog, and eventually capture? As stated earlier, this domestication of the East through early educational and commercial cartography had more to do with deeply rooted scriptural and historical assumptions than with the expansion of the British world system in the mid-19th century. Yet the East was made more accessible, intellectually speaking. This had a lasting consequence: the possibility of a more direct intervention in the area became less theoretical to both the policy makers and the emerging public opinion. By the late 1860s, the transitional East had begun to be located within the imperial architecture, not only for the advocates of a more interventionist policy but also for a wider public. It was not yet envisioned as the vital hub it would eventually become for the British Empire, but maps as well as a wide array of instruments of knowledge contributed to situate it in the position of a vital connector.

PART TWO

A SHIFTING EAST IN THE AGE OF HIGH IMPERIALISM (1870–1895)

CHAPTER FOUR

ORIENTAL DESIGNS

H. H. Kitchener's 1885 map of Cyprus was compiled from the first triangulation of the island, which was annexed by the United Kingdom in 1878 (fig. 4.1). Before we look at what the document reveals, we must go back to the late 1870s to understand the sequence of events that led to its elaboration. From 1877 to 1878, Pan-Slavist Russia waged war against the Ottoman Empire in the Caucasus and the Balkans, more than twenty years after their confrontation in the Crimea. The Treaty of San Stefano (1878) marked the end of the conflict. Turkey in Europe had shrunk, giving way to new independent states: Bulgaria, Romania, Serbia, and Montenegro. The agreement did not solve everything. A conference was held in Berlin between the European powers during the summer of 1878, with a view to finding a new balance in the Balkans. Britain stuck to its policy of vigilant support for the Ottoman "sick man." The threats posed to India and Suez, as well as the prospect of growing Russian influence in the Balkans and the Caucasus, justified this position. A diplomatic alliance between the British and the Osmanlis was secretly negotiated shortly before the Berlin Congress opened. Benjamin Disraeli was the architect of the resulting Anglo-Turkish Convention. He wanted to annex Cyprus to the British Empire. It

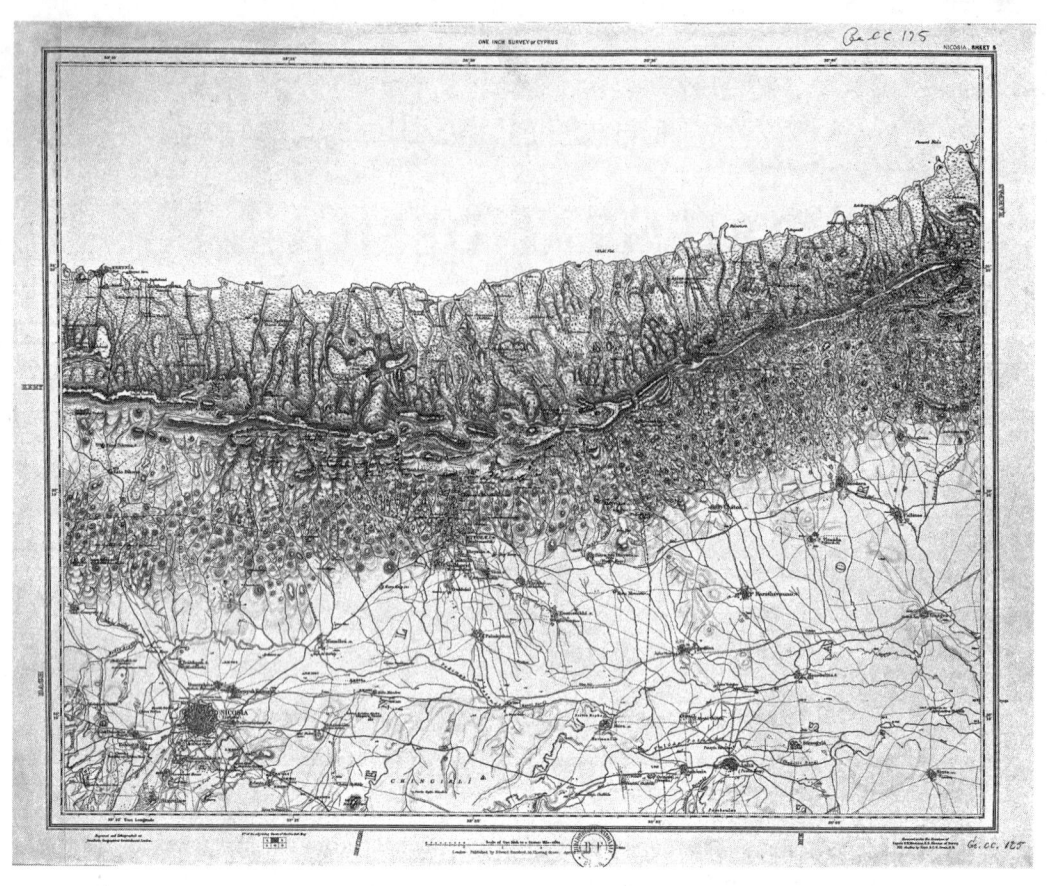

FIGURE 4.1. Herbert Kitchener, *A Trigonometrical Survey of the Island of Cyprus, pl. V Nicosia*, 1:63 360 (London, 1885). Size of the original: 49 × 61 cm. Photograph courtesy of the BNF (GE CC-125).

was a way for Britain to gain a foothold in the East. This decision opened a new frontier of imperial expansion in the area for Britain. In a broader context of intensifying competition with new powers such as Germany or Italy, Cyprus was now the key of a transitional semicolonial periphery which extended from the eastern Mediterranean to Persia.

Why Cyprus? Maps played a significant part in the decision that was taken by the Conservative cabinet in 1876.[1] Samuel Baker noted in 1879 that "innumerable and very bad maps of the island issued on linen from the War Office" were all stamped with the year 1876.[2] The government had been looking for a strategic position in the Mediterranean for some time. The navy needed a station on the route to India between Malta and Aden. The Intelligence Department of the War Office was therefore called upon to find the most suitable location. John Lintorn A. Simmons, inspector general of fortifications, and a member of the Royal Engineers, was asked to establish and compile maps to choose an island that would consolidate the British presence in the area, while securing access to the Suez Canal. Simmons wrote a report in 1877 to determine the best site for a coaling station in the eastern Mediterranean.[3] His role had hitherto been to examine Ottoman fortifications and assess their ability to withstand a Russian offensive, especially in Erzurum. He collected many maps in the process and developed a recognized geographical expertise.[4] Austen Henry Layard, ambassador to Constantinople, simultaneously worked to identify the best candidate. Maps were a crucial part of the argument put forward by the various experts in search of the right location. Colonel Robert Home, of the War Office Intelligence Department, described the different options in a memorandum sent in June 1878 to the secretary of state, Lord Salisbury. The document reviewed Crete, the Dardanelles, and many other places to see if they could meet the needs of Britain in the Levant. It concluded that Cyprus was one of the best solutions, despite the opposition of naval experts to the proposed scheme. Holme decided to submit a sketch map of the island to bolster his views.[5]

While the cartographic representation of Cyprus did not in itself lead to annexation, since that decision had already been taken by Beaconsfield and Salisbury, maps and geographical expertise were becoming increasingly authoritative tools for decision makers in the late 19th century. When Parliament came to discuss the secret agreement by which Cyprus was annexed to the British Empire, maps were once more used as supporting evidence. Some of them were displayed on the wall of one of the committee rooms to inform MPs about the situation in the eastern Mediterranean. Francis Charteris (1818–1914), also known as Lord Elcho, was clearly convinced by the exhibit: "Take a map and study the position of Cyprus [. . .]. It is the one most commanding and best situated for the protection of Asiatic Turkey and British interests in the East; and for this reason, that Cyprus commands Iscanderoon, the best harbour on the opposite coast, and also the outlet of the Suez Canal."[6] New global political structures required seemingly more accurate tools of government than in the times of the old Foreign Office. Maps were one of them. John Gallagher and Ronald Robinson underlined this process in *Africa and the Victorians*: the expansion of British power in the world was directed from London, by leaders who were "professionally and geographically" detached from the territories under their consideration.[7] The backup of a convincing cartography was now a requirement. Improved surveying techniques, the increasing precision of the equipment, and the tightening grasp of Europe on unexplored territories modified these leaders' perception of the world.[8] Contemporary observers were well aware that the blanks on the world map were quickly disappearing. This shift was concomitant with the expansion of the European colonial empires. There was an obvious relation between the mapping of the world and its domination. Felix Driver and Joseph Conrad before him have described the transition toward "geography militant" in the 1880s that was concomitant with the advent of high imperialism.[9] The East, at least its elaboration as a region by the Europeans, provides a telling example in this regard. Its cartographic and geographical construction is an opportunity to explore these parallels between imperialism and Western knowledge.

Cyprus had thus been categorized and assessed in British maps before H. H. Kitchener started his surveys. Science had already helped to situate the island inside the imperial structure. However, accurate measurements of the new dependency were lacking. Triangulation was to solve the issue. The aim of this change of scale was to ensure a more precise delineation of property boundaries, a more thorough approach to the raising of taxes, and also an effective control of the population, thanks to the identification of the various communities. The surveys began in September 1878 with the assistance of a team of royal engineers. The measurements were completed by February 1883.[10] Kitchener's triangulation of Cyprus could be taken as yet another reflection of a wider process of imperial expansion. The process proves to be far more complex.

Cypriots did not always welcome the royal engineers with great hospitality. Kitchener was shot at by a farmer near the village of Pissouri while surveying the surroundings.[11] The incident added to a long list of difficulties. Kitchener spoke a little Turkish, but no modern Greek. He had to rely on an interpreter.[12] The lack of funds became so serious that the survey was brought to a halt after a few months. Kitchener resigned before completing his work, due to disagreements between administrations. Lord Wolesley, high commissioner in Cyprus, wanted a rough survey, whereas the FO sought a genuine geodetic survey.[13] Kitchener was caught between these conflicting aspirations. A map at a scale of one inch to one mile, in fifteen sheets, was finally printed by the Stanford firm in 1885. The map was compiled from data which were mostly planimetric. Relief information was lacking. One of Kitchener's assistants, Lieutenant S. C. E. Grant, sketched the hill features. He resorted to shading instead of the more up-to-date contour lines. While it had a picturesque dimension to it, accuracy was wanting. Ethnographic information nevertheless enriched the map. A cross indicated an Orthodox village, a crescent moon a Muslim settlement. Boundaries between districts were authoritatively shown on the map. The uncertain Ottoman grasp on the territory was apparently giving way to a more technocratic management of the land. This was merely an illusion. In practice, Kitchener was never able to reach a sufficient degree of accuracy to establish a land register and truly modernize the Ottoman tithe, despite his vision of a highly accurate scientific mapping of Cyprus. The map was nonetheless used to place boundary stones to delineate properties in the early 1880s. The stones were repeatedly vandalized, and while a specific offense was introduced to try to counter this trend, it was to little avail.[14] There were multiple impediments to the delimitations implied by the survey in the following years. The power of European science had met its limits. Kitchener's trigonometric survey offered an impression of control. In truth, low-intensity resistance from the locals and the "weapons of the weak" partly defeated his purpose.[15]

The gap between an apparent panoramic control of the empire from London and the actual elaboration and uses of the map was often wider than generally imagined. The 1885 map of Cyprus is one of the many documents from the imperial archive that raises questions regarding the very nature of the intellectual grip of the British on the area. Were maps and geographical reports unequivocal illustrations of European domination? True, they were components of an assemblage of commissions, reports, and statements, which created a flow of data that seemed reliable. They fitted into the teleological narratives of progress which Europe fashioned for itself in the late 19th century. However, not only was their power contested by local forces, but their relation to the reality of the territories at hand was also distorted, or even inconsistent. Thomas A. Hull, superintendent of the charts at the Admiralty, pointed to these illusions in 1875: "People look at the beautiful maps issued by Keith Johnston, and others, and they think that we know the world perfectly: what more can be wanted? They do not know that the outlines given on such maps are to a great extent mere guesswork; the proportion of what is real knowledge and of what is imagination is not known even by well educated men."[16]

This chapter sets out to assess these discrepancies, as a new British empire was on the rise. It also delves into the competition between the various institutions and individuals involved in the British definition of the area. As the word "imperialism" was becoming less of a neologism in Britain, from the 1870s onward, the mapping of the East reveals the polycentric nature of the British Empire. As we will see, new instruments and improved techniques contributed to a questioning of old legitimacies, as far as the collection of information was concerned. The following chapter will also underline the indigenous attempts to resist, or manipulate, the knowledge and the power British maps conveyed.

Previous sections have shown how conflicted and hybrid the production of spatial knowledge on the region had been in the mid-Victorian era. In this last section we shall show how cartographic contradiction rapidly developed within the region itself to try to counter Britain's perspectives precisely at precisely the moment when it seemed to be tightening its grip.

4.1 MAPPING EASTERN QUESTIONS: BRITANNIA'S RULE OR THE RULE OF THUMB?

A great deal of misapprehension arises from the popular use of maps on a small scale. As with such maps you are able to put a thumb on India and a finger on Russia, some persons at once think that the political situation is alarming and that India must be looked into. If the noble lord would use a larger map—say one on the scale of the Ordnance map of England—he would find that the distance between Russia and British India is not to be measured by the finger and thumb, but by a rule.

LORD SALISBURY, June 11, 1877

Lord Salisbury was well known for his mistrust of maps. He believed "the constant study of maps" to be "apt to disturb men's reasoning powers."[17] However, like his colleagues, he was confronted, first as secretary of state for the Foreign Office, and then as prime minister, with an ever-increasing number of cartographic documents. He was also known for his habit of hanging large-scale maps on the walls of his house in Hatfield, where the map collection of his ancestor was carefully kept. Even he could not overlook the example of other European leaders. Germany in particular impressed the other European powers with its political and military uses of the map. As Robert Morier (1826–93) once noted, Bismarck never carried out "his studies otherwise than with an open map before him, and with careful statistical synopses of the military forces really available by the various pieces on the chess-board within easy reach."[18] Maps were one of the many tools of modernized governance that European states experimented with in the last decades of the 19th century. Alongside more and more inventive modes of graphic representation of statistics, quantitative research in various forms, and expert reports, they became customary instruments of government. A detailed study of how the East was pictured by the central government in London reveals both the limitations and the challenges brought about by these evolutions.

IN THE OFFICES: NEW DIVISIONS

The FO was one of the first major departments to reform its intelligence in the late 1860s. It was completely reorganized in 1881–82. In 1857, the FO was divided into six departments. Its structure mirrored the classical divisions of the world described by Aaron Arrowsmith's *Compendium*: a consequence of the Etonian education of most senior officeholders in the department.[19] The Near East division was thus in charge of the Orient. By the mid-1860s, the Turkish Department was merged with the Russian Department. It encompassed Turkey, Russia, Greece, and North Africa. Julian Pauncefote (1828–1902), then assistant permanent undersecretary of state for the colonies, eventually created eight geographical divisions in 1881.[20] A large Eastern Department included Eastern Europe, Turkey, Persia, Central Asia, and Egypt.[21] These attempts to modernize the geopolitical categorizations of the FO, despite the old clannish structures, were combined with other reforms and evolutions. A new building was inaugurated in 1868. It accommodated a larger library than the old Foreign Office. Supplementary clerks were recruited. The collection and cataloging of knowledge became more effective. The late 1870s and early 1880s showed an increasing awareness of the need to protect and organize information in the FO. Changes were made to the indexing system in the mid-1880s. This even involved the prompt introduction of a culture of confidentiality in the wake of the Marvin Scandal. Charles Marvin was sometimes employed as a clerk by the FO. His low wages certainly induced him to find better pecuniary opportunities. He sold the text of the confidential Anglo-Russian convention of May 30, 1878, to the *Globe*. Fortunately for Marvin, he was released of all charges, since the very notion of confidentiality was not enforced by statute.

Edward Hertslet (1824–1902), the FO librarian since 1857, was pivotal in the reorganization of the documentation. He took specific interest in maps, which were gradually becoming indispensable tools for the British diplomat, despite Salisbury's apprehensions. Hertslet supervised the compilation of several maps and published a series of volumes on the political divisions of the world. His three-volume work *The Map of Europe by Treaty* was published in 1875. He wrote a similar sum on Africa in 1894.[22] He regularly wrote memoranda on boundary disputes and British influence in various areas of the globe. Most of them included maps. Hertslet wrote, for instance, a detailed paper on the competition between the powers in the Red Sea in 1874.[23] The report begins with a meticulous chronology of the evolution of foreign interventions in the area. The comprehensive memorandum was based on the most recent data. It refers, in particular, to Richard Burton's accounts of his travels in the area. A map was inserted at the end of the document; a feature that was becoming conventional in the 1870s. Hertslet could also issue reports at the request of other departments. One of his memoranda, published in September 1880, was written at the request of the Board of Trade, which wanted to facilitate the installation of beacons along the Red Sea shores. The validity of Ottoman sovereignty over a few islands was also discussed. Maps were beginning to shape the decision makers' perceptions.[24] The establishment of an international framework to minimize tensions resulting from the late 19th-century race for empire made it impossible to overlook cartographic tools. The 1878 Berlin Conference is a blatant example of this. When Alexander Gorchakov, the Russian representative, insisted on talking with Disraeli alone, Salisbury answered: "Lord Beaconsfield can't negotiate; he has never seen a map of Asia Minor."[25] A global grid of reference was instrumental to the rise of this new international legal order. Maps, more than ever, became conflict-mediation tools. For example, the Treaty of San Stefano (1878) included four maps as an annex to the agreement. The translation of mapmaking into a legal framework was therefore one of the many ways by which cartography actually shaped territories.

Filtering the data in a knowledge-based department such as the FO raised many issues. There was no ad hoc intelligence department. Dealing with the growing tide of documents was not an easy task, since the structure was a small one. Most of the staff did not have any experience of foreign countries, for lack of resources. Hertslet and his colleagues relied on consular reports and various sources of information to write their memos and supervise the compilation of their maps. Foreign maps were copied or purchased when opportunities arose. Other departments sometimes provided the FO with updated documentation on an informal basis. Charles Wilson, from the WO, sent one of his copies of a Russian military map of Persia to Hertslet in the mid-1870s.[26] The FO was dependent on other major institutions, such as the Admiralty or the government of India. This meant, to a certain extent, that the FO documentation was already tainted with the postulates of other institutions. This was specifically true of the East. A March 1888 memorandum by Francis Bertie (1844–1919), who had entered the FO as a junior clerk in 1863, illustrates how the quasi monopoly of the geographical and strategic information gathering in the Persian Gulf and Red Sea could foster Anglo-Indian views in London.[27] Bertie first discusses the conflicting positions between the government of India—which wished to develop alliances within Arabia to secure the position of Aden and weaken Ottoman influence—and the metropolis. Gladstone was reluctant to implement an active policy toward the Arabian hinterland, for fear of openly encroaching on Ottoman interests. Bertie's concluding remarks interestingly underline the need to build alliances with the tribes of the interior and the necessity of establishing a boundary-making commission. This was precisely the Anglo-Indian position, which leaned toward a forward policy. The metropolis relied heavily on data collected by a few agents on the spot. It gave their perspective substantial weight in the designs of Whitehall on the East. The fact that the Anglo-Indian world enjoyed a unique expertise in the area conferred legitimacy to its information, even if many, Salisbury being one of them, as shown by his comments on small-scales maps, were somewhat distrustful of this input.

The FO was not the only department to review its organization in the late 19th century. Clements Markham had laboriously built up the Geographical Department of the India Office, but it was not yet a fully fledged separate department. It was adjoined to the Registry and Record Department in 1885.[28] It nevertheless benefited from the consistent flow of cartographic production stemming from the British administration in India.[29] The Colonial Office (CO) had little to do with the East at this point, except for Cyprus, which came under its supervision in 1880. However, C. P. Lucas (1853–1931), a prominent figure of the department, showed a keen interest in the Suez Canal.[30] As in many other offices, cartographic documentation had an increasingly large place in the records.[31] Chewton Atchley (1850–1922), librarian of the CO from 1881, eventually published a catalog of maps in the early 20th century.[32]

The need for better intelligence was the main object of the commission established in 1883 to consider the creation of an Intelligence Department for the Admiralty. A Foreign Intelligence Committee was founded in December 1882. It developed into a fully fledged Naval Intelligence Department under the supervision of William Henry Hall (1887–89).[33] The Admiralty's hydrographic charting of the East was updated. Several expeditions were sent to the area to gather more data, after the coordinated campaigns of the first half of the 19th century. Admiralty hydrographers were mobilized to establish new nautical charts. Measurements were made near the shores of Aden in 1870 by officers from the metropolis. Perim Island, in the Gulf of Aden, was surveyed by Lieutenant Francis John Gray in 1874. *Pilots* for the Persian Gulf and the Red Sea were updated.[34] The area of Shatt al Arab came in for the specific attention of the hydrographers, at the request of the government of India.[35] Cartographic control of the region was completed by expeditions to the islands of the Persian Gulf and the Red Sea, including Socotra. In the Mediterranean, William Wharton (1843–1905) joined the surveying ship *Shearwater* in 1872. He charted the Mediterranean for three years, including the Sea of Marmara and the Bosporus.[36] The straits attracted additional attention after the 1877–78 Russo-Turkish War. New nautical charts of the coasts of Egypt were drawn up in 1882, in connection with the British intervention against Ahmed Urabi's revolt.[37]

At the WO, Charles Wilson took over the Topographical and Statistical Department in 1870, at a time when Edward Cardwell, the secretary of state for war, was implementing a series of wide-ranging reforms in the department. Wilson was well aware of the importance of cartography and geography, as shown by his participation in the surveys in Palestine under the PEF. He had been staunchly critical of the gaps in the map collection of the department. He was an advocate of the creation of an Intelligence Branch, with Cardwell's approval. It was established in 1873 under the direction of Patrick MacDougall (1819–94). The new branch was able to print memoranda and précis in a very short time to meet strategic needs. The effective use of maps by the Prussians during their war against France convinced the WO that topographical intelligence was pivotal in a modern war.[38] The department was divided into four foreign sections, one of which dealt with Turkey and Africa. Section F was in charge of collecting topographical data. A complete map catalog was achieved in 1877 as a consequence of the reforms. Gaps in the documentation were progressively filled. A systematic numbering and indexing of the cartographic production was put into effect in 1881.[39] Further strengthening of British intelligence capacities was intended under Henry Brackenbury (1837–1914). He was appointed head of the Intelligence Division of the WO in 1886 as director of military intelligence. Brackenbury was very active in trying to liaise with other departments such as the CO or the FO.

A quantitative assessment of the WO's cartographic output shows that the Ottoman Empire, Persia, and the Nile regions attracted increasing attention. As shown in figure 4.2, the number of maps of the East in the WO maps collection increased from seventeen in 1881 to sixty-eight in 1882. British intervention in the area meant more operations on the ground. Topographical and large-scale maps were in greater demand as a consequence. One of the first major developments in the area to favor more surveying and mapmaking was the 1882 British offensive in Egypt. The British intrusion and its aftermath helped materialize a strategic area under British influence that was centered on the Suez Canal. French colonization of North Africa led to the further separation of the Maghreb from the Mashrek. The boundaries of the Arab East were pushed a little further south in Africa, as the Sudan was attached to the block of states whose fate was linked to Suez. The Sudanese coast, which controlled part of the Red Sea, had an obvious strategic significance in securing the route to India and in the suppression of slave trade.

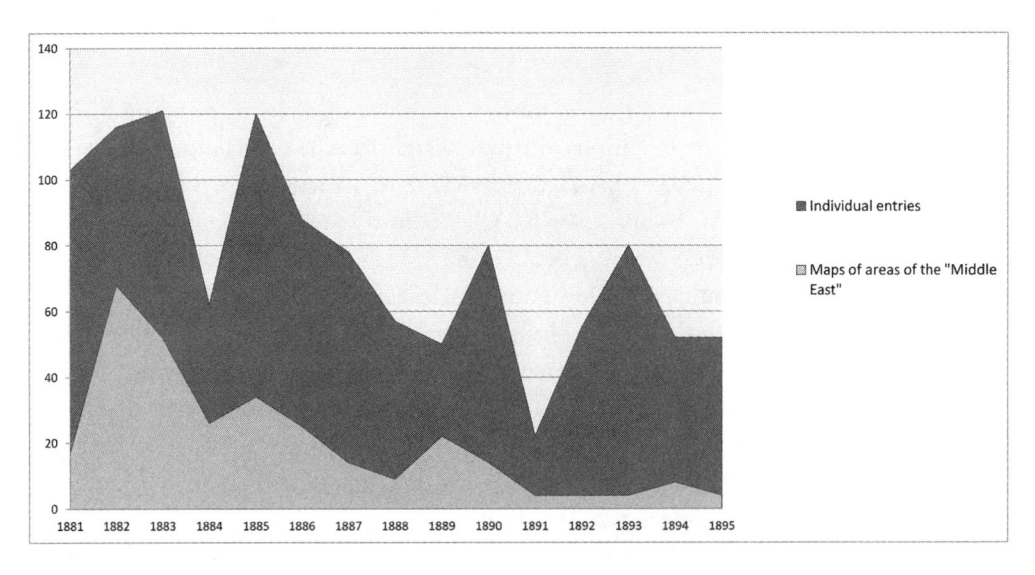

FIGURE 4.2. Proportion of maps of the Middle East in the War Office numbered maps (1881–95), based on A. Crispin Jewitt, *Maps for Empire: The First 2000 Numbered War Office Maps, 1881–1905* (London: British Library, 1992). © Daniel Foliard.

The WO had many other reasons to look east. Crises in the Balkans and the instability of Ottoman Turkey intensified Whitehall's attention to this part of the East. By the late 1870s, the WO had updated several maps of Turkey in Europe. Watchful concern for the straits, a crucial location, entailed increased data collection. Further East, Russian influence in Central Asia raised growing concern, both in India and in London. In Britain, most believed the integrity of a decadent Ottoman empire to be the most effective, and cheapest, solution to the problem of containing Russian expansionism. The WO paid more attention to the borders between Persia and Turkey on one hand and the Russian Empire on the other hand.[40] The possibility of war with Russia was not ruled out in the plans of the WO, thus prompting the establishment of a topographical database which was far more complete than what was available at the time of the Crimean War. Robert James Maxwell (Intelligence Branch) wrote a first memorandum entitled *Russian Advances in Asia* in 1873.[41] It was complemented by many maps. A report addressed to Lord Salisbury in 1876 contained a map showing the area bounded by the Sea of Azov, the Black Sea, and the Caspian.[42] Another, printed in 1877, focused on Armenia.[43] A map showing the border between Persia and the Russian Empire, to show "the probable concentration (troops) and lines of operation in case of occupation of Asia Minor by the Russians" illustrated a paper on a potential war in Asiatic Turkey and Transcaucasia.⁴⁴ The supposedly imminent Russian threat in the East fostered the production of maps.

The significance of cartography for late 19th-century strategic planning is confirmed by the many inclusions of rough sketches in the official documentation of the WO. The fact that Henry Brackenbury often attached his own sketch maps to support his argument is a salient illustration of this practice. He included a map of the Gulf region in one of his reports for 1889.⁴⁵ It underlined how dangerous it would be to allow Russia to develop its influence in northern Persia. By threatening the border, the tsarist armies could disrupt trade routes in an area where British interests ran deep. The accompanying report insisted on the fact that "commercially every road of Eastern ground in the East that falls under Russian influence is so much lost British commerce." The sketch stood as proof to the interconnections between Tehran and the Gulf which were vital to British trade in the area.

The strategic focus of the WO in the East fostered new perspectives, as evidenced in Brackenbury's cartographic projection on the Gulf and its adjoining areas. The old territorial divisions reflected in the architecture of the different sections of the Intelligence Branch gave way to new geostrategic constructions. The career of Herbert Chermside (1850–1929) is a salient example of how changing strategic imperatives delimited new terrains. Chermside had been schooled at Eton. He then trained at the Royal Military Academy in Woolwich to become a royal engineer. The curriculum included a strong focus on mapping and surveying, after Charles Wilson had successfully advocated better training in these subjects. Chermside was to remain faithful to the principles he learned from his mentor. He was sent to the Balkans in 1876–77. He was then appointed military attaché in Constantinople in 1877. He became British vice-consul in Sivas (Anatolia) in 1878. He worked for Egyptian intelligence from 1882 to 1883. He then served in the Sudan in 1883 and as governor general for the Red Sea littoral from 1884 to 1886. After an assignment in Kurdistan, he went back to the Ottoman capital as a military attaché from 1886 to 1896. He also toured Armenia in 1888. In 1896 he was sent to Crete to act as British commissioner.⁴⁶ For each of its appointments, Chermside proved to be a very prolific cartographer. There are at least fifty-three map references under his name in the official archives.⁴⁷ Some of them were included in Layard's reports when he was ambassador to Constantinople from 1877 to 1880. Others are in the archives of the WO. Chermside produced a vast number of sketches and rough surveys, particularly during the Russo-Turkish War (1877–78). He also compiled, on behalf of both the FO and the WO, maps of Kurdistan at much larger scales. Chermside's life and cartographic output mapped out the transitional East. His career was per se an illustration of the emergence of a new strategic entity in the East. It was deeply enmeshed in the Eastern Question. The territorial extension of his expertise reflected the new geopolitics of the East in the 1870s and 1880s. Mid-Victorian fragments of the East were starting to crystallize into a unified geopolitical entity from a late 19th-century British viewpoint, but it still lacked a name.

Major William R. Fox's map of the Egyptian Sudan is yet another illustration of how a modernized British intelligence could apply new projections and categorizations to the area (fig. 4.3). The document was first compiled and lithographed under Fox's supervision at the Intelligence Branch. In an emblematic illustration of how the WO publicized its cartographic achievements with little regard for confidentiality, the map was printed by G. W. Bacon, a commercial mapmaker, to be sold to the general public. One of the inserts shows a sketch map of Egypt and the Sudan. It is centered on the Red Sea itself. It materializes a strategic area stretching from northwestern Arabia to southern Sudan, thus amalgamating the Rea Sea, Aden, Egypt, Sinai, and the Sudan into a new assemblage. Interestingly, the color applied to delineate the area was the very red in which the dependencies of the British Empire were rendered on contemporaneous maps. It was a message in itself. Britain, confronted with French encroachments in East and northern Africa, was unifying the Nile basin under its rule. This commercial map, designed by a British official, was one of the many ways of proving this point.

Was the renovation of British intelligence an effective means by which to foster influence in the East? Or was it just an illustration of the "illusions of supremacy" that were often constituent of Whitehall's outlook on the world?[48] The picture is mixed. Despite the reforms, and regardless of the inflation of geographical and topographical knowledge in the WO, the Intelligence Branch remained mainly "a depository of facts."[49] Lack of planning and ad hoc intelligence continued to characterize military action. Both observers and critics of the British war efforts during the Second Boer war (1899–1902) were still astonished by the lack of proper topographical intelligence in the army. Most interventions in the late 19th century relied on local informers, travelers, and informal intelligence. And if there was a quantitative shift, the aggregation of large-scale maps in the newly created intelligence departments in London did not constitute a coherent collection. Large areas were left uncharted. Technical factors were not the only reason why a systematic cartography of new terrains was not always diligently implemented. More complex phenomena slowed down the resistible ascension of a policy-making bureaucracy. Too systematic a prospective planning was believed to be a potential limitation, and even a

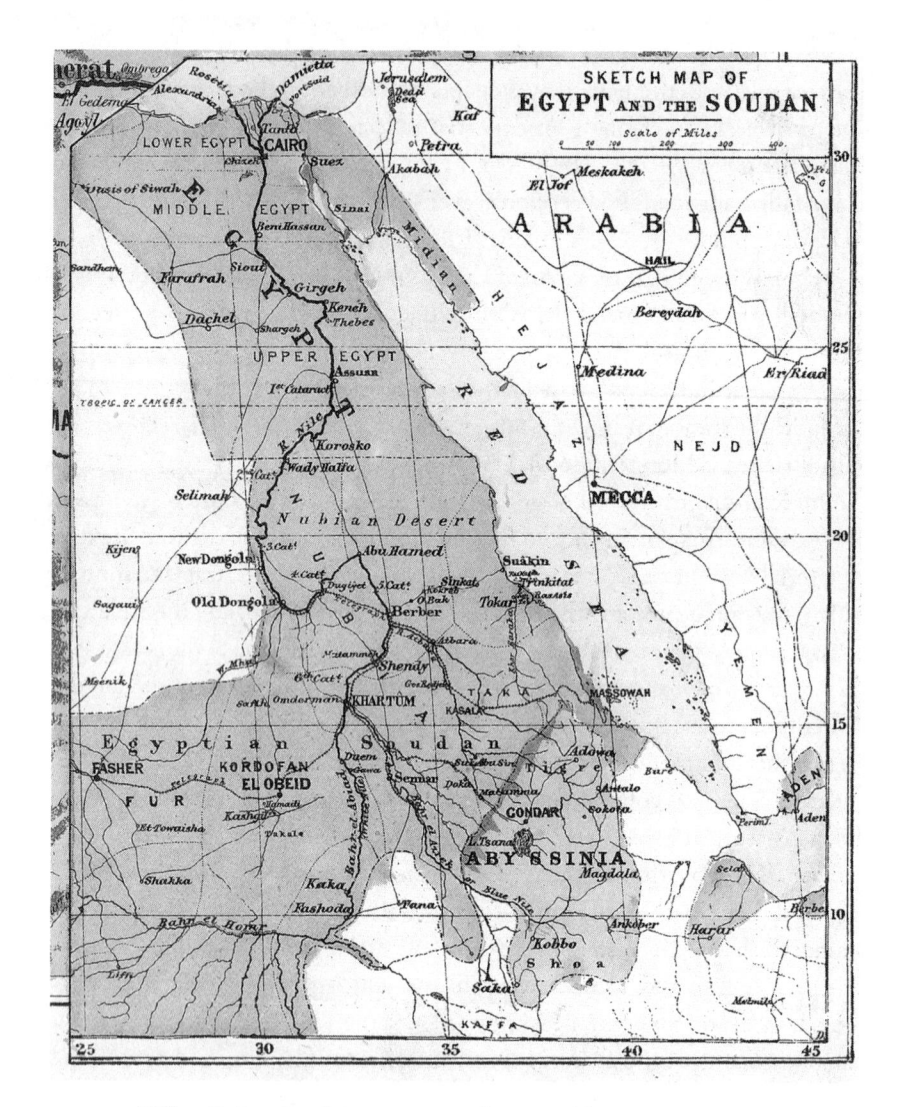

FIGURE 4.3. William R. Fox, *Bacon's Large-print Map of Egyptian Sûdan*, 1:2,217,600 (London, ca. 1885). Size of the original: 48 × 72 cm. Photograph courtesy of Société de Géographie (BNF, Cartes et Plans SGY F-42).

threat, to the decision-making capacities of the political leadership.[50] On a daily basis, little use was actually made of the précis and documentation outside expert circles, however much the signs of a "changing nature of governance," to use Jeremy Black's words, were multiplying.[51]

This survey of official maps of the East would be far from complete if it did not look beyond the traditional delimitations of what was strictly official, state-managed, and what was not. As we have seen in the first chapter, some of the components of the British cartographic grasp on the region were elaborated on the fringes of officialdom.

EXPERIMENTING OUTSIDE WHITEHALL: THE PALESTINE EXPLORATION FUND AS AN INCUBATOR

Between 1872 and 1877, the PEF implemented systematic surveys in western Palestine. The financial difficulties faced by the organization were such that the WO thus provided personnel and funds. Most members of the expedition, such as R. W. Stewart, Claude Reignier Conder, and Horatio Herbert Kitchener, were royal engineers sent to the Holy Land on "special duty." The WO's increasing involvement in the PEF's activities translated into more invasive supervision.[52] Not only were the surveys a welcome addition to the department's information on the area neighboring the Suez canal; they also provided an opportunity to develop and assess exploratory procedures in an environment that was both partly unknown and not too much of a diplomatic issue, despite renewed competition with the French in the field of biblical archaeology.

This campaign, like the first surveys of the 1860s, had an experimental side to it. Charles Wilson and Charles Warren, two prominent members of the fund, were particularly interested in promoting new surveying practices. Palestine offered a perfect terrain for the implementation of new techniques. In the 1860s, for instance, James McDonald had already used photography in a pioneering way for the survey of the Sinai Peninsula. Metropolitan topography seemed to lag behind the Anglo-Indian expertise. The partly unofficial surveys directed by the PEF were an opportunity for a new generation of military topographers to experience improved methods. This was an occasion to narrow the gap between the Anglo-Indian surveying traditions and the still uncoordinated metropolitan practices.

Laying out a baseline was the first step in the quasi-geodetic survey planned by the PEF. It was measured by R. W. Stewart in 1871 in the Plain of Esdraelon. From there, a reference frame, that is, a network of triangles, could progressively be established to cover the whole of Palestine. A map showing the trig points and a reference grid detailing the triangulation net used by the surveyors are still visible today in the British Library manuscript collection.[53] The royal engineers established procedures that did not call for any considerable workforce. The camp was set in a location from which specific points, hilltops as a rule, could be reached by the party. Once settled, British surveyors cut trees and plants to get an unobstructed view of the surrounding lands. One of them would then calculate the coordinates of salient objects, such as large trees, mosques, or ruins in the field of vision by means of a theodolite, while another filled a notebook with the relevant data and even started to plot. When a grid of fixed points had been set out, the party divided into subgroups to fill in the details inside each triangle. The royal engineers' tasks at that stage included hill sketching. The officers used chronometers, astronomical observations, and also barometric readings to refine their measurements. Buildings and ruins were measured by compass and chain. The explorers gathered a vast assemblage of botanical, archaeological, and toponymic information. The usual team of surveyors was as small as possible. Conder's account emphasized the cost-effective organization of the PEF's surveys. It comprised three noncommissioned officers and eleven local inhabitants, most of them Maronites, a typical manifestation of the royal engineers' mistrust for the Muslim population. One of the locals was a scribe who had to write down and cross-check place-names with the villagers.

Opposition from the locals was one of the most significant obstacles impeding the work of the surveyor. Yet Palestine was not as hostile a territory to surveyors as the Indian frontier. Undercover pundits were not vital to the undertakings. British officers were only partly disguised. They wore a Bedouin kaffiyeh that, from afar, made them look like Turkish officials.[54] It did not mean that everything went smoothly. In July 1875, Conder was gravely wounded in Safed by an angry mob. Some of the whitewashed cairns that were set up by Conder or Kitchener as reference stations visible from a distance were destroyed by the neighboring villagers.[55] It took a painstaking five years to cover the entire territory from "Dan to Beersheba and from the Mediterranean to the Jordan." The surveying methods developed in the Holy Land did not require a considerable labor force while offering the cartographers a respectable degree of accuracy. The experience marked a break with the techniques previously used in Britain. Metropolitan surveyors had demonstrated their ability to achieve the survey of an unknown terrain under relatively unfavorable conditions. This gain was crucial in the development of surveying techniques used in the exploration of Africa a few years later. What started as a scientific and religious enterprise in a strategic sideshow unintentionally became an experimental terrain for cost-effective imperial mapmaking.

After the end of Conder's and Kitchener's surveys in 1877, the PEF considered conducting surveys east of the river Jordan, an area then named "Eastern Palestine" in Britain. The fund had received the support of the American Palestine Exploration Society to achieve this goal. For a few months in 1881, Conder's expedition carried out measurements on nearly five hundred square miles between the Jordan and Amman. The WO eventually redeployed its efforts toward the Nile regions because of the instability in Sudan. Since the PEF could not fund the project on its own, the survey came to an abrupt end. The Mid-Victorian fascination with the biblical East was starting to wane.

The maps were photolithographed by the Ordnance Survey in 1880 in twenty-six sheets at a scale of 1:63,360. Volumes on Palestine's fauna and flora, its geology, and its history and ethnography were published between 1882 and 1889. This work constituted an extraordinary source of information about Palestine. Even the most Anglophobic catholic French people acknowledged that "Englishmen, who are to be seen wherever material interests arise, wherever new expectations of wealth and trade begin to surface, do not neglect purely scientific or religious investigations [. . .]. These insular practitioners teach us some valuable lessons."[56] The compilation of data from the surveyor's sketches was mostly performed by the staff of the Intelligence Department of the War Office.[57] The maps showed comprehensive indications about the vegetation, landscape, densities, and agriculture. The cartographers experimented with new graphic representations to depict a variety of nontopographical information. From this perspective, the cartographic representations compiled from Conder and Kitchener's notebooks innovated by showing a more diversified geography than was usually displayed by the WO's maps. The PEF's publications became references for biblical scholars for the following decades.

Zincography added to the technical possibilities.[58] Printing in color extended the scope of the experimentation. The visible variations in the cartographic depiction of Palestine contributed to modifying British perceptions of a territory which was no longer lifeless and frozen in its sacred past. Representations of a depopulated Holy Land of the early 19th century slowly gave way to maps displaying the human geography of Palestine, such as the spatial structure of its agriculture.[59] Was this a tool to reclaim the Holy Land, a cartographic expression of the British colonial gaze? The modernity of the PEF's surveying and mapping accomplishments should not be overestimated. It is too often a matter of conventional wisdom that these British experiments in the East represented the indisputable expression of an imperial grasp. The fact that General Edmund Allenby used some of the maps printed in the 1880s by the fund during the Palestine campaign in 1916 is regularly taken as evidence of the imperialistic nature of the PEF's endeavors. Yet, for lack of accurate elevation measurements, the surveyors and cartographers working for the PEF significantly used shading, which was far from offering the most up-to-date graphic option for the representation of the relief. Contour lines and hypsometric tints were already widely used in Europe. The various publications which were to follow from the survey made use of contour lines.[60] Conder himself privileged this solution whenever the data were accurate enough to allow it: "The only method by which general results can be obtained from isolated observations of level is by the use of contours, or lines of equal level, the tracing of which indicates the relative positions of the features of the ground."[61] The rough altimetric observations gathered between 1872 and 1877 were far from accurate enough to substantiate an updated representation of the relief. General Allenby, as we will see later, quickly realized the defective nature of the PEF's maps in that respect. As impressive as Conder's achievements were, they never amounted to an effective intellectual control of the area by the British.

The surveys of Western Palestine and the failed attempt to survey Eastern Palestine were therefore not as imperial as has sometimes been argued.[62] It is undeniable that the competition between the British and the French in the area was one of the underlying motives of the PEF's mapping of the Holy Land. However, Conder's and Kitchener's trigonometric measurements were more than just another example of imperial competition applied to a scientific field. They were fields of experimentation, from the gathering of data to their publication. More important, they participated in the development of a network of British agents which would prove to be instrumental in the elaboration of British influence in the East.

Charles Wilson was at the core of this nexus. A leading figure in the survey of Jerusalem in 1864, he was appointed executive officer of the Topographical Department in 1869 and promoted to the position of director of the Topographical Department and assistant quartermaster general at the Intelligence Department in 1870. He took over the direction of the Ordnance survey of Ireland in 1876 while keeping an eye on the PEF, of which he was a member until his death. He then served in the Servian Boundary Commission in 1879 before being appointed military consul general in Anatolia, when H. H. Kitchener, whom he already knew through the PEF, was military vice-consul. He toured Asia Minor with Professor William Mitchell Ramsay (1851–1939) and Chermside in 1881.[63] Ramsay was to become one of the most influential scholars on the history of ancient Asia. He counted Gertrude Bell and David Hogarth among his students. Wilson traveled from Aleppo to Smyrna with Wilfrid Scawen Blunt and his wife, Lady Anne Blunt, who had visited the interior of Arabia in 1878. In 1883, he founded another society, the Palestine Pilgrims' Text Society, with Walter Besant. From 1884, he was the chief of the Intelligence Department in Egypt. Wilson's career and expertise were chiefly built on his Palestinian experience. It was here that he had begun to develop a network of friendships and collaborations with some of the most influential British agents in the East in the late 19th century and early 20th century. Some of them, like Kitchener, were directly connected to the PEF. From Ramsay to the Blunts, Wilson's acquaintances encompassed a British nebula which is central to the prehistory of the Middle East as a geographical concept. His experience in Palestine legitimized his expertise and his authoritative position on the East.

One of the reasons why Palestine became a training field for the metropolitan military topographers was the realization that London was less advanced than other imperial centers, such as India, in the field of intelligence collection. This helped establish a specific British expertise on the topography of Palestine, which had wide-ranging consequences on the elaboration of the "Middle East." Despite the fact that the area was, from the 1870s, an integral part of the corridor between London and India, Eastern experts such as Chermside, Wilson, and Kitchener became trusted interlocutors for policy makers who were sometimes at a loss when facing a sometimes undecipherable area. Their expertise, categorizations, and conceptualizations provided a grid that seemed useful to many. Their networks also started to form an embryonic administrative framework for the area.

4.2 THE INDIANIZATION OF THE CARTOGRAPHIC GAZE

The unmatched expertise of the cartographers and surveyors trained in the Raj was a feature of the Anglo-Indian system of government. Every officer in the field had to supply topographical intelligence to the best of his abilities. A specific surveying and mapping culture developed into a system that was sometimes exported outside India, as we shall see later. Various manuals, such as R. Smyth and H. L. Thuillier's *Manual of Surveying for India*, republished in 1875, and Lowis d'Aguilar Jackson's 1880 *Aid to Survey-Practice* illustrate the comprehensive approach to surveying which was devised in India.[64] They testify to the advance taken in India in the different stages, from surveying to printing, relating to the production of maps. Anglo-Indians played a key role in drawing the contours of the East thanks to this comparative advantage. Several expeditions beyond the borders of the Raj provided the Anglo-Indian surveyors with a wealth of experience.

Three distinctive features of their practices must be underlined. First, they were quick to adopt, and sometimes even developed, what were the most innovative instruments. A position of inspector of scientific instruments was created in 1862 with the aim of examining the quality of the equipment used. Thomas Cushing, the inspector from 1876, tested an average of ten thousand instruments a year.[65] The portability, design, and precision of the surveying appliances to be used in an expeditionary survey were improved. In the 1880s, Jackson privileged a small set of instruments, including an aneroid barometer, less bulky than the traditional mercury barometer, a six-inch theodolite, a plane table, and small chronometers.[66] Most of these instruments had not yet been widely popularized in Britain. They made it possible to standardize the survey of large stretches of new terrains, with an acceptable degree of accuracy. Jackson estimated his error in longitude for a 160-mile route survey to less than 2 minutes 30 seconds.[67] Some of these instruments could easily be hidden, which brings us to the second notable characteristic of the Anglo-Indian school of expeditionary surveying. Indian assistants, trained in specific schools, accomplished most of the work.[68] This cost-effective solution also had the advantage of putting British lives less at risk in the sometimes unfriendly environments that were to be explored beyond the border. The pundits were the vanguard of the Indian surveys outside the Raj. The third feature which we should underline derives from the preexisting station bases on which the Trigonometrical Survey of India was elaborated. The surveyors could rely on properly localized reference points to project themselves within territories where no systematic triangulation could be envisaged. Persia, from Baluchistan to the Persian Gulf, was surrounded by these reference points. No one was in a better position than the Anglo-Indian officers to delineate the easternmost limit of the transitional East. Many Anglo-Indian experts, such as Thomas Hungerford Holdich (1843–1929) of the Afghan Boundary Commission of 1884–86, and Robert A. Wahab (1855–1915), who was in charge of the Himalaya party in the mid-1880s, would later be sent to Egypt and the Arabian Peninsula. Another network of British agents which proved to be instrumental in the invention of the "Middle East" thus revolved around India.

Not only was British India a pioneer in the standardization of data collection in uncontrolled terrains. It also developed a very modern organization by which to collate and print this knowledge. The Department of Surveyor General was divided into six branches. The Surveyor-General's Office was in charge of surveys, compiling, and engraving maps. A Lithographic and Photographic Office dealt with the task of printing. An entire office was dedicated to the design and construction of instruments. The Trigonometrical Survey Office and a Revenue Survey Office complemented the structure. A new building housing most of these officials was built in Dehradun (now capital of the state of Uttarakhand) and began to be operational in early 1880s. These departments responded to requests from other branches of the Indian government. In 1879–80, they compiled and printed maps of Afghanistan based on the most recent surveys to serve the military campaign. The efficiency of the department was such that some of the maps produced for the intervention of 1882 in Egypt were compiled and printed in India. In 1886–87, the Russian advances toward Persia made the publication of an updated map of Baluchistan an urgent necessity. The task was swiftly performed. Through the adoption of photogravure, these services seem to have become more and more efficient in the provision of maps to a nebula of military and diplomatic personnel.

FROM INDIA TO LONDON

The scope of the Indian surveys extended far beyond the Raj itself, moving into areas which seemed to fall under Whitehall's supervision, such as Mesopotamia and Persia. London could not ignore the data collected by Anglo-Indian agents, who were often acting well beyond the reach of the central government. The specificities of the Anglo-Indian outlook on the transitional East were not fundamentally affected by the post-1857 reforms. Changes induced by the Government of India Act 1858 did not radically disrupt the internal logic of the Indian subempire. Resistance to the growing involvement of the metropolis in the daily management of Indian affairs continued well into the early 20th century. The amount of trust to be placed in the Ottoman ability to withstand Russian expansion remained a key point of contention between the FO and the government of India. The former considered the stability of the Ottoman Empire to be essential to British interests, while the government of India cared less about the sultanate's sovereignty. In the early 1870s, India pushed for the creation of a protectorate in southern Arabia, a move entirely in contradiction with the Ottoman interests. Advocates of a forward policy in India, and even in the IO, such as Lord Randolph Churchill, were numerous. Many of them feared Russians would disrupt the overland route to Britain or the Persian Gulf. They started to threaten northern Persia from the 1880s onward, when they established their ascendancy over the Caucasus. Anglo-Indian officials pressured London into protecting the Indo-European axis and the Persian Gulf. The propagation of maps participated in the debate and in a redefinition of the British view of the region. The IO at home received copies of the documents produced in India and partook in what Arnold Kaminsky called the "Indian ethos."[69] The India Council, the interface between London and the Indian government in the late 19th century, still enjoyed strong connections with India. Most councilors had direct ties with the Anglo-Indian world. The resilient autonomy of the IO was reinforced both by a lack of supervision from Parliament, which did not get involved much in the affairs of India, and its independence from the Treasury. Like other departments, the records of the IO were reformed in the 1870s.[70] Part of British India's map production was thus easily accessible to decision makers in London.

The most interventionist Anglo-Indian views were conveyed to Britain through various channels. The Royal United Service Institute (RUSI) was one of them. This early example of a British strategic think tank was one of the places where defense policy of the empire was devised, sometimes with impressive results. Its members were among the first in Europe, in the 1860s, to predict the advent of trench warfare. Many of those who participated in discussions and published in the *Journal of the RUSI* were in fact Anglo-Indians. Henry Rawlinson, Lord Wolseley, and Frederic Goldsmid were all members of the institute. In the 1870s, it functioned as an echo chamber for those who opposed the tenets of a "masterly inactivity" policy, which advocated as little intervention as possible in Afghanistan or Persia.[71] After the conquest of the Khanate of Khiva in 1873 and Merv in 1884, Russia apparently had India within reach. Britain's geographical and cartographic attention to the bordering regions of Afghanistan, Persia, and the Russian Empire increased. Anglo-Indian experts thus became key stakeholders, producing several documents to specify the locations of the boundaries and the area of contact between Persia and the Russian Empire.

Geographical accounts and surveys were frequently put forward to contradict this cautious perspective. The Anglo-Indian officers and strategists used maps as a crucial tool by which to disseminate their views in the metropolis. They filled the blanks, changed scales, and surveyed the chessboard of the Great Game, to their specific ends. Valentine Baker's expedition in 1873 is a perfect illustration of the influence gained by the proactive Anglo-Indian circles. Baker, accompanied by William Clayton and William Gill, a fellow of the RGS and a royal engineer, explored northern Persia. His account was published in 1875 in the *Journal of the RUSI*.[72] Shortly afterward he published a popularized narrative of his adventures with the aim of warning the British public of the "proximate danger of a Russian conquest of India."[73] It included a "Sketch map of Central Asia, showing the Advances of Russia," thus forcing the issue even more. The protection of the North-West Frontier of India became a growing concern for prominent strategists such as Edward Bruce Hamley, who presented a paper at the RUSI in 1878.[74] The lecture was chaired by Henry Rawlinson. The War Office showed an interest in Hamley's intervention.[75] A map based on Gill's measurements was also compiled at the FO in 1878.[76]

There was a strategic divide between the metropolitan government and the Indian government as to the reading of the data. Interpretations of the Russian threat differed. If cartographic emphasis was common in India, London did not always heed the expertise of the Raj for institutional reasons. However, while some, such as Gladstone or Salisbury, resisted the flow of data, preferring to form their own opinion as to what was to be done in Persia, the authority stemming from the quantity and quality of the maps and knowledge produced in India undoubtedly facilitated the circulation of views from British India. Interestingly, the Indian government itself was confronted with the same challenges.

RESIDENTS AND POLITICAL AGENTS: SUBEMPIRES IN THE SUBEMPIRE

James Onley has shown how the influence of British India far exceeded the traditional definitions of the Raj.[77] An ever-growing network of dependencies, protectorates, protected states, and diplomatic representatives in Mesopotamia, Arabia, Persia, and the Gulf safeguarded the overland and sea routes in the late 19th century. Piracy, which impeded the expansion of British trade, was another concern. The Raj gradually sought to cut off the pirates from their havens on the coasts of the Arabian and Red Seas. Moreover, the decline of the Ottoman sovereignty over local emirates facilitated the transformation of what was an essentially maritime influence into an involvement of Britain with the hinterland. The seaman's view gave way to that of the landsman. The patient elaboration of an Indian subempire in the Gulf transformed what Persians called the "Persian Gulf," the Arabs, the "Arabic Gulf," and the Ottomans, simply "our Gulf" into a British lake whose shores were to be better controlled, and therefore inevitably surveyed and mapped out.[78] A nexus of Anglo-Indian agents was instrumental in the process. The Indian Political Service (IPS) employed scores of officers, most of whom reported back primarily to the Indian Foreign Department in Calcutta, to manage this sphere of influence. These agents and residents enjoyed a certain degree of independence. Their freedom of action, in comparison to that of other officials in the formal empire, was a distinctive feature of their action in the region and was to have long-term consequences. With few dissenting voices to cross-check their analyses, they enjoyed a quasi monopoly on the production of official information on the area. Some in London, such as Arthur Arnold (1833–1902), a Liberal MP for Salford and an Oriental traveler, voiced their concern: "India should not bear the charge of the Consular and Agency expenditure on the Persian Gulf, and upon the Tigris and Euphrates [. . .]. The concerns of British trade and commerce in Western Asia should be in the hands of officers more completely responsible to the Home Government."[79]

Given the specificities of the Anglo-Indian outlook on the functions to be performed by British representatives in the field, and, above all, given the focus on the need to survey and collect data as often as possible, it is not surprising to see most of the members of the IPS zealously explore the territories under their scrutiny in the late 19th century. Operations grew more methodical than in the 1850s. With the protection acquired from the sheikhs of the Gulf or from the tribes, the British often enjoyed a local sanction enabling them to recruit larger parties and to bring more equipment. They could also stay longer in the field. These exploratory enterprises had the advantage of inducing a form of soft confrontation with the Ottoman authorities, even though Whitehall, and specifically the FO, rarely supported moves that appeared to countenance any attempt to dismantle the theoretical Ottoman rule in Arabia. Surveys were not only disinterested schemes to collect ethnographic knowledge; they were also means to promote and shape forward policies in the area. Were they part of an overarching imperial project? Dissenting views between the British government in India and Whitehall did not amount to a coherent perspective on the region at the time. The saturation of data produced by Anglo-Indian agents in the area was part and parcel of a wider debate.

Samuel Barrett Miles's career gives an opportunity to understand how the Anglo-Indian personnel shaped the perceptions of the territories they administered in the Gulf. Miles (1838–1914), the son of a major general, served in the Bombay Native Infantry for years before he was assistant resident in Aden from 1867 to 1869. He was appointed assistant political agent for Gwadur (Persia) in May 1873.[80] He became political agent in Muscat in 1874. He occupied various positions in the area, holding office as consul general of Baghdad from 1879 before returning to India in 1887. After years spent between Turkish Arabia and Arabistan, Miles was fluent in Arabic. When he took his post in Oman in 1874, the political residency of Bushire, which was previously attached to the Bombay Presidency, had become a department of the central government in Calcutta. As stated earlier, these reforms participated in the broader expansion of Anglo-Indian-styled modes of government to the Gulf. British agents were commissioned to translate the area into a geographically understandable territory, a space to be reordered. This reorganization prompted the production of much more regular and detailed reports, as well as the development of a network of informers in the Gulf. The British agent in Muscat also coordinated his work with Bushire. A stream of data from the Persian Gulf, including maps and ethnographic reports, fostered the Anglo-Indian influence on the area. Miles strenuously participated in this collection of knowledge.

He first explored Aden. Along with Werner Munzinger (1832–75), a Swiss-born explorer who worked for the British government, he penetrated into the hinterland of the southern Arabian Peninsula in July 1870. Their accounts were published in the *Transactions of the Bombay Geographical Society*.[81] They were also read by G. P. Badger to the RGS in London.[82] The publications included a map and several geological sketches. In 1873, while in Gwadur, Miles explored the Mekran coast, in order to complement the surveys made in 1865 by one of his colleagues, E. C. Ross.[83] However, it was in Oman that he proved himself to be a prolific surveyor.[84] He accumulated a wealth of information which he only belatedly published.[85] It seems that he kept the most sensitive information for his official correspondence.[86] Miles, a fellow of the RGS from 1872 on, also drew original maps of the area.[87]

A look at his exploratory practices gives an impression of the multifarious skills of the Anglo-Indian agents. Miles embarked on his journeys with a great variety of instruments. He had a thermometer, of which the readings were recorded in his notebooks.[88] He also used an aneroid barometer and a telescope.[89] A typical feature of the work of these highly autonomous British agents was their capacity to adopt quickly the newest techniques. Miles was no exception, since he was equipped with a camera when he surveyed Oman in the 1880s. He was probably the first photographer of the hinterland.[90] Most of the pictures are now lost, except for those published by his widow in 1919. Like his predecessors, he traveled undisguised, seeking as much protection from local sheikhs as he could find.[91] Unlike the previous generation of British explorers in Arabia, he distrusted local information, drawing on it "with much reluctance, because experience shows that natives may give distances and names, but have no geographical instinct, a proof of this is Colonel Chesney's map of Arabia."[92]

Besides the architectural, botanic, geological, archaeological, or even meteorological data collected by Miles in his surveys, the most salient feature of his explorations was his ethnographic and anthropological perspective on Oman. He was always very careful in his descriptions of the tribal networks and in identifying the more loyal supporters of British influence in the region. Ross eventually published a tribal map of Oman which was based on part of Miles's observations.[93] In Oman, the British could rely on the allegiant Al Said dynasty. Sultan Thuwaini bin Said al-Said (1821–66) signed a first agreement with the British government in 1861. His successor, Sayyid Turki bin Said (1832–88), signed a treaty to suppress the slave trade in 1873 under the auspices of Bartle Frere. Miles also cataloged the feuds which might endanger the position of the ruling family. He was particularly concerned about the incursions of more religiously radical tribes from the Arabian hinterland. His maps and explorations were a significant step in the delimitation and redefinition of Oman. This taxonomy of the tribes, sheikhs, and ethnic groups became a key characteristic of the IPS's intelligence in the region for decades. With his surveys and reports, Miles, along with many of his colleagues, participated in the elaboration of the nexus of family states of the Arabian coast, a crucial component of the late 19th-century Indian subempire.

The Anglo-Indian map of the East transformed the region into an intermediate space that was bound to fall under British, that is, Anglo-Indian, influence. A sometimes reluctant metropolitan government was encouraged to consider Arabia, the Red Sea, and the Persian Gulf territories which were at once threatened and vitally strategic. The experts of the British Raj had a clear grasp of the apparent inability of the Persian and Ottoman governments to ensure the stability that was vital to British communications.[94] The elaboration of an official map of the transitional East was not a centralized process, as we can see here from the decisive role played by the Anglo-Indians. The agents who surveyed the terrains and compiled the maps could often be far removed from the metropolitan center, in which case they promoted a vision of the territories in question which could be different from that of London. Miles's surveys reflected the wider process of Anglo-Indian expansion in the area. They figured an ordered region, where British rule and expertise had managed to stabilize local chiefdoms and eradicate slave trade. They disciplined the land that stood between Suez and Bombay; a hub of the British world system that was henceforth within the reach of the empire.

4.3 MAPS FROM EGYPT

The quasi protectorate in Egypt developed into a subimperial center as well. An Anglo-Egyptian outlook was rapidly consolidated among the many voices then describing and analyzing the area. British experts developed a distinct perspective on regional issues, promoting their specific understanding of boundaries and geographical classifications. British agents enjoyed a greater liberty to develop their own approach of intelligence than in the metropolis, where political scrutiny could dampen the technocrat's enthusiasm. Reginald

Wingate, director of military intelligence for the Egyptian army, designed a prototypal articulation between intelligence and topography in Egypt.[95]

REGISTERING THE LAND AND TAMING THE NILE

While military uses of the map were perfected, British protectors also initiated the creation of a land registry between 1878 and 1888. Auckland Colvin (1838–1908), an Englishman born in Calcutta, became the director of these early surveys. The experience began in three provinces (Gharbia, Fayoum, and Qaliubia). A baseline of more than a mile was established. In what was an imitation of standard practice in the Raj, crews of locals supervised by British inspectors were recruited. Part of the team was responsible for the triangulation, while the others made detailed measurements of the properties and buildings, using steel tape. The hiring of Egyptian workers, nearly 30% of the employees, facilitated the relations with the villagers. The result was eventually rather piecemeal, since none of the three provinces had been fully surveyed a few years later. The villagers resisted these attempts to map out their properties by displacing the marks left by the topographers.[96] The experimentation nonetheless proved valuable for future projects.[97]

Another attempt at a better cartographic control of Egypt was organized under the auspices of the hydrographic survey of the Nile, beginning in 1889. The need for accurate topographic maps and surveys in the lower valley of the Nile was pressing in view of the many irrigation projects planned by British advisers in Egypt. These attempts, once again inspired by the Anglo-Indian experience, constituted a first in the Arab East. The preface to R. H. Brown's work on the irrigation of the Fayoum (Brown was employed by the Egyptian Department for Public Works) leaves little doubt as to the sources of inspiration of the Anglo-Egyptian experts: "During the last nine years it has fallen to the honorable lot of a small band of English engineers, most of them trained in India, to effect a revolution in the irrigation system of old Egypt, and thereby materially to improve the wealth and agricultural prosperity of the country."[98] The precise definition of property boundaries, the establishment of a land registry for taxation purposes, and the delimitation between the desert and the cultivated land were at odds with local uses. The British trigonometric perspective on Egypt, despite the inconclusive nature of these first surveys, began to produce what was a new conception of the Egyptian territory. It was rendered more amenable to modernization.

In the Egyptian cradle of hydraulic despotism, controlling the Nile became one of the most symbolic schemes of the British experts in Egypt. An agreement with the French, who were given control over the Antiquity Department in compensation, provided that the British would supervise irrigation works. The khedive Ismail had considered transforming the river into a waterway well before the 1880s. He financed the expedition of Samuel Baker in 1865 and hired a number of American cartographers under the direction of General Stone to explore the Nile. It was Colin Scott-Moncrieff (1836–1916), the "prefect of the Nile" in the Ministry of Public Works in 1883, who was to to make Egypt fertile again. He was a former member of the Indian Survey Department and, as such, another example of the Anglo-Indian influence in Egypt. He participated in the construction of several civil works. Another specialist, with a similar background, William Garstin (1849–1925), joined him in 1884. A group of engineers from British India was thus central to the development of an irrigation plan for the Nile Valley. The construction of a dam at Aswan, and a project to use Lake Moeris, near the oasis of Fayoum, as a reservoir, were two central features of the project. Topographical surveys were, in all logic, an indispensable first step toward irrigation planning. To that end, the Anglo-Indian irrigation expert William Willcocks (1852–1932) carried out several expeditions and surveys. In his role as director general of reservoirs for Egypt, he published a report with several maps in 1889.[99] The exploratory appetite of these engineers, who were obviously inhabited by a lofty conception of their mission, did not stop in Egypt. Command over the Nile's upstream was a necessity to ensure a sustainable irrigation for the veiled protectorate. The map of the Nile was to extend to include the Sudan and even Uganda. Only there was the Nile to be effectively and entirely supervised. Scott-Moncrieff formulated a very Anglo-Egyptian rhetorical question in 1895: "Is it not obvious that the Nile, Lake Victoria Nyanza to the Mediterranean should be under a single government?" The map of Anglo-Egyptian Nile logically led British protectors further south, to the Sudan and beyond.

Most of the agents involved in these surveys were also members of the Khedivial Geographical Society and the Egyptian Institute, where they could meet with explorers such as Samuel Baker. They propagated their analyses in Britain. Willcocks led the offensive.[100] London and British agents had, thanks to them, firsthand knowledge of the investment opportunities in Egypt. While the connections between the surveyors and the private investors who wanted a share in the promised prosperity of the land of the Pharaohs are not well documented, there is little doubt as to their existence. One connection can be partly accounted for in the case of the Aboukir Company. The firm was established in London and Cairo in 1887 by Robert Lang Anderson and was to invest in a scheme of land reclamation by drainage near Lake Mariut in the Nile Delta.[101] It had obtained a government concession. Willcocks was present at the inauguration of the pumping station.[102] Other traces of these overlapping interests are to be found in the Aswan Dam project. Willcocks's conclusions on the scheme were finally approved by the Anglo-Egyptian government in 1895. Three years later John Aird's firm became the main contractor on an enterprise which was awarded without any tender bid. The ethical contradictions of this cluster of British private interests and Anglo-Egyptian schemes were pointed out in a 1909 book: "Government participation in reclamation work in Egypt is confined, strictly speaking, to the conservation and distribution of water in the public benefit [. . .]. The Egyptian Government is not concerned with the actual operations of land reclamation further than granting concessions [. . .] to various private reclamation companies [. . .]. These are somewhat numerous, incorporated land-development companies having multiplied rapidly in Egypt during the last ten years, and are chiefly English joint stock companies having their head offices in London, the most important of these being the Aboukir, the Kom Ombo, and the Behera companies."[103] Surveys and maps could serve many interests. British mapping in Egypt provides many examples of how maps could become powerful agents of change.

Richard Burton, then a famous explorer, convinced Tawfiq, the khedive of Egypt, to support an expedition to explore the Arabian coast of the Red Sea in 1877. Burton carefully recorded geological, meteorological, and topographical data.[104] An insert representing the northern part of the Red Sea in an appendix map contains one of the earliest mentions of oil in the region.[105] Burton's passing remarks about oil attracted the attention of the Anglo-Egyptians. American and French companies were already extracting oil in the region in the 1880s. While oil was then mainly refined to be used in the lamps, experiments on motors with special burners were already underway. Circulation between the Anglo-Egyptian and the Anglo-Indian worlds favored an early awareness of this issue. British administrators in Egypt were familiar with oil extraction, thanks to their relations with engineers from the Raj who had observed, with great curiosity, the first Russian experiments on oil for the motorization of their fleet in the Caspian Sea. The potential offered by this new use spurred the members of the Department of Public Works into ordering a report on the matter from the Anglo-Indian expert Charles Stewart (1836–1904), who had been consul in Resht on the Caspian from 1866 on. He had drawn the attention of the British government to the matter on repeated occasions. Stewart and a geologist explored the Egyptian coast in 1888. They published a report with various geological maps. It underlined that "the importance of settling whether petroleum exists in large quantities beneath the surface becomes one of very great importance indeed [. . .]. If oil was struck in any quantity on the Red Sea coast, Baku would become a dust heap, and I think he was right."[106] While geological mapping of the area was still in its infancy, converging reports of travelers were beginning to attract the attention of British agents in Egypt. The large-scale extraction of oil took place much later, but Anglo-Indian maps of the Red Sea undoubtedly accompanied the emergence of raw materials on the Eastern stage.

The Anglo-Egyptian administration had its part to play in reshaping British views of the East. The specificities of the Egyptian and Sudanese situation distinguished its outlook from the Anglo-Indian standpoint. Anglo-Egypt gazed toward Arabia and the tributaries of the Nile. British Egypt conceived its role as a pivot between Asia and Africa, an acknowledgment of the geographical centrality of Egypt in the new equilibrium of the British Empire. At the crossroads between the Nile as an African axis and the Suez Canal it represented the crucial segment on the road between Europe and Asia. This vision, combined with Anglo-Indian geographical representations, favored a recentering of the metropolitan perspective, which until then had been primarily focused on Asiatic Turkey and its border with Russia, from the Bosporus and the Black Sea to the Indian Ocean, the Red Sea, and the Gulf.

British mapmaking in Egypt must also be understood in relation to domestic politics. Some in London doubted the merits of an intensifying British involvement in Egypt. They feared that the costs of semicolonial expansion in the Nile Valley would far exceed its alleged benefits. Cadastral surveys and mapping projects in general participated in the imperial persuaders' rhetoric of expertise to counter these views. Alfred Milner, a staunch advocate of systemic empire-building policies, was one of the first to take Egypt as a model for the colonial policies to come. With its figures, political analyses, and appendix maps, his book *England in Egypt*, published in 1892, inaugurated a new genre of Orientalist writing.[107] Egypt as shown in surveys, reports, and graphic modes of visualization was rationally represented as a laboratory for a new type of empire. This specific Anglo-Indian perspective had far-reaching effects on the Egyptian territory. The mapping of Cairo or the Nile Valley was performative. Maps were not anymore the passive registering of traces of the past or adjuvant to "scripting Egypt," to use Derek Gregory's expression.[108] They were one of the instruments to reform and transform the Egyptian land. As we will see later, this new regime of cartography in the East was to become a key feature of the British outlook on the area as a whole, Egypt being but a starting point.

4.4 "NO MILK FOR BABES": FIELD OBSERVATIONS, INSTRUMENTS AND LEGITIMACY[109]

Thomas Cook's *Handbooks* often advertised scientific and optical instruments from well-established firms such as Negretti & Zambra in London.[110] Pocket barometers, binoculars, and telescopes were no longer the monopoly of the explorer. Even tourists were expected to take account of the "scientific" dimension of their journeys. As the blanks on the map receded, the professionalization of the surveys left less room for the amateur. The "Eastern traveler" had to adapt to new requirements. If travel accounts without an appendix map were common in the first half of the 19th century, the same was no longer true in the 1870s and 1880s. Scientific maps became an indispensable accessory to the travelogue. While it had always been the role of the educated traveler to provide data, the last decades of the centuries saw the generalization of what were more demanding standards. The former monopoly of the adventurers in the East was threatened by more rigorous and efficient surveyors from India or the Royal Engineers. In order to foster a more scientific approach to exploration, the RGS provided the aspiring explorer with one of its most popular publications in the second half of the 19th century: Its *Hints to Travellers* went through updates and reissues for decades.[111] Significantly, surveying techniques and map plotting were emphasized in the editions of the 1870s and 1880s. The publication testified to major evolutions in the representation of what the traveler was to accomplish.

By the late 19th century, the mid-Victorian combination of literary construction and scientific accomplishments was considered less and less satisfactory. While fieldwork remained essential, metropolitan institutions played an increasing part in data validation. The RGS and other learned societies, such as the British Association for the Advancement of Science, were instrumental in disciplining exploration.[112] Every amateur could turn into a potential topographer. Learned societies provided instruments, as well as medals and grants to promote stricter scientific standard, and to spread an instrumental culture among British explorers. A well-received paper at the RGS meeting meant consecration.

Did these evolutions reshape the British map of the East? They certainly strengthened a positivist outlook on the area. The rhetoric which started to prevail figured territories that were scientifically disciplined. The Orient was not different from other divisions of the globe in that regard. Yet its specific shade of 18th- and early 19th-century exoticism consequently waned. The discovery of Arabia, Syria, and Persia in the 1870s and 1880s provides telling examples of the changing nature of exploration in the late 19th century.[113] Anne and Wilfrid Blunt's exploration of central Arabia is emblematic in this respect. Wilfrid Scawen Blunt married Lady Anne Blunt in 1869. Her personal wealth provided the funds for several travels in the East, from Anatolia in 1873 to Egypt in 1876, then to Mesopotamia and Persia in 1877–78. The couple set out to explore Arabia and the Nejd in 1879. They published an account of their journey in *Pilgrimage to the Nejd* in 1881.[114] As shown by Ali Behdad, the contrasting parts played by husband and wife during the voyage testify to the progressive implementation of the scientific requirements of exploration put forward in Great Britain.[115] Wilfrid Blunt followed the "Byronic script" traditionally devolved to the Eastern traveler. He somewhat despised the RGS's attempts to fill in the blanks on the map of the East. He was eventually to describe the learned society as "the instrument of Europe's penetrations and conquests against the wild races of mankind."[116] Blunt wanted to experience an Orient he believed was still untouched by modernity.[117] By contrast, Anne Blunt took pride in collecting as much accurate information as she could.[118] It was Anne Blunt alone who was in charge of the instruments used during their explorations. She had a compass and barometer, and she was familiar with chronometers.[119] Her readings were carefully reported in pocket books "on horseback or whenever I could manage it."[120] She also gathered a wealth of sketches. Their scientific contribution was underlined in one of the reviews of *A Pilgrimage to Nejd*: "Various illustrations from her drawings are given, of which some illustrate the physical formation of the more striking parts of the country, and are acceptable as its only existing truthful representations."[121]

Anne Blunt's participation in the extension of a technical culture in the service of the exploration of Arabia illustrates the specific part played by women in the representation of the Orient.[122] She was the daughter of Ada Lovelace (1815–52), who invented the analytical engine, a revolutionary calculating machine, with Charles Babbage (1791–1871).[123] Ada had a clear-cut vision of what her daughter's education should be: "If she will only be kind enough to be a metaphysician or a mathematician instead of a silly manikin dangling Miss."[124] Her grandmother, Lady Byron, was a great admirer of pioneer pedagogues such as Johann Heinrich Pestalozzi (1746–1827) and Philipp Emanuel von Fellenberg (1771–1844). Her influence was felt in the early education of Anne Blunt, who was instructed in the very unladylike (by 19th-century standards) disciplines of mathematics and natural sciences. This exceptional emancipation explains how she developed her abilities in geographical and topographical surveys. In an era when the *Times* could express doubts concerning "the general capability of women to contribute to scientific knowledge [. . .] their sex and their training [rendering] them equally unfitted for exploration," Anne Blunt's accomplishments were remarkable.[125] Even the RGS, which held out against the membership of women for decades, acknowledged her contributions.[126] When Wilfrid Blunt read a first paper on their explorations to the RGS in November 1879, a startled Anne was invited by Lord Northbrook, the secretary, to sit next to her husband on the platform.[127]

Explorers such as Anne and Wilfrid S. Blunt could ask the RGS to lend them instruments. Many artifacts from this period are now kept in the society's archives, such as the aneroid barometer used by Charles Doughty in Arabia between 1876 and 1878 and the one used by General Charles Gordon in the Sudan in 1874–76.[128] Not only did the metropolitan learned societies promote the best uses for the better preparation of the putative explorers, they also asserted themselves as "centers of calculation," to use Bruno Latour's expression.[129] They controlled instruments before and after the exploration. They collected field books to compute calculations. The recurrent inconsistency of some readings was often underlined.[130] Learned societies contributed to the elaboration of a new regime of legitimacy.

Charles Doughty's explorations provide a further illustration of this process. He arrived in Damascus in August 1875 to join a hajj caravan to Arabia. He unsuccessfully sought funding from the British Association and the RGS. Both organizations declined his request. The reasons underlying the RGS's refusal were straightforward: "The country which Mr. Doughty proposes to examine will be regularly surveyed by the English and American exploration Funds during the course of the next two years."[131] The age of exploration by dilettantes was coming to an end. Doughty embarked on his travels nonetheless. He wandered the deserts of Arabia for nearly two years. In trying circumstances, he developed strategies to achieve the most accurate measurements possible. He had brought a prismatic compass, which he generally kept hidden in a box, and an aneroid barometer. He insisted on making barometric readings on a regular basis. This was "no milk for babes," as he wrote in his accounts, since the task of manipulating these paraphernalia could well put his life at risk. While drawing sketches of Ha'il, the Wahhabi capital of the Nejd, from the rooftop of a palace, he was caught with his papers by a passerby and narrowly escaped being lynched by an angry mob.[132] He treasured his topographical observations so much that when he was exiled in the desert far from the capital of Nejd, his first thoughts were for his pistol and maps: "I returned and armed myself; and rent my maps in small pieces, lest for such I should be called in question, amongst lettered citizens."[133]

When he returned from his tribulations, both financially and physically ruined, Doughty found that very few publishers or scientific bodies were actually interested in his accounts. He made a first presentation before the Bombay Branch of the Royal Asiatic Society in 1879 and found continental Europe to be more interested in his findings than his motherland. Kiepert's *Globus* published Doughty's own version of his travels with sketch maps in 1881–82.[134] Ernest Renan, whom the explorer met in Paris in 1883, funded the publication of part of his notebooks.[135] Doughty was eventually invited to an RGS meeting in November 1883, where he was able to make a short report.[136] The *Proceedings* printed a map showing his route to go with the article.

The title of the map, *Sketch Map of Part of Itinerarium North Western Arabia and the Negd*, folded in the second volume of Doughty's *Travels*, is in itself proof of the highly unusual outlook of the author on his experiences.[137] The term *itinerarium* referred to a long tradition of Roman road maps which also comprised medieval and Renaissance guides for pilgrims en route to the biblical lands. The document echoed the religious concerns that pervade Doughty's account. It also mirrored his stylistic research aimed at the revival of a language and rhetoric belonging to the past. Doughty was very specific about what the map should be. He corresponded, sometimes bitterly, with Ewery Walker, who was drawing a reproduction of the *Itinerarium* in one of the rooms of the RGS for Stanford, imploring him: "In a word please keep wholly and only to my Map; as is only fair and right since it is to illustrate my Wanderings."[138] The conflicting construction of the map is reflected in the end result. It is a singular combination of a very modern instance of chromolithography applied to maps and an archaic representation of the relief. Doughty/Walker's attempts at chorography and geological mapping produced a singular document (fig. 4.4). The density of visible information on the document is a distinct trace of the direct interventions of the author on the draftsman's work. There are comments on ethnographic and archaeological findings. There are elevations, a multitude of place-names, layers of color showing geology, lines tracing his itinerary: the information overload is striking and is in stark contradiction to the standards of the RGS's publications. And even if Doughty could justifiably claim that every "chartographer of those parts of Asia has founded upon my labours," his map was an oddity which he himself tended to reject.[139] Doughty's cartography of Arabia was caught between graphic visualization and the written word, between the travelogue and the modern colored map.

Doughty was eventually granted a gold medal by the RGS in 1912, a very late recognition of his accomplishments. What Anne Blunt's measures and readings had brought to the couple's expertise on the East was lacking in Doughty's

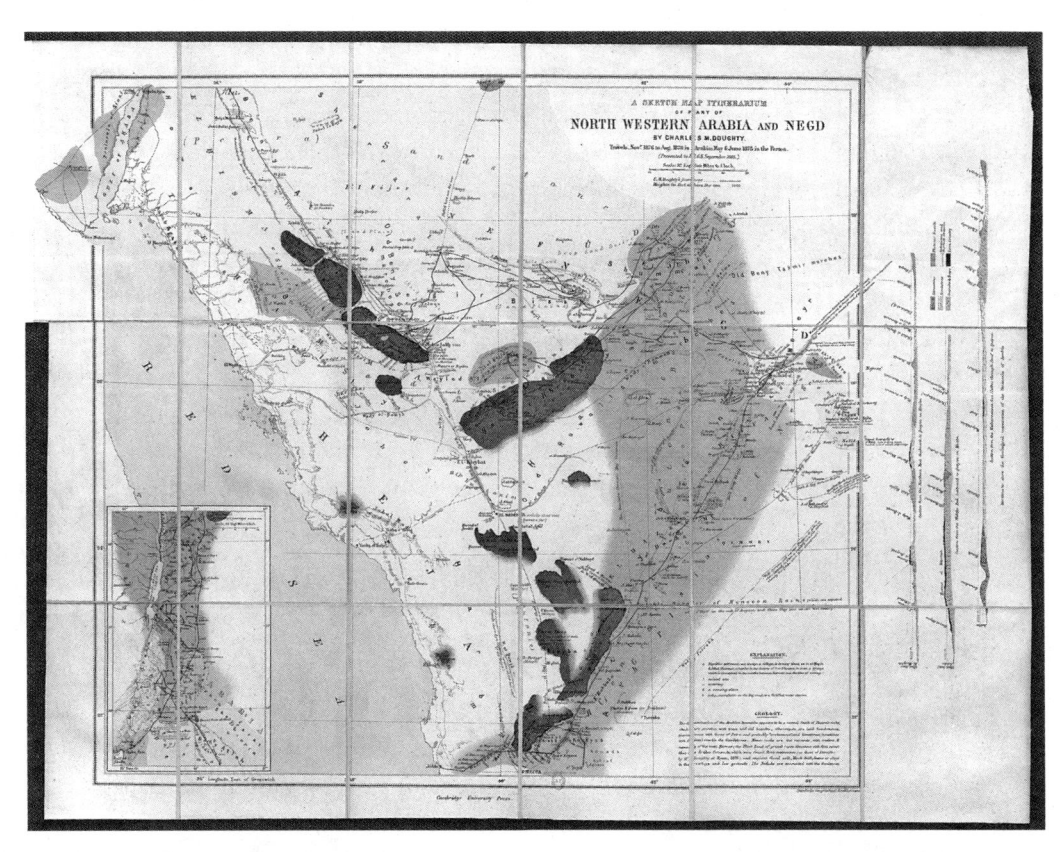

FIGURE 4.4. Charles Doughty, *A Sketch Map of Itinerarium of part of North Western Arabia and Negd*, 1 inch = 32 miles (London, 1884). Size of the original: 49.5 × 67.3 cm. Photograph courtesy of the BNF (Cartes et Plans GE FF-2006).

accounts. He did not comply with the new demands of the metropolitan scientific bodies. His unusual and almost pictorial map, regardless of its innovations, was to become an exception, as the British cartographic production evolved toward standardization. Neither his colorful accounts nor his multilayered representation of Arabia fitted into this new geographical world. It took twenty years to sell the five hundred prints of Doughty's volumes with their folding map. To a certain extent, this commercial failure illustrated the end of an age of romantic and selfless adventure. Joseph Conrad, looking back on decades of evolution of Western geography in 1924, acknowledged the transformation: "No doubt a trigonometrical survey may be a romantic undertaking, striding over deserts and leaping over valleys never before trodden by the foot of civilized man; but its accurate operations can never have for us the fascination of the first hazardous steps of a venturesome, often lonely, explorer jotting down by the light of his camp fire the thoughts, the impressions, and the toil of his day."[140]

In the late 1870s, there was the hope for a stricter scientific coordination in geography and cartography. The example set by the Indian expertise, the experiments of the PEF, the emphasis on better intelligence collection, and the militarization of the mapping of the East all converged as factors in the decline of the figure of the lonely and adventurous Eastern traveler of the mid-Victorian age. It seems that the balance was struck in favor of larger institutions, such as scientific bodies or ministerial departments, whose procedures and critical outlook sometimes questioned the legitimacy of the observer in the field. The rejection of Doughty's still-enchanted East and the overlapping approaches of the Blunt couple illustrate this shift from a mode of introspective research, often coupled with a design of atonement, to a more systematized grip on the region's geography.

The old information order gradually made way for a new one as the identities and status of the explorers changed. The increasing standardization of the collection of knowledge favored a homogenization of the views on the region compared with the somewhat chaotic cartographic translation of the area in the mid-Victorian period. Despite this, the East remained elusive. Many of its territories still offered opportunities, even if Africa could prove a more profitable choice to make a name for oneself. Even though the explorations and maps we have examined participated, during the last decades of the 19th century, in a wider project of codification of the world, some terrains in the East still eluded the attempts at taxonomy, for lack of resources, interest, or abilities. Resilient blanks on the map questioned the actual degree of control and knowledge exercised by Western powers in the region. We thus need to add one further shade in order to understand the complexity of the geographical construction of the East. The still-imperfect survey of the Orient left a margin for the locals to intervene in their own cartographic portraiture, to adopt and use the rhetoric of the British map to resist it.

4.5 MAPS AND COUNTERMAPS IN PERSIA

British India's role in the delimitation of Persia's western boundaries in the late 19th century is well documented.[141] One of the main factors in this projection of the Raj outside its borders was Russian expansionism in Central Asia. Boundary making and the consolidation of buffer states was a way to temper tsarist ambitions. Maps were central to the enterprise. In a region where the very notion of territoriality did not fit into Western categories, the cartographic language profoundly modified the rules of the game. The very notion of boundaries seemed partly irrelevant to local rulers, who traditionally relied on what were poorly defined borders from a European point of view. The Indian surveying drive helped locate and define divisions that were meaningless to local rulers.

The Raj-Persia border in Baluchistan was one of British India's primary concerns. This arid region extending from the Arabian Sea to southwestern Afghanistan was explored from the 1860s to the early 20th century.[142] Frederic Goldsmid, who had become an expert on the region thanks to his role in the construction of a cross-continental telegraph line, headed an arbitration commission in 1871 which was to survey the southern part of the border.[143] In the late 1870s, Thomas Holdich explored Afghanistan and the Perso-Indian borderland. He eventually became the head of the Baluchistan Survey Party from 1883. In 1896, a Perso-Baluch Boundary Commission surveyed and demarcated the remaining part of the boundary north of Goldsmid's arbitration. The second disputed delimitation was between Afghanistan and Persia in the Seistan and Khorasan area. The indefatigable Goldsmid became chief arbitrator of the commission which surveyed part of Seistan-Afghanistan border in 1872. In Khorasan, another segment of the border between Persia and Afghanistan was settled by General Charles S. Maclean's arbitration (1891).

Anglo-Indian surveying and mapping expertise gave the Raj an obvious vantage point in negotiations in which British stakes were high. Most of the arbitrations showed unambiguous signs of the power of the Western mapping processes and techniques in shaping Persia's eastern border. British India's influence on these settlements could often be felt. In most cases, the role of a small group of military officers from the Indian army was decisive. When in 1871, at the request of Persia, the British government set up a commission to delineate the border between India and Persia, Goldsmid appeared to be the best option on account of his experience in the installation of the Indo-European telegraph.[144] Goldsmid had to negotiate his way between the Persian commissioner, Mirza Ibrahim, and the delegate of the Khanate of Baluchistan, Sardar Faqir.[145] The discussions took several months. They hinged on the arduous convergence of the conflicting views and interests of Whitehall, the viceroy in Calcutta, the Bombay Presidency, and the British mission in Tehran. Captain Beresford Lovett compiled a map of the border area. It served as the reference document for the partition of Baluchistan, which was divided into two without proper consideration of local conditions. Goldsmid's assigned mission, namely, to preserve the stability of Persia and the protection of India, was successful. The same logic applied to the division of Seistan in 1873. The commission consolidated the newly created entity as a buffer state between India and the Russian Empire by giving Herat to Afghanistan against the opinion of the Persian commissioners.[146] Maclean's 1891 line in Khorasan was once again a very generous settlement for Afghanistan. The Hastadan Plain was given to Afghanistan, thus enlarging the territory which in theory was to fall into British India's sphere of influence. The Russian outcry against this demarcation was proof enough of the Raj's intervention in the arbitrations. London was often kept at a distance from proceedings that were monitored by the Indian government. A letter from Edward Hertslet dated November 6, 1889, testifies to the ability of the government of India to select carefully the information that was to be sent to London. The FO librarian realized that he was unable to get his hands on a version of the Goldsmid's 1871 map. No one seemed to have a copy of the document in Whitehall.[147]

British maps of the eastern limit of the transitional East seem to offer a standard imperialist tale in which British agents, unchecked by local powers posing as passive spectators, single-handedly explore a territory. To subscribe fully to this narrative would be to ignore the various instances of counter-mapping which punctuated these decades of arbitrations. Local rulers, and the shah himself, were able to use such negotiations for their own purposes. An anecdote told by Holdich is particularly relevant in this regard. It involved a subsurveyor nicknamed the Munshi.[148] He was one of his most trusted plane tablers of the border surveys. Like many of his colleagues, Abdul Subhan, a Kashmiri, was instrumental in collecting topographical information in terrains that would have been deadly for the British surveyor. The Munshi was trusted with secret missions. He sometimes traveled as a native doctor to avoid detection.[149] Holdich never disregarded the part played by these Indian agents in the exploration of the northwestern frontier. He described them as "climbing over the hills like ants, and leaving no peak, no gully unvisited or unrecorded."[150] Their names are often erased from the imperial archives and the tales of adventure of the Great Game. Without them, the British map of the East would have been full of blanks. The subsurveyor Ahmed Ali charted 19,000 square miles of western Baluchistan in the 1880s.[151] Imam Sharif surveyed about 1,100 square miles in one month (July 1889) in the same area.[152] Hira Sing and Ata Mahomed plane-tabled the Birjand region during the summer and autumn of 1885.[153] Abdul Subhan, the "Munshi," was certainly one of the most respected pundits. He had moved up the ladder, from accountant to draftsman and explorer. He participated in the Yarkand mission and was one of the best authorities on the Oxus region and Afghanistan. Hence Holdich's surprise when he met the Munshi in Kabul. Abdul Subhan had resigned from the Survey of India and was working for the amir of Afghanistan in Kabul. The underpaid pundit had turned to the other side. The Indian government apparently refused to raise his wages, and Abdul Subhan decided to sell his expertise to the amir of Afghanistan, Abdur Rahman Khan. He became surveyor general in Kabul, until he was executed for some obscure reasons. For Holdich, who thought of him as a traitor, it was a satisfactory ending. The Munshi's career would be purely anecdotal if it were not for what it reveals about the amir. Holdich underlines in his book "the remarkable acumen which the Amir has always displayed about maps [. . .]. He not only knows the meaning of a map when he sees one, and can decipher its topography, but he has shown a shrewd appreciation of the part which a map may be made to play in the political arena of boundary agreements."[154] How many maps and documents the amir was able to collect we will probably never know. Archiving state papers is often the privilege of stable states. The variable quality of the historical records, if not taken into account, may distort our perception.

Fortunately, some examples of countermapping are still to be found in Western archives, such as the exceptional Persian map from 1883 kept in the manuscript collection of the Bibliothèque Nationale de France (fig. 4.5). It was stamped by Amir Heshmat Al-Molk of the Khozeimeh dynasty.[155] The cartographer, Mohammad-Reza Mohandes, was a tutor of mathematics at the Military College of Tabriz. It shows the suggested boundary between Persia and Afghanistan. The notes stated the Nasseredin Shah's recriminations against the Goldsmid Arbitration Commission which divided his lands. The map was addressed to Malkam Khan (1833–1908). Mohandes, the Persian cartographer and surveyor, had already compiled a map of Tabriz with Colonel Qarajadagi in 1880.[156] The emergence of this local cartography using Western standards reflects what Nancy Peluso calls countermapping, that is, the use of maps to subvert the dominant narrative.[157] The 1883 map of Seistan proves that the Persian elite knew how to use the Western cartographic vocabulary against its European practitioners. They relied on an emerging Persian surveying and mapping expertise in order to attempt to counter the British arbitrations, when they believed their interests to be jeopardized. The reforms implemented by Mirza Taqi Khan Amir Kabir, Nasseredin Shah's grand vizier, in the 1850s, included the foundations of several colleges. August Křiž taught surveying and mapping at the Dar al Fonun (Polytechnic College) in Tehran. He compiled a map of the capital with his students in 1858. The input of Western specialists

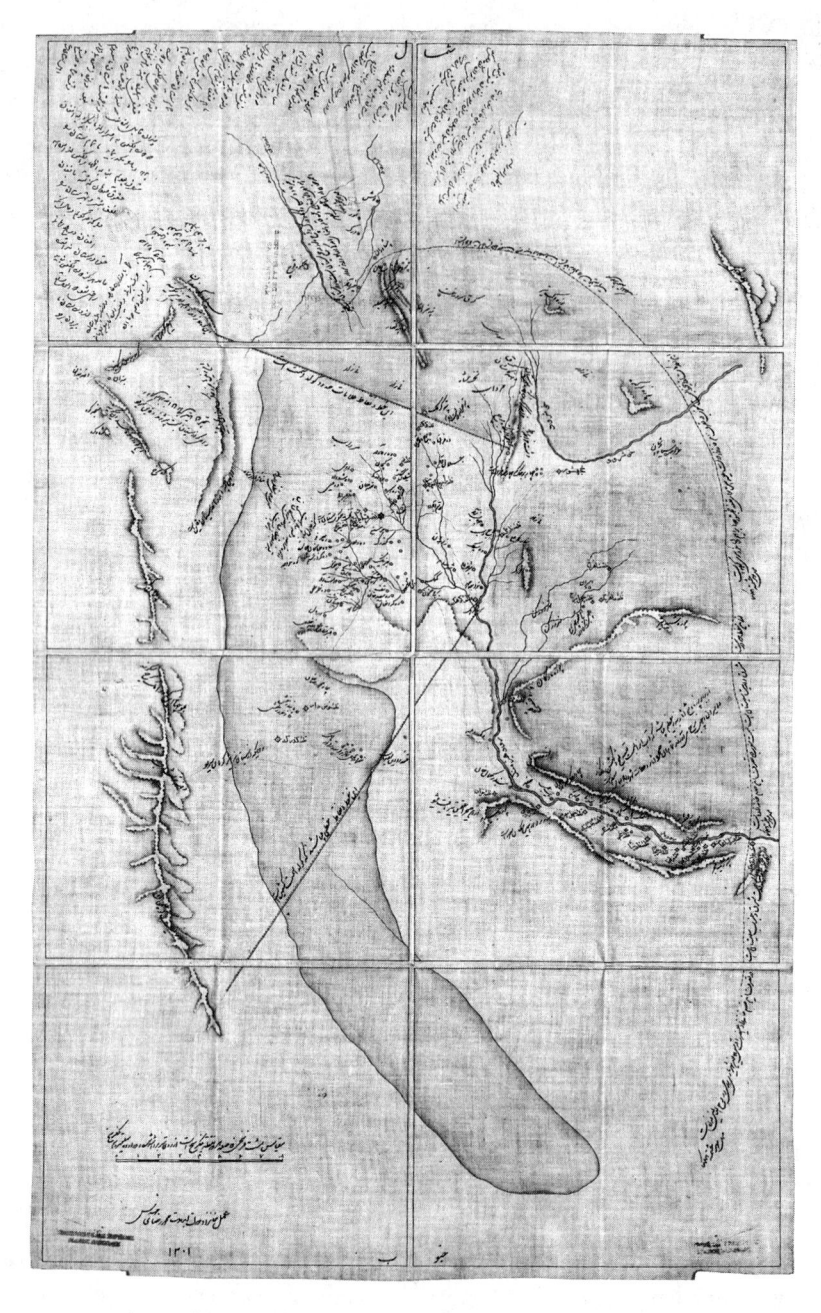

FIGURE 4.5. Mohammad-Reza Mohandes, *Map of Seistan* (1883). Photograph courtesy of the BNF (Manuscrits SUPPL PERSAN 1994).

FIGURE 4.6. Sebah and Joailler, *École Turque*, photographic print (Istanbul, 1890). Size of the original: approx. 20 × 27 cm. Photograph courtesy of the Library of Congress (catalog number: 2010648429).

facilitated the development of a local cartography. Once again, as must have been the case for the amir of Afghanistan, most traces of these uses of the map are now lost.

Such an incorporation of the European cartographic rhetoric was also a feature of the reforms of the Ottoman Empire in the late 19th century. The authorities were actively developing a network of collaborations, the aim of which was to create their own maps. European cartographers and surveyors were recruited to instruct a new generation of local administrators and experts. A specific topographical and cartographic department was set up in 1880. Ottoman cartography drew on its own resources and surveys. It could also buy, translate, and copy French, German, and British maps. An early 20th-century example of these practices, a series of maps of the Ottoman Empire copied from Francis Maunsell's work, is in the RGS archives.[158] As an example of the ample historical evidence of these hybridizations, a 1890 photograph by Sebah and Joaillier provides an insightful perspective (fig. 4.6). The photographers owned a studio in Istanbul and worked under the patronage of Sultan Abdul-Hamid II. The picture belongs to a well-represented genre in late 19th-century

Ottoman photography: the illustration of the renewal of the educational system. Interestingly, in this instance, maps and globes are displayed as outward signs of the modernity brought about by the sultan's reforms. Some of them, such as the wall map shown in the top right corner, though written in Arabic, display a vision of the world in two hemispheres that was conventional in Britain and Europe at the time. What might appear as mere propaganda should not hide the fact that, as the Ottoman elite was becoming more and more familiar with European geography, surveys, and maps, a local expertise was being developed, inducing a more balanced dialogue with Western maps. To British eyes, since the Ottoman Empire was the "sick man of Europe," the true consequences of these evolutions were overlooked. As we will see later, this would transform into a hard lesson learned in Gallipoli in 1915.

Mohandes's 1883 countermap and Ottoman reforms both reflect how European cartography and its universalizing epistemology could very well be recycled and directed against foreign influence. This illustrates how maps and mapping are part of a dynamic process. The relative paucity of records to document these contentious uses must not hide the undeniable existence of competing knowledge systems in the Middle East.

CONCLUSION

Topographical and geographical knowledge was elaborated in a complex of collaborations and oppositions, both inside and outside the British sphere. There was a dual process of interaction, between the metropolis and other imperial centers and also between these subempires and their own agents in the field, where everyone tended to downplay the crucial input of local information. Such multiplicity did not always materialize in a clear-cut division between the metropolis and the rest of the empire. Each actor believed in an idealized cluster of British interests. They formed a nebula of agents who were working in the direction of a supposedly shared objective. All in all, the British "mental map" of the transitional East was a multiple-sourced construct, which involved Anglo-India, metropolitan Britain, Anglo-Egypt, and a range of other stakeholders. It did not form an imperial system as such. The conception of what "empire" or influence meant for the area enormously varied. Yet the multiplying, and sometimes erratic, mapmaking enterprises of the 1870s and 1880s testify to a common drive to explore and domesticate.

In this chapter, I described the explosion of large-scale maps and the multiplication of increasingly scientific surveys that followed the intensifying involvement of Britain in the East. Were these fragmented parts of the East, like Cyprus, Arabia, or Palestine, geographically interconnected from a British standpoint, or was the picture kaleidoscopic? As stated in the first chapters, political, scriptural, and historical presumptions that conflated these entities together predated the invention of the "Middle East." Britain's direct intervention in the area from the 1870s on added a new layer to the embryonic division of Asia. The developments of the Eastern Question, its legal codification through international treaties, and the growing likelihood of an Ottoman collapse, which unified the apparently disparate territories studied here into a single entity, also called for new geographical categorizations. This codifying process was still underway. The 1878 Treaty of Berlin, one of the first international agreements to encompass an area that was very similar to what the inventors of the "Middle East" had in mind two decades later, was vaguely designed for the "settlement of the affairs of the East."[159]

Significantly, the apparent lack of accuracy in the cartography of the world, the many gaps of Whitehall map catalogs, and the very variable uses made of these documents by the political leaders were not just the consequences of a yet-unfinished journey toward progress. The 1870s and 1880s saw the rise of massive systems of data collection—some resisted it, showing little enthusiasm for this new mode of governance. They expressed defiance toward the modernized sources of information out of fear that they would lead to a shift in the power balance in favor of the military expert and the technocrat. Finally, while we must distrust the narrative of progress elaborated inter alia by the annual addresses of the president of the RGS, British surveys and maps of the 1870s and 1880s displayed a very different East from the one that was elaborated during the mid-Victorian era. Late 19th-century maps of the area showed newly delimited borders, ethnographic figurations, colorful geologies, and projected railways. Its surveyors and explorers were no longer the lonely travelers of the first half of the century. Most of them took part in what were coordinated projects. The exotic pictorial map gave way to a modernized cartography that was yet to come to fruition.

What of the inhabitants of the East in this process? The reactions of local populations to British mapping and mapmaking are one of the most difficult aspects to document. Doughty bluntly stated that the "wide wanderers" he met in Arabia had "little understanding of the circumscriptions which we easily imagine, and set down in charts."[160] But traces such as Mohandes's countermap of Seistan are testimonies to what certainly was a widespread phenomenon of resistance and contradiction. This contestation was far more developed than narratives of European exploration and official reports supposed. Local powers and populations were not passive bystanders in the construction of spatial knowledge on the region. The fluid nature of the interplay between self-styled European dominance and indigenous spatialities cannot be underlined enough.

CHAPTER FIVE

VIRTUAL TRAVEL IN THE AGE OF HIGH IMPERIALISM

Charles Gordon's adventures and their catastrophic outcome in Khartoum in 1885 were a media event that became a focal point of the British imperial narrative. They inspired the British printing industry into publishing a remarkable array of maps and accounts. The apparent craze for news and graphic representations of Britain's doomed efforts to relieve Gordon led illustrated papers to print a stream of cartographic illustrations. Maps of the Nile basin punctuated the campaign. G. W. Bacon's 1884 *Bird's-eye view of Egypt and the Sudan* is a striking example of the graphic representations of the imperial stage which multiplied in the wake of Gordon's failed attempts to retain control of the Sudan (fig. 5.1). This imagined aerial view is emblematic of a popular genre that combined cartography with artistic license. The late Victorian period witnessed a burst of enthusiasm for this type of representation. Standard maps were often too austere to compete with the vivid illustrations printed in the weeklies. By contrast, bird's-eye views offered a seductive scenography and a system of visual recreation that could immerse the viewer in the battlefield.

In Bacon's 1884 *Bird's-eye view*, embattled Sudan is viewed from the south. The region is shown as the key to the Nile basin. In the distance, the fate of

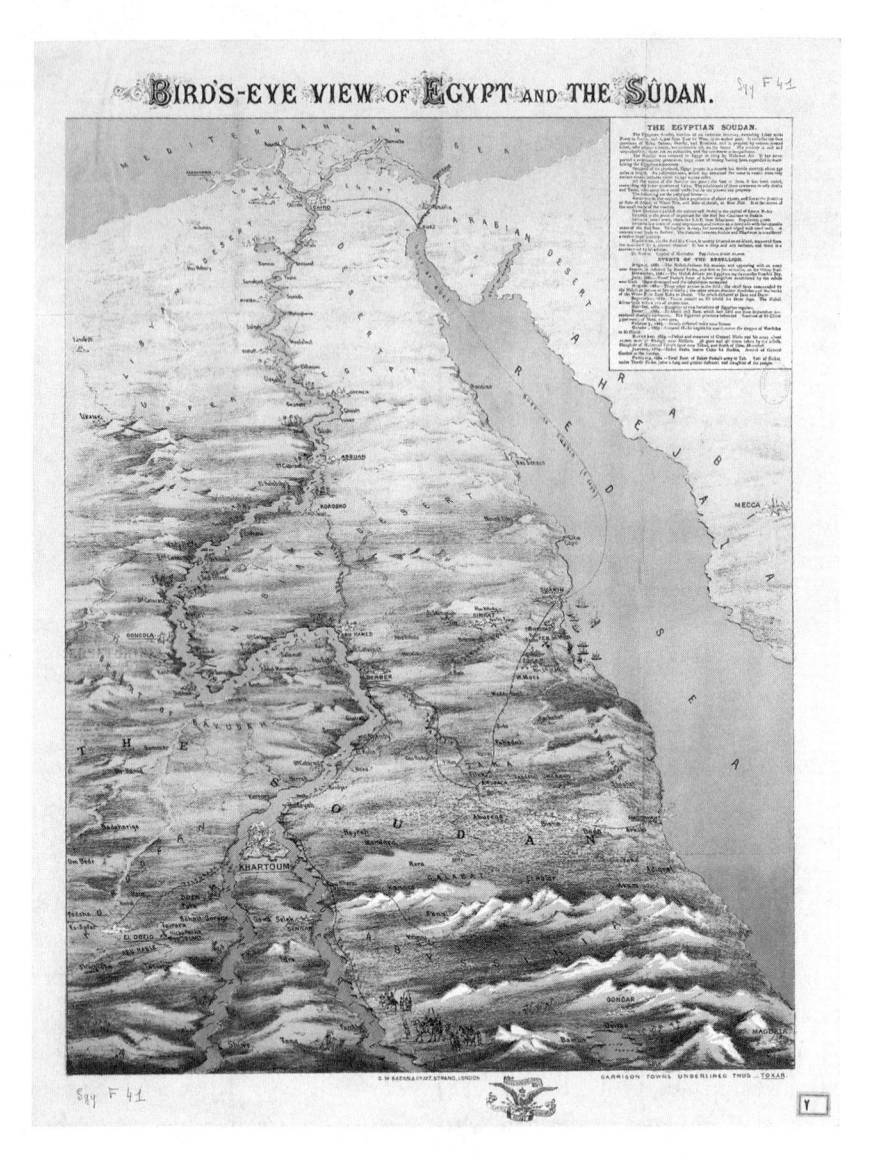

FIGURE 5.1. George W. Bacon, *Bird's-eye View of Egypt and the Sûdan*, scale not given (London, 1884). Size of the original: 63 × 49 cm. Photograph courtesy of Société de Géographie (BNF, Cartes et Plans SGY F-41).

Cairo, the fertile delta, and the Suez Canal appears ineluctably associated with Gordon's fate in a larger-than-life Khartoum. Here and there horsemen are to be observed ominously wandering around in the desert. British boats landing at Suakin draw attention to the vital role of this port on the Red Sea. An insert lists battles and describes the most significant features of the geography of the Sudan. Gordon's stand is pictured as the key to Egypt and the Red Sea. No scales, or any of the technical indications commonly added to the scientific map, were added: the onlooker was there for the thrill. The cartographic document had become accessible: Bacon's bird's eye view sold for one shilling.

This image is one of the many chromolithographed illustrations published by Bacon in relation to the Madhist rebellion. The publisher seized the opportunity to sell aerial perspectives of the battles of El-Teb (February 1884) and Tamai (March 1884), where General Gerald Graham successfully fought back Osman Digna, a Madhist military commander.[1] He also printed other more conventional cartographic documents, such as his 1884 *War Map of Egypt*. Many of the well-established mapmaking establishments seized the opportunity. W. & A. K. Johnston published a *Large Scale War Map of the Eastern Soudan* in 1885.[2] G. Philip released a *Special Large Scale War Map of the Soudan* in 1885.[3] The editor at the publishing firm of Stanford went one step further. Gordon's sanctification as an imperial martyr fostered a market for memorabilia in the mid-1880s. The firm even printed a facsimile of one of the hero's sketch maps, transforming the general's own surveys into a relic.[4] The illustrated papers in turn proved innovative in their uses of the map. In March 1884, the *Graphic*, for instance, published a hybrid document, a composite of cartography and travelogue, which further enhanced the geographical continuities between the Egyptian delta and the Sudan.[5]

Bacon's print exemplifies a turn from the written word to a more graphic figuration of the world's geography. Maps, like photographs, transformed the position of the metropolitan observer.[6] The very nature of traveling and travel imagery evolved as a consequence. Cheaper atlases, photoengraved images, and bird's-eye views turned the metropolitan observer into an armchair explorer, a stationary beholder of the world. The East was a vehicle for the purveying of salable news and information. The attraction of instantaneity began to challenge the traditional lure of the exotic and biblical Orient. Cartographers, mapmakers, and the imaginations of editors competed to supply dramatic cartographic documents. They intensified their efforts to attract a broader audience, which often felt concerned with the events that shook the region, such as the Bulgarian atrocities of the late 1870s.

This chapter assesses the effects of the growing imperial involvement in the region on the cartographic production of the 1870s and 1880s. In particular, it describes how the biblically tinged Orientalism of the mid-Victorian era was eventually challenged by new conceptualizations. It shows how the confrontation with distant lands and the empire profoundly changed the way British people located themselves in the world. It also analyzes how representations of the region were shaped by innovative printing processes and mapmaking techniques. In doing so, it demonstrates that cartography was progressively exported out of its context of origin to become a widespread type of illustration.

5.1 THE MAPMAKER'S EAST: AN IMPERIAL GAZE?

A PROLEGOMENON: AUTHORSHIP AND AUTHORITY

Many different actors were involved in the process of mapmaking. There were those who gathered the data. This was the part played by the explorer and the surveyor. Others compiled, designed, and drafted the map. Very often, there was an author, not always able to draw a map, who oversaw the creation of the document. He guided, amended, and commented upon the work of the cartographer himself. In the later stages of the creation of the document, the mapmaker was the one who printed the map, but he was not always the publisher. Stanford, Bartholomew, and Philips prepared documents that were printed by others. The long process from manuscript to print involved a succession of interventions that might sometimes leave the observer wondering who the author of the map really was. From data collection to its compilation by a specialist, mapmaking was the product of a complex dialogue between several agents.

The increasing complexity of the production chain forced major publishers into structuring their activity on an industrial scale. Major firms moved away from the map *d'auteur* to adopt a cost-minded division of labor. Stanford, for instance, employed nearly one hundred people in the 1890s in its London offices, which were specifically designed to streamline processes. Another illustration of these productivity-oriented reorganizations is given by the collaboration between J. G. Bartholomew and Thomas Nelson beginning in 1888. The partners, then known as the Edinburgh Geographical Institute, built new and larger premises on Park Road, Edinburgh, in 1889.

The industrialization of mapmaking was concurrent with an even greater emphasis on the scientific legitimacy of commercial cartography. Map reviews published in geographical magazines emphasized the significance of technical credibility. In a context of competition between European mapmakers, concerns expressed by the various geographical societies as to the relative quality of British mapmaking in comparison with German prints became recurrent. In an increasingly competitive market, atlases which did not incorporate the latest data could encounter sharp criticism. Updating antiquated material no longer sufficed. In the *Scottish Geographical Magazine*, the author of a review of the very expensive *London Atlas of Universal Geography* by Stanford regretted that "the really disappointing feature of the atlas is that so many of Mr. Arrowsmith's old maps are retained [. . .]. We do not in the least depreciate the work of John Arrowsmith, for no better or more thorough map-work has been done in England since his day; but he had not the advantages of modern surveys."[7] Stanford had bought part of Arrowsmith's stock in the early 1870s. Some of the maps dated back to the 1840s. The inclusion of the most recent information became a commercial imperative. Mapmakers had to strengthen their relationships with surveyors and explorers to gain direct access to new material. As a consequence, the most successful firms cultivated privileged connections with major figures of British exploration around the world. They also tended to build up their own reference libraries.[8]

The most productive suppliers of topographical knowledge therefore had an instrumental role in shaping what the cartographic figuration of Eastern places would be. Anglo-Indian cartography of the East was more likely to be transcribed into the commercial sphere in Britain thanks to a comparative advantage given by its surveying expertise. The multiplication of documents highlighting the critical connections between the Raj and Britain was an echo of this influence. Maps illustrating the communications between the mainland and India were found on a regular basis in 1880s and 1890s atlases. This is epitomized by a map of the "Road to India," centered on the Nejd, which was published in 1887 by John Walker in an inexpensive atlas entitled *The British Colonial Pocket Atlas*.[9] In Bartholomew's 1891 version of his *Popular Atlas* a specific map was devoted to the illustrations of India's communications with Egypt and the Mediterranean. Concurrently, Red Sea–centered maps became commonplace in late 19th-century atlases. Input from the many surveys that were organized after the 1882 intervention in Egypt allowed publishers to provide much more accurate maps than in the 1850s and 1860s. A map of the Red Sea from Johnston's *World Wide Atlas* (in its 1892 edition) was partly compiled from the data collected by Burton on the Midian, or by British officers stationed in Suakin.[10] The region now under British influence in the Middle East was, as a result of the congruence of various elements, not least of which were those from commercial publishers and their markets back home, more prevalent in cartographic output than in the Palmerstonian era. A stream of data from the Anglo-Egyptian and Anglo-Indian world favored the elaboration of maps that advertised and justified the extension of British power in the region.

NEW TECHNIQUES AND NEW CONCEPTIONS: THE EAST IN COLORS

The fall in printing costs accelerated in the 1870s with the constant improvement of presses. Color printing became affordable and allowed experimentations. Chromolithography became increasingly common for well-established firms such as Stanford, Bacon, Bartholomew, and Johnston. As a consequence, colored maps became frequent at every level of the printing industry, from the specialized journal to the illustrated press, both in Britain and in Europe. Karen Pearson's surveys on the "color revolution" in 19th-century cartography testify to that fact.[11] The first irresistible evolution involved the generalization of a new representation of the terrain. Hachuring, which had been the main device for the representation of relief before the mid-19th century, was gradually replaced by other techniques. Hypsometric tints and the use of contour lines spread in the late 19th century, even in the cheapest maps. A. K. Johnston's map of Syria and the Suez Canal is an early illustration of the use of contour lines in what was a reasonably affordable atlas.[12] Never before had such an apparently modern figuration of the relief been at the disposal of the general public. Sacred Palestine and the Dead Sea were suddenly given a relief that only James Wyld's georama or the PEF's relief maps had previously showed.

Chromolithography broadened the thematic possibilities of the commercial map. Geology, the distribution of populations or religions, and even densities were more easily included in the representation of the East. This had consequences for how the traditionally biblicized and exoticized Orient was displayed. The first traces of modernity which punctuated the cartographic representation of the region in the mid-Victorian era became increasingly significant. New geographical categories emerged as the interconnections of Britain, its empire, and the rest of the world became more obvious. E. G. Ravenstein's maps of Asia for George Chisholm's *Handbook of Commercial Geography* (1889) confirm this impression.[13] Chisholm was a member of the RGS from 1884, and the Royal Statistical Society from 1886.[14] He was a staunch advocate of economic geography. His book was a very early illustration of how maps could explore the potential value of a territory. It contained a map of the railways in Turkey as well as indications on the geography of commodities in Asia. In reaction to the emergence of this new genre, the Bartholomew firm designed the twenty-seven maps of the *Atlas of Commercial Geography* published by Cambridge University Press in 1889.[15]

These pioneering maps were a reflection of a fast-evolving cultural environment. The last two decades of the 19th century witnessed a profound shift in the way Britain related to the outside world. The words and ideas used to describe the British Empire and the nature of its global influence were distinctly different from the rhetoric and concepts of the mid-Victorian age. The 1880s witnessed the first manifestations of constructive imperialism. The Imperial Federation League, which, as its name indicates, advocated the advent of a Greater Britain inclusive of the white dominions, was founded in 1884. The supporters of the confederation promoted other central ideas. They explicitly insisted on the notion of development, that is, on the need to plan more expertly for the economic growth of the British Empire and the territories under its influence. What better vehicle than maps to convey these views? Arthur Silva White, a major figure of the Royal Scottish Geographical Society and an enthusiastic supporter of the Britannic Confederation, a topic on which he edited a book in 1892, designed a series of cartographic documents to further this cause.[16] He first invented contour lines of a new type, which he named "chrestographic curves."[17] They were meant to show the "progressive value of a given region." He applied his invention to a map of Africa in 1891. This early version of a human development index was a combination of six indicators (climatic phenomena, natural communications, natural resources, exterior trade and commerce, indigenous political conditions, and foreign political conditions, including the capacity for development of European institutions). Egypt's and South Africa's potential for development were thus enhanced.[18] Silva White designed similar maps in a later work on Africa, again with the help of Ravenstein, the cartographic editor of the *Geographical Magazine*.[19] They included information about Arabia, Persia, and the Turkish Empire. In a book released in 1899, he then focused again on Egypt, a field of experimentation of constructive imperialists such as Alfred Milner.[20]

As with other evolutions of the map industry in the late 19th century, these cartographic innovations were not specific to the representation of the Orient. Given the singularity of some Eastern territories in the European and British imaginations, these new economic maps, coupled with the apparent modernity they displayed, conflicted with the more traditional and stationary figuration of the early decades of the century. In a context of growing concern over foreign competition, and specifically German involvement with the Ottomans, the increased number of economic maps was one of the many ways to raise the public's awareness of the ties that bound the British world system together. The transitional East was incorporated into an international environment characterized by a first globalization of trade, in which the Suez Canal played a significant part. Not only were these regions, which were traditionally remanded to the periphery, visually connected to the Western world; they were also given a more utilitarian value and a potential future.

Of the many colors used by British cartographers, one bore a symbolic weight: the red on the maps became the natural figuration of the British imperial influence. It occurred in the overwhelming majority of atlases and individual maps of the empire in the late 19th century.[21] Its use is particularly interesting in the case of the region under study here, which was mainly located in Britain's informal empire. Except for Aden and Cyprus, one would not expect to see any territories between Egypt and India painted in red. British mapmakers were nonetheless often liberal in their use of the imperial red. A map published in Bartholomew's 1891 *Popular Atlas* illustrates these overlapping uses (fig. 5.2). Aden is logically represented in a red surface. The coloring of the southern coast of the peninsula, shown in red from the east to Ras Fartak, is rather more surprising. The map gives the impression that Britain legally controlled the entire coast and several miles into the hinterland. This was not the definition of British sovereignty in southern Arabia, according to the FO's official position at the time. It is an echo of the Anglo-Indian position on the geostrategic importance of the area. Some in India wanted British influence to extend around Aden to secure the coaling station. Surveys were organized to study the tribes in order to assess who was to accept British protection. A map from an official memorandum, written by Francis L. Bertie in 1886, a supporter of Anglo-Indian intervention in Aden, depicted the distribution of the various populations.[22] An earlier account on the same subject, by F. M. Hunter, assistant to the resident in Aden, was also published in Bombay the same year.[23] As J. G. Bartholomew compiled this recent information in order to design his 1891 map for Nelson's atlas, his work legitimated the extension of the British influence in southern Arabia. His use of the red color to figure Egypt is even more remarkable. Egypt's apparent sovereignty was respected. It is painted in a neutral yellow, though a glance at the ill-defined border of Egypt in the Libyan Desert shows a surreptitious use of the imperial red: an illustration of what the expression "veiled protectorate" truly meant.

PURCHASING MAPS

The average price of individual maps and atlases evolved very slowly. The one-guinea atlas became a reference price in the late 19th century. Individual maps published by G. W. Bacon to illustrate the 1882 intervention in Egypt were, for example, sold for one shilling. In Johnston's range of atlases published in 1895, prices ranged from thrupence to six pounds six shillings for the overpriced *Royal Atlas*. The new pocket and school atlases were sold at affordable prices

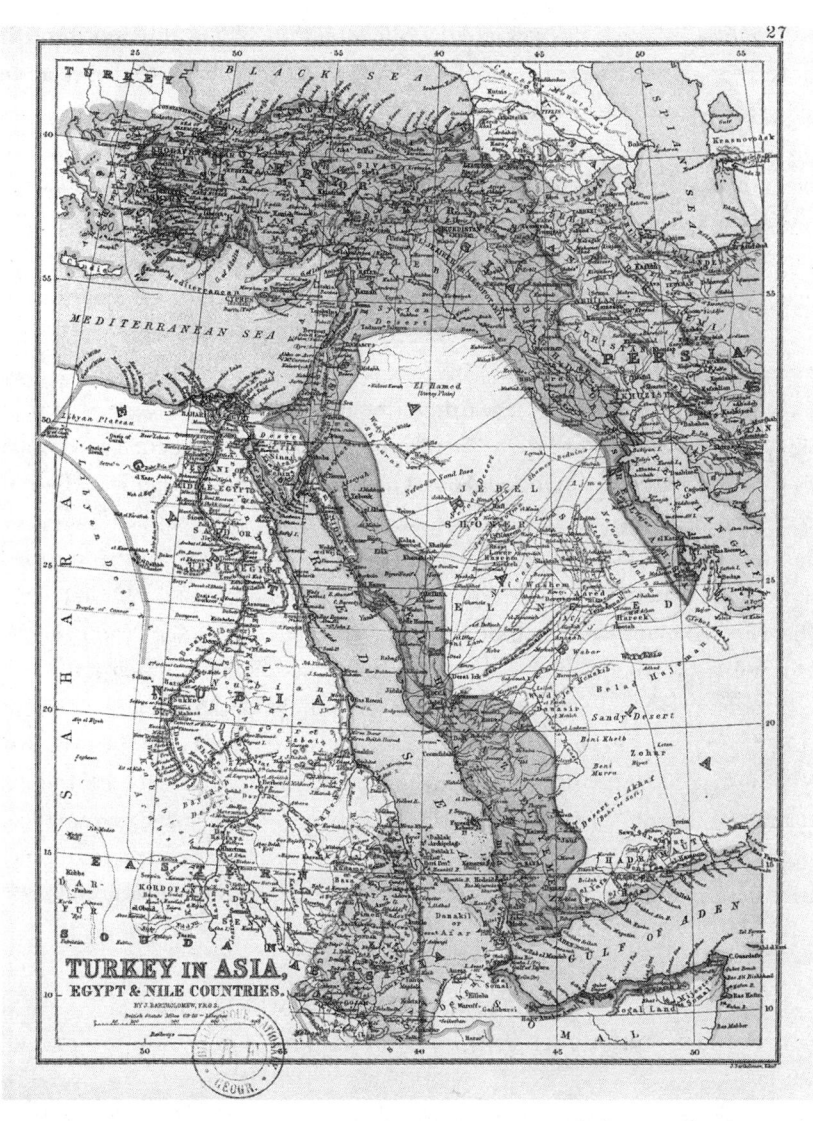

FIGURE 5.2. John G. Bartholomew, *Turkey in Asia* (detail), in John G. Bartholomew, *The Popular Atlas of the World* (Edinburgh: T. Nelson, 1891), map no. 27. Photograph courtesy of the BNF (Cartes et Plans GE DD-402).

for the middle-class buyers. Commentators still deplored the fact that quality atlases remained too expensive. While the best atlases kept their aristocratic standing, most mapmakers published cheap versions of their reference books, such as J. G. Bartholomew's *Royal Shilling Atlas*.[24] The gradual decline of prices of cartographic products, such as individual maps, combined with growing interest in geography, expanded the market.

The East in particular attracted public attention. The archives of the Bartholomew mapmaking firm include two families of records that give an incomplete but relevant view of the production and sales of maps in the late 19th century. The *Day Books* and *Printing Records* provide a rough overview of the number of individual maps printed by Bartholomew from 1869 onward. A look at the production of single maps of Egypt and Palestine demonstrates the level of interest for these two territories shown by the firm's clients. In absolute terms, a large quantity of maps was printed by the Scottish firm: tens of thousands for Egypt, nearly two hundred thousand for Palestine. A more detailed review of the data reveals that, in the case of Egypt, the demand increased during the intervention against Ahmed Urabi in 1882. Bartholomew printed thirty thousand maps of Egypt for various customers. He opened a new market for his maps of Egypt in the 1890s when H. Gaze and Sons, one of Thomas Cook's competitors, asked the mapmaker to print tourist maps of Egypt. On December 21, 1891, Bartholomew printed 10,100 sheets, each containing eight maps of the Holy Land under the title *Palestine to Illustrate the New Testament*. These 80,800 maps were intended for the National Society for Promoting the Education of the Poor, which advertised the collection in its catalog at the price of six shillings, hoping that the useful map would end under many Christmas trees. The *Printing Record* also lists the original maps designed by the firm from 1877 (fig. 5.3). The relative importance of the Orient compared with India or Australia is salient. These indications testify to the existence of a genuine market for maps of the East. This feature of late 19th-century British geographical culture is particularly reflected through the uses of maps in periodicals.

5.2 GEOGRAPHICAL IMAGINATIONS AND THE NEWS: "OUR LITTLE WARS TEACH US GEOGRAPHY"[25]

Legal and technical developments in the 1870s and 1880s changed the way dailies and weeklies were edited, printed, and sold. The reproduction of maps for the press was automatized from the 1870s. When a map was urgently needed to illustrate an article, editors could rely on the technical expertise of Mr. Shanks's Patent Type Foundry established at Red-Lion Square in London. The newspaper would ask for a prominent mapmaker to design a simple outline map. Once the map was ready, it was sent to the foundry, where it was mechanically reproduced and reduced with an engraving machine. Words and letters were added with templates. Once the initial autographic block was ready, it was copied to produce stereotype blocks, which were sent to the newspaper's

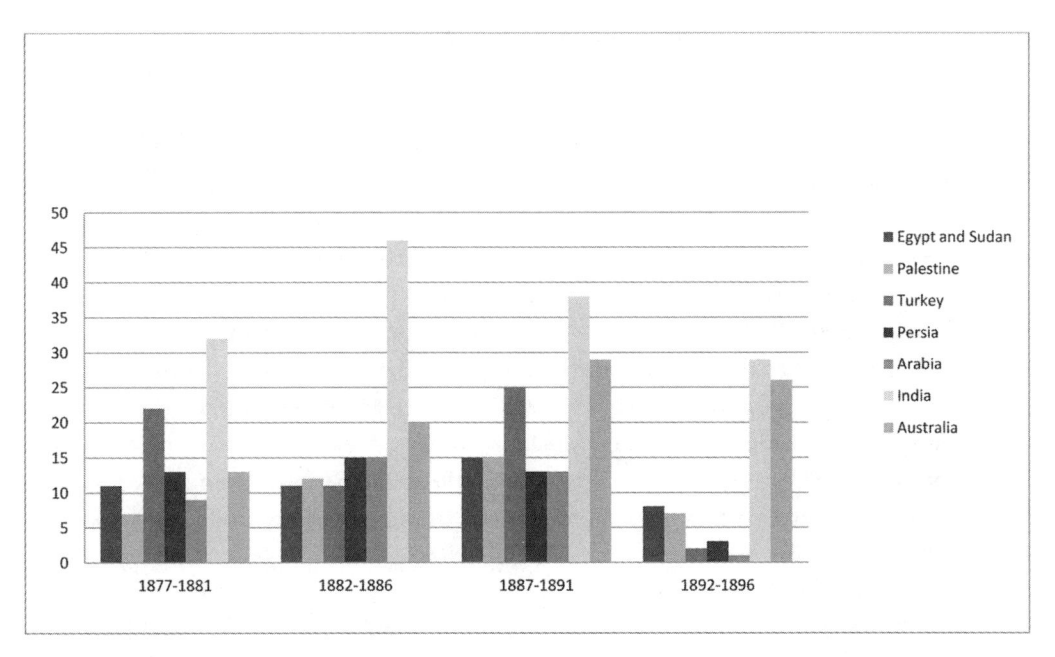

FIGURE 5.3. Original maps listed in the Bartholomew print records. © Daniel Foliard.

presses and thus included in the type as easily as any other paragraph. The process was simple and cheap. Even provincial newspapers such as the *Leeds Mercury* did not hesitate to use Shanks's services. The regional daily published a map of Turkey in Europe in July 1876 to illustrate the Serbo-Turkish War.[26]

The Nile Valley attracted extensive news coverage in the 1880s. From 1882 on, the suppression of Ahmed Urabi's revolt triggered a flow of maps in the popular press. The *Penny Illustrated Paper*, which targeted the working class, illustrated Lord Wolseley's campaign from Ismailia to Cairo with a series of rough maps that were part of a *Pictorial Chronicle of our War against Arabi Pacha*.[27] The process used to print the graphics was developed by Alfred Dawson in 1872. It was derived from Edward Palmer's glyphography. Dawson's firm, the Typographic Etching Company, was one of the first in Britain to offer phototyping solutions to everyone, from book publishers to newspapers. Maps, images, or plans were copied on blocks that were sent to the printer. The traditional engraver's work was replaced by mechanical processes. Even the cheapest papers could afford to include a cartographic document to attract readers thanks to lower production costs.

Journalistic maps were often sketchy. Despite their rough outlines, they were an adjuvant that strenghtened the legitimacy and authority of the written word. One of the most active promoters of British "new journalism," the flamboyant W. T. Stead, quickly realized the potential of this category of graphics when he became editor of the *Pall Mall Gazette* in 1883. Stead is famous for his investigative and sensational articles on prostitution and white slavery.[28] He was in the vanguard of a journalistic revolution that has been well researched. A somewhat overlooked dimension of his editorship was his noticeable interest in maps. From 1883, the Madhist rebellion gave the *Pall Mall Gazette* several opportunities to enlighten its readership on the topography of Egypt and the Sudan. Over a period of two years, the newspaper consistently printed multiple sketch maps of the terrain in order to document the British attempts at repressing the Islamist revolt (fig. 5.4). Most episodes of the war were cartographically depicted a week or two after the events, from Colonel William Hicks's catastrophic defeat at the battle of El Obeid in November 1883 to Charles Wilson's failure to relieve Charles Gordon in Khartoum in January 1885. Stead's *Pall Mall Gazette* played no small part in creating Gordon's legend. Maps were central to the narrative.

The combination of new printing technologies with Stead's own crusades made this pioneering coverage of the Madhist wars possible. Like many Britons, Stead developed a marked interest in the Eastern Question in the late 1870s. He was one of the strongest voices to denounce the Bulgarian atrocities, a central issue in British politics of the late 1870s. He was particularly alarmed at the fate of Eastern Christians. As a consequence, the highly graphic cartographic coverage of Gordon's demise was far from neutral. It was part of a larger attempt to force decision makers into action. An educated readership, provided with modern and graphic representations of the unfolding crisis, was to put democratic pressure on the British government. Maps of the East turned into arguments to be marshaled in a fierce political debate. As noted by Stéphanie Prévost, Stead made relentless use of journalistic cartography to promote his views.[29] He used maps to express his indignation when the 1894–96 Armenian massacres made the headlines in Britain. One of his articles on the subject included a groundbreaking map showing the locations of the Ottoman crimes, and the casualties.[30] Stead thus developed in the 1880s and 1890s a new genre in the cartographic figuration of the East: the humanitarian map. As early as July 1883, the *Pall Mall Gazette* published a map of the cholera pandemic in Egypt.[31] It echoed the emergence of medical cartography.[32] It was also a symptom of the increasing importance of the humanitarian outlook, then a facet of the British public understanding of the East.

The new journalistic cartography did not supersede the earlier, mid-Victorian uses of the map in the press. The large foldout map, with its many

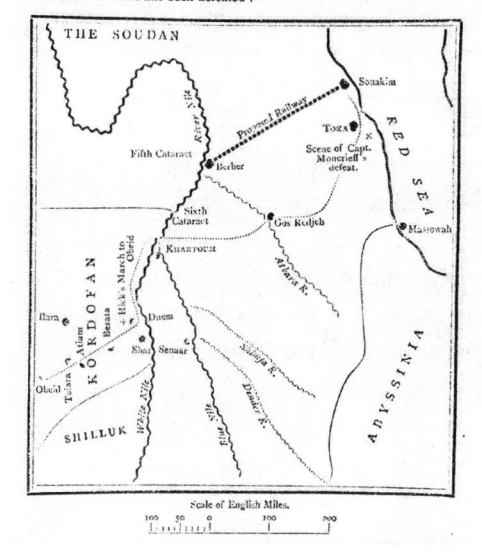

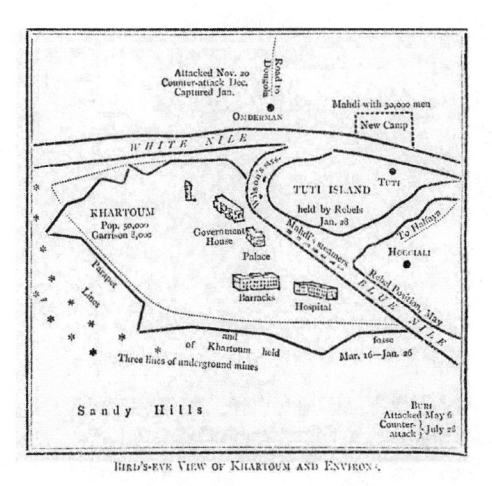

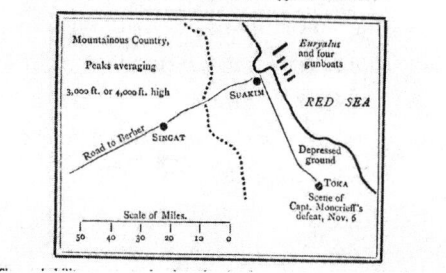

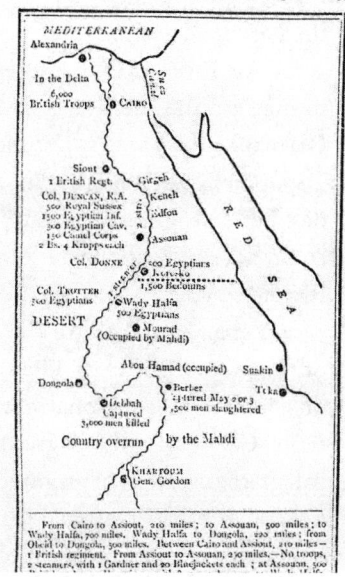

FIGURE 5.4. William Stead, *Mapping the Mahdi rebellion in the Pall Mall Gazette*, various scales (London, 1883–85), in *The Pall Mall Gazette* (clockwise from top left): November 22, 1883, p. 11; July 4, 1884, p. 7; December 10, 1884, p. 4; February 6, 1885, p. 8). Image © The British Library Board. All Rights Reserved. Images reproduced with kind permission of The British Newspaper Archive.

details, kept its consumer appeal. It was still printed as an exceptional object, on offer in the supplements of the great illustrated papers. These spectacular cartographies looked as if they were designed to be framed and hung at home. Such graphics were of course much more expensive to make and print than the sketch maps of the dailies. That is probably why even the distinguished *Illustrated London News* did not hesitate to reprint a twenty-year-old map of the Turkish Empire in its April 23, 1877, extra supplement.[33] Most of the time, given the very competitive market for illustrated periodicals, the editors provided their readers with a range of highly elaborate and inventive maps to foster public enlightenment and supply a form of amusement. With the mastery of chromotypography from the 1870s, the duplication of spectacular colored maps became more common. In 1877–78, while the Russo-Turkish War energized the jingoists, the *ILN* offered its consumers various eye-catching illustrations, such as a view of the Balkans in 1877 (fig. 5.5). A map of the Sea of Marmara was then printed on March 9, 1878. It staged the Russian threat to the straits with an innovative point of view.

In the attempt to stay ahead of the competition, the *ILN* employed the best talents available. One of its main cartographers to illustrate articles on the Eastern Question in 1877–78 was actually one of the experts in the field. A survey of the weekly in the late 1870s shows that most of the maps relating to the Eastern Question were designed by E. G. Ravenstein (1834–1913). This German cartographer, a student of Petermann, worked for decades for the Topographical Department of the WO. In 1872, he was offered the position of chief cartographer of the RGS, which he apparently declined because he could not smoke in the building. He started working for the *ILN* in the 1870s. Ravenstein could use the RGS map room and network to design up-to-date documents. His work for the *ILN* testifies to an often overlooked dimension of the journalistic map. The cosmopolitan Ravenstein was actually a pioneer of scientific internationalism. As early as 1871, he attended the first International Geographical Congress, which was in held Antwerp, hoping to participate in the creation of an International Geographical Association. He was also a member of the National Olympian Association, further evidence of belief in international cooperation. Ravenstein's career and work are illustrative of the denationalizing forces that often characterized European scientific communities in the late 19th century.[34] A look at the very professional and purely descriptive maps of the Balkans he designed for the *ILN* in 1877–78 leaves little doubt as to the fundamentally educational nature of his journalistic cartography. One should not systematically exaggerate the extent to which imperialistic assumptions underpinned late 19th-century cultural representations. It is

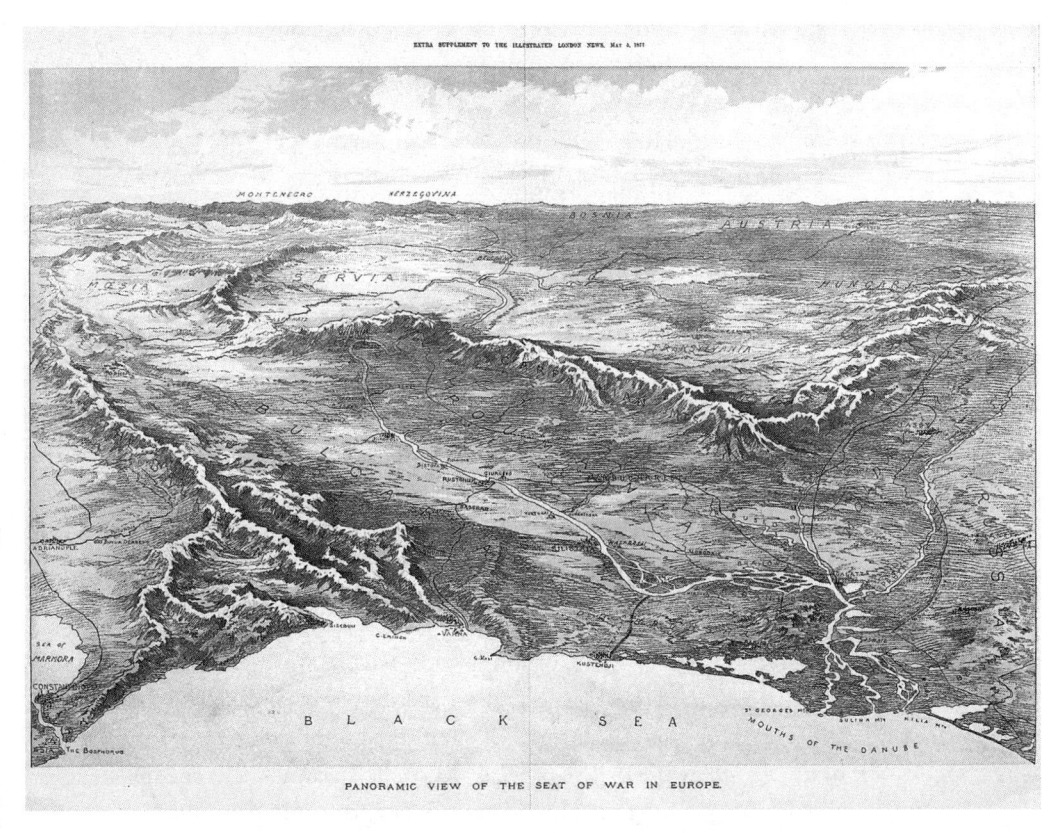

FIGURE 5.5. *Panoramic View of the Seat of War in Europe*, in *The Extra Supplement to the Illustrated London News*, May 5, 1877. Size of the original: 40 × 54 cm. Thinkstock standard license.

however true that the nature of the relations between the press and the experts cast doubt on the neutrality of the cartographic figuration of the East in the popular press. While Stead was obviously printing maps aimed at questioning governmental action, other editors were much less critical of the nature of British influence in the region.

The *Graphic* printed what, in this respect, was a very significant map in 1884.[35] It was entitled *The Russian Approach to India*. It manifested the advances of the tsarist regime toward the Indian Ocean. The region, deprived of its mountains and deserts, looked like an inviting prospect for a Russian army trying to invade the Raj. Neither the Himalayas nor the extremely inhospitable deserts of Baluchistan seemed to stand in its way. The rudimentary nature of the new printing methods contributed to this impression. Interestingly, the author of the accompanying article referred to Charles Marvin's *Russian Annexation of Merv*, published weeks earlier. Marvin, a Russophobe, who had sold the text of the secret agreement between the United Kingdom and the Ottoman Empire to the press in 1878, was one of the leading advocates of a more aggressive policy.

It is therefore indisputable that journalistic maps had a substantial part to play in the public debates regarding the Great Game of the Eastern Question. Did they shape the readership's views on the region? To arrive at any accurate assessment of the reception of this kind of documentation is, in reality, a complex and ultimately insoluble question. To acknowledge an obvious evolution in the quantity and quality of maps in the press from 1870 to 1895, both of the East and of the rest of the world, is one thing. To assess the effects of their popularization on the readership's awareness of international affairs is quite another. One thing seems clear. The millions of buyers of British newspapers were not "absentminded" or indifferent to the destinies of the United Kingdom in the world.[36] The cartographic competition between the various illustrated papers is in itself a confirmation of the existence of a growing demand. The substantial circulation of the leading periodicals, combined with the steady interest in the Eastern Question, nurtured by the Bulgarian atrocities, news of discoveries in Palestine, Arabi Pasha's revolt, Gordon's death, Russian threats in Central Asia, and the Armenian massacres, certainly contributed to placing the East on the map, at least for a significant part of the British public. Surveys of the coverage of the region by the British press, by Michelle Tusan and Roger Adelson, corroborate what is suggested by other statistics presented here: the East held a preeminent position in the British late 19th-century imaginings of the world.[37] Neither the commercial map nor the journalistic map was the only vector through which the region came to achieve such a pervasive presence in British imaginations, but cartography nonetheless played a crucial role in framing the area for the British public.

5.3 A SMALLER WORLD

The already-mentioned *Hints to Travelers* offered an enticing isochronic map in its fifth edition.[38] It helped the aspirant explorer to measure the actual travel time between London and any place in the world. While part of the Orient still remained unreachable, Smyrna, Cairo, and even Jerusalem, appeared to be as near to Britain as Paris, Rome, or Berlin. In the 1890s, the P. and O. weekly service from London could take travelers to Port Said in twelve days on luxurious steamships. The birth of tourism indubitably shaped cartographic and geographical imaginations. Thomas Cook's schemes to embark European travelers to Egypt and Palestine are well known. The popularization of tours from the 1870s onward transformed the perceptions of the eastern Mediterranean. The area was then partly under British control. Egypt and Palestine were opened to the middle class, and not just to the aristocrats on their Grand Tour. The emergence of modern tourism cannot in itself stand as sufficient explanation for the improvement of the British public geographical knowledge. A proliferation of travel guides, cartographic artifacts which prolonged or even replaced the actual journey, followed the development of tourism in the East. The best-known series of guides were published by Thomas Cook, George Bradshaw, and Karl Baedeker.[39] Maps became increasingly common in these works. Such documents could develop into an adjuvant to the preparation for the planned travel, a discursive mode for its anticipation, or even a substitute for the actual "in situ" enterprise. Sales of these handbooks largely exceeded the number of "real" travelers. Travel guides were generally low-priced books, costing between four and six shillings in the 1870s, approximately ten pounds in 2014. They offered a geographical and cartographic perspective to an audience far wider than the readership of the mid-Victorian travelogues. The advent of mass tourism directly and indirectly helped to place the East on the map of the world's fast-developing communication routes, spurring the adventurous explorer in search of distinction and authenticity to journey greater distances, into the deep Orient of Arabia and southern Persia.

The emerging market of armchair traveling quickly fired the imagination of entrepreneurs. Maps were displayed through a range of new artifacts. One of the most popular media of the precinema era, the lantern slide show, helped to foster the population's geographical interest. Among the slide shows touring Britain in the late 19th century, itineraries and travels were quite popular. There was an increasing demand for photographic tours in Egypt from 1882, for instance.[40] The shows were not designed for a limited audience. Some projectors, such as the Sciopticon, were bright enough for very large theaters, which could host more than 1,600.[41] These new technologies participated in a profound transformation of late 19th-century visual culture, thanks to a mass consumption of projected images. An example of these journeys through the magic lantern was published by York and Sons. Frederick York (1820–1903) was a renowned lantern slide manufacturer, established in London from 1884. He produced more than a hundred thousand slides a year in the 1890s.[42] His series on Egypt and the Holy Land took the observer on the traditional tour, from his hotel in Cairo to Abu Simbel. Most of the photographs were from negatives by Frank Mason Good (1839–1928). One of the introductory slides displayed a map of the Nile copied from Samuel Manning's *Land of the Pharaohs*

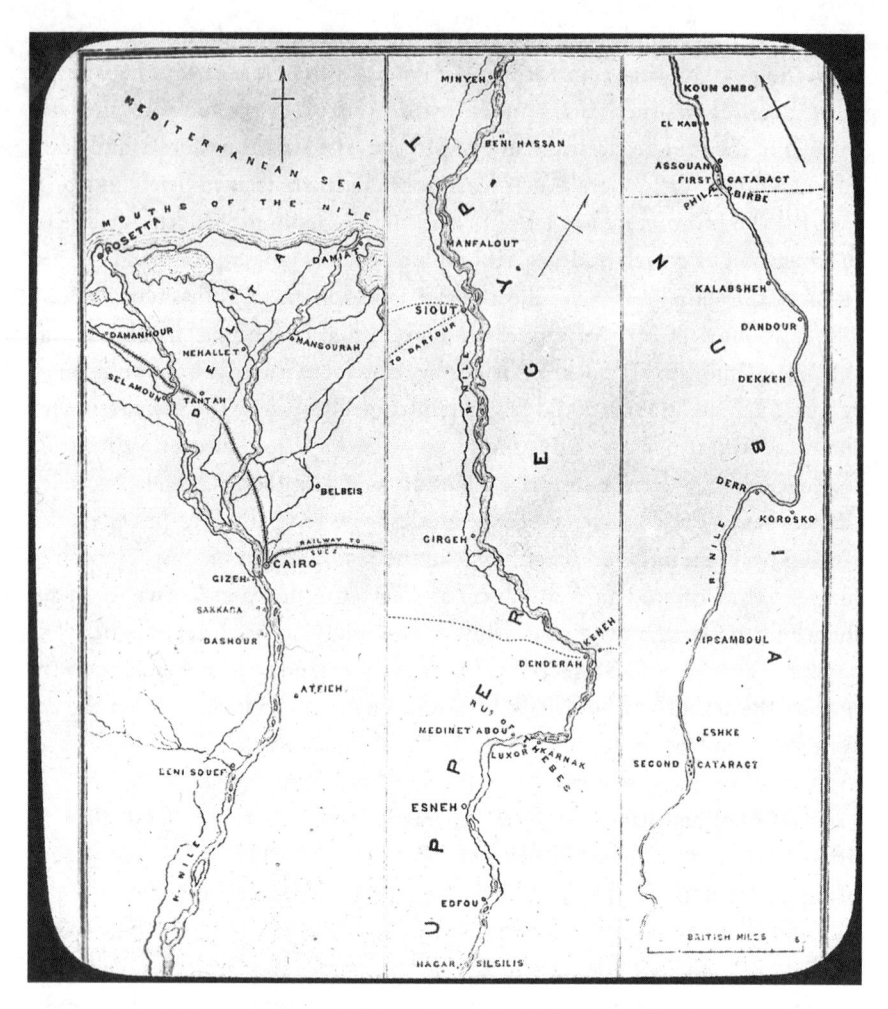

FIGURE 5.6. *Map of Egypt on a Lantern Slide*, manufactured by York and Sons (London, ca. 1880) from Samuel Manning, *Land of the Pharaohs: Egypt and Sinai: Illustrated by Pen and Pencil* (London: Religious Tract Society, 1875), 81. Size of the original: 8.2 × 8.2 cm. Collection of the author.

(fig. 5.6).[43] The lecturer could thus familiarize his audience with the topography of Egypt and the highlights of an armchair journey, experienced in the atmosphere of a lecture room, one that was documented in the accompanying books sold by York.[44]

Stereoscopic views were another vehicle for the cartographic representation of the Orient. They became highly popular in the second half of the 19th century. The London Stereoscopic Company was able to sell more than five hundred thousand views in two years. According to William Darrah, stereoviews were "the first visual mass medium."[45] Underwood and Underwood printed more than ten million views a year in the early 20th century. For a Victorian audience, the tangibility of three-dimensional images was an advantageous substitute for the discomforts of an actual journey. The process was ideally suited to the diffusion of a very popular type of cartography, the relief map. The opportunities offered by the combination of stereoscopy and model maps were noted as early as 1864 by Francis Galton from the RGS.[46] Best sellers in that category were the stereoviews showing H. F. Brion's relief maps. The Society for Promoting Christian Knowledge published one of his works, a photorelief map of Asia in 1884.[47] He also sold a relief map of the Sudan which illustrated British attempts to save Charles Gordon in Khartoum.[48] Another popular cartographic document in stereo was the raised map of Palestine by the assistant secretary of the PEF, George Armstrong. It took him seven years to produce the model, the PEF subsequently selling the resulting map in various different versions.[49] The *Palestine Exploration Fund Quarterly Statement* praised the achievement in no uncertain terms: "There are the seas, the lakes, the mountains, and valleys, all so perfect and distinct that one can travel over the ground and visit the cities and towns. With the Bible in hand the holy sites can be inspected, the historical events of the narration can be followed, the movements of the various tribes can be traced, the operations of war can be grasped and easily understood. With this Raised Map before him a Moltke could sit and plan a campaign as if it were a chess problem."[50] When the Keystone View Company started to sell stereoviews of the raised map in the 1890s, the exceptional panorama was made available to a much larger audience (fig. 5.7). The map was still popular in the early 20th century.[51] Lantern slides or stereoviews were the byproducts of the wider popular culture, which readily assimilated the Eastern Question and its geography.

Corroborative evidence shows that late 19th-century maps of Eastern lands had become fairly common objects. Between the distributions of maps of Palestine by the Religious Tract Society, the decreasing costs of spectacular individual maps, the success of the cheap pocket atlases, and the increasing number of maps printed in the newspapers, cartographic representations were multiplied exponentially. On a larger scale, both the outward migratory flow from Britain to the empire and the development of international communications fostered a growing awareness of the outside world.[52] Even though the transitional East was outside the formal empire, it was an object of curiosity and attention. A wide range of motives, such as a greater awareness of the unparalleled interconnectedness of the globe, humanitarian concerns, religious

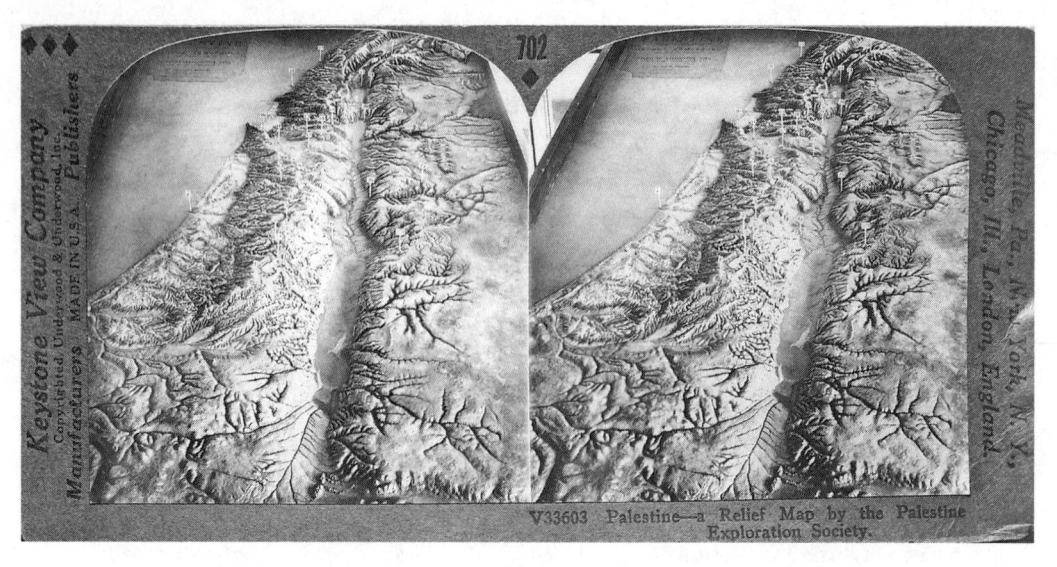

FIGURE 5.7. Keystone View Company, *Steroscopic View showing Relief Map of Palestine by the Palestine Exploration Society*, late 19th century. Size of the original: approx. 17.4 × 9 cm. Collection of the author.

anxieties, and straightforward jingoism, shaped the British mental map of the region. The educational uses of geography and cartography testify to these multifaceted constructions.

5.4 "ENOUGH OF JERUSALEM"?

> I fancy we've had enough of Jerusalem, considering we are not descended from Jews. There was nothing first-rate about the place, or people, after all—as there was about Athens, Rome, Alexandria, and other old cities.[53]
>
> THOMAS HARDY, *Jude the Obscure*, 1895

Did the setting up of a modern educational system in late Victorian Britain answer Sue Bridehead's brash remarks? The biblical ennui of Hardy's character, on a visit to Christminster to see a raised map of Jerusalem, might not be a mere literary device. Contradicting trends redefined the position of geography and maps in British schools. Raised models of Palestine certainly enjoyed an enduring popularity well into the 20th century, and with 50% of children attending Sunday schools in the late 19th century, the *geographia sacra* of the East was still alive and well.[54] Yet, the proponents of a complete overhaul of the nature of geographical education, whose voices were first heard in the mid-Victorian period, enjoyed a far larger influence than before. John Scott Keltie's 1885 report on the state of geographical education, which compared the British situation with Continental educational systems, is one of the many illustrations of this growing concern.[55]

The RGS was one of the most active pressure groups to advocate a reform of geographical education in Britain. Clements Markham, its secretary from the 1860s to the 1880s, laid emphasis on the issue from the early 1870s. He led a movement which continuously argued the need for drastic evolutions in this respect.[56] Many other learned societies, such as the Manchester Geographical Society and the Royal Scottish Geographical Society, went along with the campaign. The Geographical Section of the British Association published various articles on the question.[57] Prizes were even invented to promote geographical education.[58] In 1886, the RGS organized an exhibition of appliances in order to demonstrate how teaching and learning material could be used in schools to foster geographical literacy. Underlying all these efforts was the feeling that Britain was lagging behind other European powers in the field. The largest empire on earth did not know the globe well enough. Keltie witnessed a telling conversation between two visitors to one of the RGS's exhibitions. An English lady asked a German lady "if she could read a map, as she herself could not read at all [. . .]. The German lady was quite astonished to hear this, as it had been part of her education to read maps."[59] One of the staunchest advocates of a comprehensive reform of the curriculum, Halford J. Mackinder (1861–1947), summarized his overall impression of the situation in 1895 thus: "We have had in England good observers, poor cartographers, and teachers perhaps a shade worse than cartographers. As a result, no small part of the raw material of geography is English, while the expression and interpretation are German."[60]

Both the contents and the relative position of geography in the educational system were targeted by the reformers. The hitherto unchallenged domination of physical geography, the infamous "capes and bays" approach, and of sacred geography, was for instance questioned by Sir Scott Keltie's report. His conclusions were highly critical of the pedagogical practices he observed during his inspections. Mackinder urged a redefinition of the discipline itself. One of his most famous works, "On the Scope and Methods of Geography," put forward the idea of a geography that should be the science of man in the territory.[61] While the updating of the curriculum in any educational system is never an easy task, the first signs of a modernization of geographical education are nevertheless to be found in the 1880s. Oxford and Cambridge Local Examinations, an influential test which set the tone for other institutions, was progressively updated to follow the new rationale. In *Gill's Oxford and Cambridge Geography*, the reference book written to help the students work for the exam, only two pages focused on Palestine, a far cry from the mid-Victorian geographies. More pages were devoted to the study of other Eastern territories, such as Persia, Arabia, and Egypt. Significantly, the few paragraphs describing the Holy Land were no longer about its sacred geography. They accounted for the population and economy of actual Palestine. In the other passages on the East, the reader was provided with an updated depiction of the area, including reference to the Mahdist revolt, information on trade, and remarks on transportation.

With the Forster Act of 1870 and the creation of a coordinated primary education, geography became compulsory. Official guidelines in 1871 and 1875 formally installed it in the educational landscape. From 1875, class subjects were introduced from standard 2 to 6. Pupils had to pass an examination in two subjects, from a list which included grammar, history, elementary geography, and plain needlework. The schools would then be granted four shillings for each successful pupil.[62] A continuous strain of official recommendations insisted on the importance of proper geographical courses. In the 1880s, for example, the Department of Education for England and Wales revised the instructions to school inspectors, requiring them to pay more attention to the emphasis laid on the British Empire and its geography in the schools. Geography made its way into the timetables.[63] The implementation of the recommendations progressed in the late Victorian period. From the 1890s, the new geography was taught in most elementary schools.[64] The 1894 *Report of the Committee of Council on Education* shows that a majority of the pupils had geographical classes. Fourteen thousand schools out of approximately twenty thousand, about 70%, had chosen to teach geography as a class subject.[65]

Not only was the place assigned to the subject greater than before; there was also an evolution in its contents. The 1871 New Code emphasized the importance of maps as pedagogical tools. This became the case, too, in all the four nations of the British Isles. J. Gordon's general report on the Church of Scotland schools in the Edinburgh area for 1872 underlined the reality that most teachers in the district proceeded "by application of large schoolroom maps, sketches on the black board, and by frequent exercises in map drawing," in accordance with the new official requirements.[66] The 1884 Education Code, which set the curricula for the seven standards, stated that pupils were to be taught the "meaning and usage of a map."[67] In standards 5–7, cartographic illustrations were mandatory. In 1894, the Education Department reiterated the instructions: "Objects should always be present, and in sufficient numbers; and the chief aim should be to call into activity observation and the construction

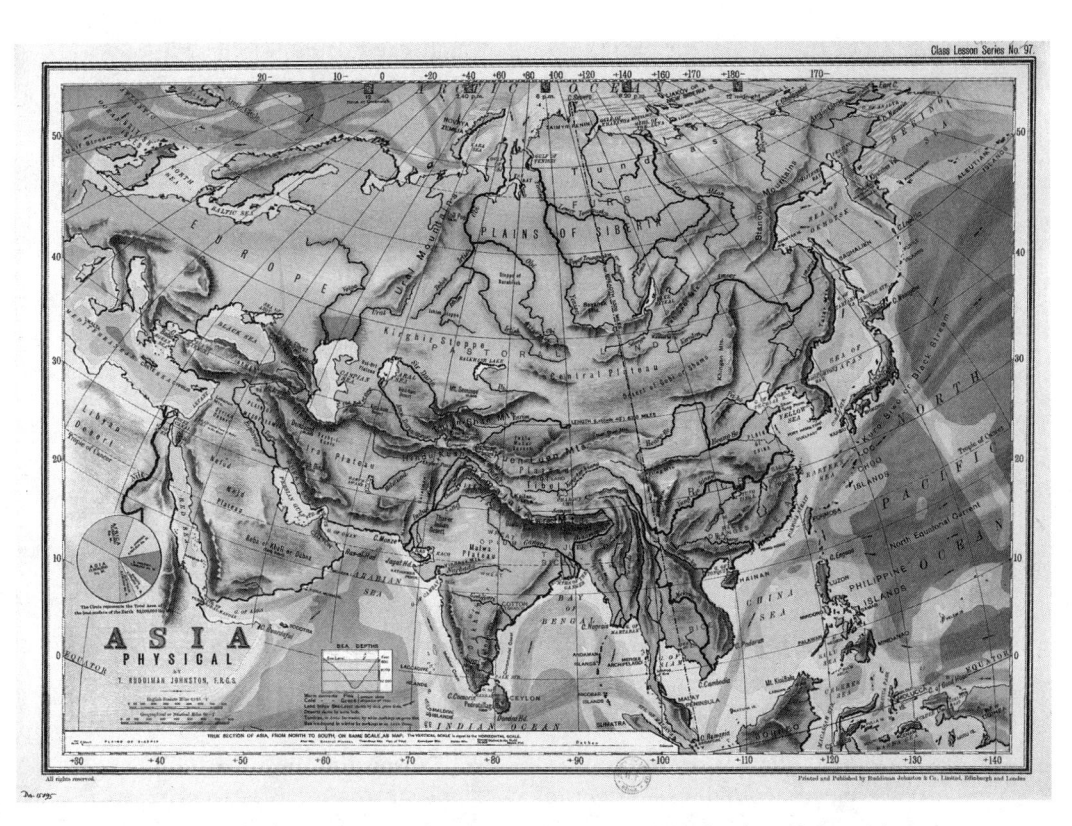

FIGURE 5.8. T. Ruddiman Johnston, *Class lesson series, Asia Physical* (Edinburgh, 1885). Size of the original: 86.3 × 61 cm. Photograph courtesy of the BNF (Cartes et Plans GE CC-929).

of clear mental pictures [. . .]. Geography, where it is a class subject, should be treated in a similar way and should be taught by visible illustrations and by actual modelling in sand and clay."[68] Grants allowed schools to buy materials to meet the official requirements. Maps became a familiar sight in the classroom. Publishers invested in the market. T. Ruddiman Johnston's "new geography" wall maps were praised by the RGS, for instance. They included diagrams, dramatic figuration of the relief, and data on human and economic geography. Johnston's map of Asia thus hinted at Egyptian cotton. It also located the Djebel Shammar, an unusual feature for a school map and an echo of Doughty's and Blunt's explorations in the area (fig. 5.8).

Were teachers able to put into practice the ambitious requirements drawn up in London in the wake of the RGS's offensive for a better geographical education? Corroborating evidence appears to confirm it, even if the disparities between schools and regions were often wide. While geography was an extra subject for training teachers in the 1870s, it was one of the most popular.[69] The syllabus of the subject for examinations of male candidates described in the 1871 New Code insisted that the training teachers be "able to draw a map of the four quarters of the globe" (first year) and be proficient in the "physical, political and commercial" geography of the British Empire" (second year).[70] In any case, the renovators of British geography took great care to spread their vision among primary and secondary teachers. The Geographical Association was established in 1893 to that end. Mackinder was one of its founders.[71] It worked both as a lobby, pressuring for the recognition of geography in all educational institutions, and as a forum for teachers to further their knowledge and pedagogical skills in the subject.[72] The association grew to become the interface between academic research and the educational world.

At the same time, modern geography emerged as an academic discipline. The RGS systematically encouraged universities to open departments, and schools to recruit teachers with specific training in geography in order to resolve what was believed to be a weakness: Britain's backwardness in geographical literacy and expertise. The first chair of geography was founded in Oxford in 1887. Its first holder was Mackinder. Once the initial enthusiasm had waned, Mackinder acknowledged the gap between the popularity of his courses and the very few students who actually passed the examinations in geography.[73] In terms of contents, his teachings reflected the tenets of the "new geography." In consequence, considerable attention was thus given to the East. An entire set of lectures dealt with Mediterranean lands, Central Asia, and India, in which he anticipated his 1904 heartland theory.[74]

The efforts of the learned societies went beyond primary and secondary schools, or even university education. Emphasis was laid on adult education in general. The Manchester Geographical Society and the Royal Scottish Geographical Society organized sequences of conferences specifically designed for this audience, other public bodies being asked to collaborate by modernizing their own training programs. The Board of Trade examinations, for instance, was reformed to leave more space for sea charting, topography, and cartography.[75]

The Elementary Education Act of 1870 did not undermine the influence of churches in the educational system. According to the New Code (1871), qualifications for pupil-teachers included detailed geographical notions on "Europe and Palestine."[76] This is a further confirmation of the resilience of sacred geography. Trainee teachers were almost inevitably asked about the geography of the Holy Land in examinations. For a woman to become a teacher in the 1890s, she had to possess knowledge of the "general geography of the continent of

Asia. Sketch-maps, such as should be drawn by a teacher in the illustration of lessons, may be required of Hindostan, Palestine, and China."[77] The British Palestinian tropism survived the birth of the educational system. The Holy Land did not cease to be holy in the late Victorian map of the East. The biblicist perception remained, thus, one possible prism through which to grasp the Orient, but alongside others, whose relevance was gathering strength.

THE PUPIL AND THE MAP: CONTRASTED ANSWERS

The official guidelines partly give a flawed assessment of what pupils and students really knew. Everyday practices in classrooms did not always live up to the experts' expectations. In many instances maps were indeed used, but with little effect. An 1894 report observed that in many cases "strings of names and disjointed facts are still too often the only stock-in-trade of many of the teachers of geography whom we meet in the schools [. . .]. The map-drawing done as a rule is not of very much value."[78] Modernized educational uses of maps took years to spread for lack of adequate teacher training.

Another hindrance to children's geographical literacy was the continuing disaffection of the poorest students. Of more than sixty-seven thousand deprived pupils in London, 50% had dropped out of the system by standard 5 in 1894. Only eight thousand actually reached the latter grade. The inspector reckoned that, by his best estimate, no more than half the children had a basic geographical knowledge.[79] And yet the generalization of primary education was concomitant with a notable increase in the demand for and supply of geographical and cartographic documents. Julie McDougall's quantitative assessment of printing and production of Bartholomew's school atlas records how the market for this type of literature increased markedly from the 1890s to reach its peak in the early 1920s.[80] However, such an analysis of educational publications provides us with a very limited picture of children's geographical literacy in this period, and we should cast our net more widely.

The development of the British educational system in Britain fostered literacy in general. A market for children's periodicals and books developed in the 1870s. This phenomenon, studied in connection with the imperial propaganda in a book edited by Jeffrey Richards, widened the geographical horizons of children.[81] The *Boy's Own Paper*, published by the Religious Tract Society, sold more than six hundred thousand copies a week in 1880. G. A. Henty (1832–1902) and Robert Ballantyne (1825–94) often sold two hundred thousand copies of their novels. The birth of a mass culture for children was accompanied by other forms of geographical education. The *Boys' Own Journal* released a series in five episodes titled *A Boy's Tour Through Egypt, the Desert, and Palestine* between August and September 1884. It was illustrated by maps and views. In fact it was not uncommon to find the small-sized child-friendly map in this literary genre. This surprising feature was remarkably characteristic of the publications of G. A. Henty. The prolific author used the East several times in his books. He had himself traveled the region. He first published an account of the Crusades entitled *For the Temple: a Tale of the Fall of Jerusalem* in 1888.[82] He then took his readers to Egypt and the Sudan. The British campaign against the nationalist rebellion of 1882 gave him material to write *A Chapter of Adventures: Or, Through the Bombardment of Alexandria*.[83] The Mahdist rebellion was the background for *The Dash for Khartoum* and his 1903 *With Kitchener in the Sudan*.[84] The Great Game was also, naturally, a gift for Henty's talents as a popularizer. His *For Name and Fame, or, To Cabul with Roberts* did more for the name and fame of its author than it did for the literary standards of its young readers.[85] Even so, they would find maps of Tell-el-Kebir and of the port of Suakin in its pages, and, in the editors' mind at least, these maps were part and parcel of its attempt to foster young imaginations.

Geographical readers were intended to make geography accessible to children. Their narratives, coupled with a profusion of illustrations, attracted the young reader more than the arid school geography textbooks.[86] The Committee of Council on Education recommended their use in its report of 1881. These books were often offered as prizes in schools. Most major educational publishing houses published their own proprietary versions. The fifth-standard reader was specifically designed to explore Asia and Africa, that is, most of the territories under study here. It laid greater emphasis than traditional textbooks on "human geography," that is to say, overused stereotypes, as shown in a passage describing Persians in Blackwood's *Fifth Geographical Reader* (1884). According to the author the Persian was "the Frenchman of the East. He has ready wit, a polite and persuasive manner, and great animation in his talk. He presents a great contrast to the Turk. The Turk is a man of few words and grave speech; the Persian is fluent in conversation and flowery in description. The Turk is a quiet farmer, a plain man, and a soldier; the Persian is a man of the world, a trader, and an artist."[87] Maps and illustrations accompanied the young reader in his Oriental itineraries. The same book included a map of Asia and a mediocre map of Egypt.

CONCLUSION

Despite the resilience of the more religiously oriented geographies and maps of the East, the region lost part of its sacrality. Palestine and the biblical lands remained an exceptional object and a reliable source of income for editors. The transformation of the curriculum, as well as the definition of geography as a discipline per se, favored the emergence of differently construed spatial imaginations. The journalistic map and the "seat of war" panoramic views fostered new perspectives, as the telegraph and the mechanical reproduction of images accelerated the flow of news. The coming of mass literacy, combined with a "tidal wave of print," brought the East closer to the industrial world, though not yet near enough to be utterly deprived of its seductive exoticism.[88] It would be overly simplistic to find in the first developments of the mass production of maps the evidence of an emerging homogeneous perspective on the region and the world in general. Commercial cartography was in no obvious sense imperialistic in the late 19th century. While Bartholomew's imperial red layers were often generously applied, in W. T. Stead's hands maps could very well become a graphic reminder of governmental failures, at the same time that Ravenstein's maps for the *ILN* probably aimed at nothing more than to educate the general public. The ambivalent British mental map of the East had several differentiated overlays, to which the advent of high imperialism was to add yet more contrasts. The broader evolution of late Victorian imperialism as well as the impact of early globalization served to place the transitional East in a strategically commanding position. Debates as to what this position would actually turn out to be, however, had only just begun.

PART THREE

THE FABRICATION OF THE MIDDLE EAST (1895–1921)

CHAPTER SIX

SEEING RED?

In March 1899, Lord Salisbury, the British prime minister, received a dispatch to which a sketch map was attached (fig. 6.1). This hand-drawn map furnished the sensation of entering a forbidden part of the world. The map compiled data gathered by a young intelligence officer, Percy Molesworth Sykes (1867–1945). It showed the roads between Birjand and Zabol in eastern Persia. The document accompanied one of the countless reports this ambitious young man sent directly to Lord Salisbury. The singular sketch wrongly suggested that, if Russia were to attempt an invasion of India, its troops would move unopposed on existing roads toward the North-West Frontier of the Raj. In fact, the region being one of the driest of the Iranian plateau, any invasion using that route was bound to fail. In the enclosed letter, Percy Sykes advocated "the establishment of an Indian Political Officer at Birjand, as well as in Seistan." This was not a first. For years, Sykes had inundated his senior officers with letters and maps, arguing for the creation of a British consulate to control both Khorasan and Seistan, provinces caught between Persia, Afghanistan, and British India.[1] He would go on doing so for the rest of his life, besides publishing a great number of books and articles.[2]

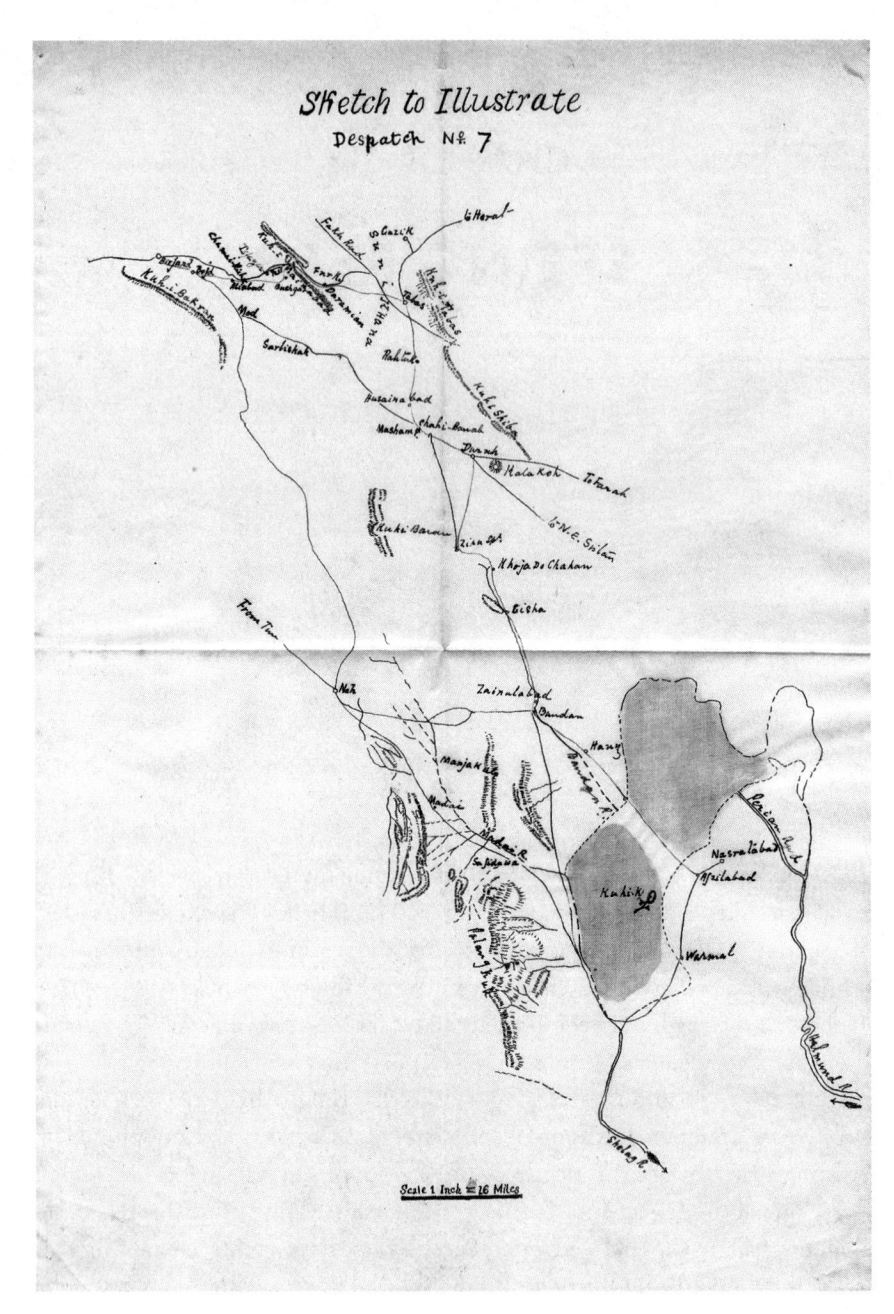

FIGURE 6.1. Percy Molesworth Sykes, *Sketch Map of Khorasan*. GB165–0276, Percy Sykes Collection, box 2, file "Foundation of the Sistan Consulate; Despatches to Lord Salisbury 1899–1900." MEC Archive, St. Antony's College, Oxford.

Did Lord Salisbury ever examine this rough map? Did he consider the potential of Sykes's routes to the Russians, should they decide to move their pawns in the Great Game? Probably not. This sketch, like so many other documents in the imperial archive, has more to say about how British field agents used their topographical expertise and about the flow of data they produced to legitimate their own careers than about their actual grip on the terrain. His reports, dispatches, and sketches were part of a strategy of making his mark, and justifying his views. They may have been expressions of Western domination of the East; they certainly served to highlight his Oriental adventures. Even his contemporaries acknowledged this fact. George Churchill, Oriental secretary to the British Legation at Teheran, sometimes glossed over the most preposterous assertions in Sykes's reports. One of Churchill's comments, dated August 1910, mocked his quixotic colleague's misrepresentations and overemphasis: "*Inshallah!* What an ass Napoleon Attila Sykes is in some ways."[3] Sykes's self-glorifying expertise did not mesmerize everyone, yet it won him a distinguished position in Persia. By 1916, he commanded the South Persia Rifles, a British-funded force that fought in the Persian campaign.

When examining British maps and explorations of the East in the late Victorian period, it is worth keeping Percy Molesworth Sykes in mind. From the delimitation of borders to geological surveys, which quickly became essential with the transition to fossil fuels, an entire range of activities required the medium of cartography in order to be effective in the eyes of policy makers. More than ever, men on the spot held a strategic position. Knowledge came at a premium. Their maps, surveys, and memoranda were part of the flow of paper that influenced and grounded the decision-making process in Britain and India. In the Middle East, only a few agents in the field were actually gathering data. Their various individual topographical and geographical skills conditioned the nature of the knowledge that was available in London. As Sykes's example shows, the data could be limited, manipulated, or omitted. We must not lose sight of the simulative nature of maps, of topographic exploration, and of the gathering of geographical data. As with many other forms of information produced by the different components of British power, not only did maps offer intrinsically limited and distorted representations of a territory, they could also give the illusion of power. Understanding how topographical and geographical information was used and processed by the various state institutions is therefore a necessary prologue to this chapter.

Parliament was denied the possibility of lengthy discussions on British strategic orientations. They were in the realm of the Royal Prerogative. While the FO did regularly prepare Diplomatic Blue Books to keep Parliament informed, they did so retrospectively. Foreign affairs were therefore seldom debated in both houses even though Radicals vainly requested the creation of a Foreign Affairs Committee to foster parliamentary supervision of the matter.[4] Geographical information and maps of foreign places nonetheless made their way into Westminster. Cartographic documents relevant to ongoing debates were often displayed in the Tea Room in the late 19th century.[5] On several occasions, maps of the Middle East were presented to MPs to palliate their lack of geographical literacy. In 1902, Joseph Walton, a fellow of the RGS and a founding member of the Royal Asiatic Society, clearly stated the limits of Parliament's expertise on the Persian Question: "I had the pleasure of travelling over the new trade route from Ispahan to Ahwaz, and I wish [. . .] that the Government could arrange [. . .] for us to have a large map hung within our sight [. . .]. When I speak of the new caravan route from Ispahan to Ahwaz I question, with all deference to hon. Members, whether 19 out of 20 of them are able actually to locate the places."[6] In 1912, Aubrey Herbert, discussing the Anglo-Russian agreement, wondered "how many people who take a strong line about Persian affairs have really studied the map of Persia."[7]

Cartographic and geographical documentation appeared more regularly in the cabinet papers. The East was not overlooked. In 1897, a memorandum on the races of Turkey was printed for the cabinet.[8] It contained reports by W. E. Fairholme and John Ardagh from the Directorate of Military Intelligence of the War Office and an ethnological map of Turkey.[9] As cartographic representations of international issues became more commonplace in British governance, civil servants familiar with maps became more influential. The geographical intricacies of the overstretched British Empire were opportunities for the experts to shine before philistine MPs or overworked office staff. Francis L. Bertie, strategically posted near the FO map room, is an example of this rising class of technocrats in the late 19th century. A senior clerk, then an assistant undersecretary in the Eastern Department, Bertie developed an undeniable expertise: the authority of his memoranda went unquestioned.[10] He frequently sketched maps to illustrate his reports.[11] He was eventually rewarded by being made ambassador in Rome in 1903 and then later in Paris. In remote places, where central government had to rely on highly autonomous agents, field operatives had a quasi monopoly over information. It offered them a great advantage in the decision-making process, as illustrated by the case of Arabia. Information mastery thus played a crucial role in the emergence of a policy-making bureaucracy. Early 20th-century reforms in various offices were specifically implemented to manage the growing flow of data that resulted from Britain's global position. The creation or the reorganization of systematic map collections, as well as the gathering of geographical data, was central to such endeavors.

A number of warnings, from war-torn South Africa to the unstable Balkan Peninsula, pushed to the fore the arguments of those in the War Office who promoted more prospective strategic planning in the early 20th century. In accordance with the recommendations of the War Office (Reconstitution) Committee, set up in November 1903, intelligence gathering was again reformed. A Directorate for Military Operations was established in February 1904 with four sections: strategy, foreign intelligence, special duties, and topography. The Topographic Section became the Geographical Section of the General Staff (M.O.4) in 1908, a section of the newly created Foreign Intelligence Section. The M.O.1 was the Strategical and Colonial Section, the M.O.2. dealt with Europe, and the M.O.3. with Asia.[12] Intelligence in general and maps in particular were central to this series of reforms. The WO attempted to accomplish its goal of providing as many accurate cartographic documents as possible on any and all potential battlefields. Ad hoc intelligence would no longer be the rule. The East was not, however, the WO's main concern; when Henry Wilson became the head of the D.M.O. in 1910, the first map he wanted on his office wall was that of the French-German border.[13] Yet the East was not neglected. The M.O.1. suggested several schemes indicating its concerns regarding the Suez and the overland route.[14] The India Office's Political and Secret Department progressively collected hundreds of maps in addition to those archived by the Geographical Section.[15]

A new generation of political leaders seized upon these new instruments for projecting power. The ambitious Joseph Chamberlain, colonial secretary from 1895 to 1903, was described by the *Auckland Star* as an "up-to-date man [who] studied the map of the world and the history of the world-wide British possessions through his searching eyeglass before shaking hands with the officialism of his many desks in Downing-street."[16] This expertise was a way to promote the extension of one's supervision over new territories. The IO, which already controlled the Persian Gulf by the late 19th century, wanted to oversee Kurdistan and Mesopotamia as well.[17] The Survey of India's prominent role in cartographic depiction gave the Raj a knowledge that proved eminently useful in widening its sphere of influence. Mastering information also provided arguments when dealing with the forces of an emerging public opinion, which could no longer be ignored by policy makers. The IO, for instance, had strong ties with Fleet Street.[18] Even the FO, traditionally more reluctant to cultivate relations with the press, entertained a special relationship with Valentine Chirol from the *Times* in the early 20th century. Both departments disseminated carefully selected information, geographical details, and official maps to foster their views.

Did this modernization of Whitehall's "mental map" of the world give birth to a coordinated perspective on the Middle East? Conflicting visions and differences of administrative opinion were part of the decision-making process in the late 19th century, but various attempts were made at the time to devise a more coherent overseas defense policy. In the wake of the Russian-Turkish war, the Colonial Defence Committee was appointed in March 1878. It designated the Carnarvon Commission in 1879, whose purpose was to issue recommendations on the organization of colonial defense. The committee accumulated a vast amount of evidence, including annotated cartographic documents, which had been partially compiled in the last of the three reports released in 1882.[19] The protection of the Suez route and the fortification of Aden were among the commission's proposals. In 1885, efforts to improve collaboration resulted in the creation of the Colonial Defence Committee.[20] This newly established entity did not fully succeed in regulating disparate positions between the War Office, the Colonial Office, and the Admiralty.[21] It was not until the Second Boer War that a new wave of reforms was set in motion. The anxieties raised by the conflict had a lasting influence on British imperial policies. Arthur Balfour, then prime minister, established a Committee of Imperial Defence (CID) in December 1902 in order to improve interdepartmental coordination.[22] The CID forthwith set out to underline the critical importance of the Suez Canal; it suggested various defense schemes for the Middle East and discussed the consequences of a hypothetical occupation of Constantinople by Russia, along with the latter's intentions in the region.[23] The Home Ports Defence Committee and the Air Committee were two of the four subcommittees that dealt with the more technical aspects of strategic planning.[24] Though everybody agreed upon some basic principles asserted by the CID, such as the protection of the Suez route, the committee did not play the significant role it had initially been given. The WO, among others, had its own agenda in the region, and the CID maintained its advisory basis. True cooperation was not yet a reality. On a more technical note, the Boer War also exposed the "long-standing neglect of mapping in several areas of strategic importance."[25] Therefore, greater cooperation between survey departments was sought in the British dependencies. A Colonial Survey Committee first met in August 1905; it was designed to supervise and coordinate surveys throughout the British Empire. The old ad hoc measurements and maps were to give way to a more systematic perspective. The first annual report dealt with the Anglo-Egyptian Sudan survey and had little to do with anything other than Africa and the

Crown Colonies.[26] The committee was not yet able to systematize surveys in the informal empire and remained mostly advisory. In the absence of effective forms of cooperation, the IO, the FO, the Admiralty, and the WO often elaborated their own approaches to the region.

In this chapter, I shall set out to unravel the interplay between the various official and nonofficial actors who produced knowledge on the East in the late Victorian period. I will focus on the imbrications of British influence as reflected in mapping and mapmaking. I will pay specific attention to the dissensions between the various imperial centers which shaped the understanding of the East and therefore question imperial core-periphery interrelations in that regard. This section also draws attention to how decision makers deployed the cartographic and geographical expertise available to them. I will examine speculative, argumentative, and prospective manipulations of maps of the East and show how mapmaking practices played their part in the crystallization of Britain's conceptualization of the area. Previous chapters have already demonstrated how the local knowledge that had contributed to their elaboration was often eliminated in the ultimate iterations of British maps. Pursuing that line of inquiry, I will consider how hybridization and resistance actually characterized the construction of knowledge in the area.

6.1 THE APPROPRIATION OF EGYPT: THE EMERGENCE OF A NEW IMPERIAL CENTER

AN EXEMPLARY EXPERTISE

Egypt and the Sudan became key elements of the British informal empire in the East in the late 19th century. Arabi Pasha's 1882 nationalist revolt and his subsequent defeat at the hands of Lord Wolseley's troops marked the beginning of what is commonly known as the "veiled protectorate." Evelyn Baring, Lord Cromer, became the British consul-general in Egypt in 1883 and did not hesitate to force reluctant ministers to resign when they contradicted his directives. Anglo-Indian administrators were given strategic positions as advisers to the Egyptian rulers, overseeing de facto the country's government. This Egyptian experience of governance strongly influenced the emerging constructive imperialist stance in Britain. Joseph Chamberlain as well as Alfred Milner viewed the development of the Nile countries as a test case to their renewed imperial designs. The British technocratic expertise was to prove exemplary in reforming a supposedly backward state. Accurate maps were crucial to the enterprise.

Consequently, the British map of Egypt and Sudan in the early 20th century could be considered an example of a cartographic achievement in a colonial context. A cadastral survey was implemented in 1892.[27] Every stage of the project was executed to the highest standards. A two-mile baseline was measured out in Giza to establish a triangulation net. Different chains were used to avoid measurement errors. A strict and cost-effective division of labor was implemented. Field surveyors used theodolites and chains to measure properties and buildings. Data were recorded in standardized notebooks that were sent to Cairo to the Computation Office, where the information was verified and computed. The organization of the Indian surveys, where the local workforce was instrumental to the cartographic project, was a frame of reference for the Egyptian scheme. As in the Raj, the accounts regularly omitted such local cooperation and input. That it was essential, however, is indicated by an unusual photograph that shows the Egyptian surveyors at work, which is to be found in Henry G. Lyons's memoir.[28]

Egypt became a reference for cadastral management and the institutionalization of private property in a colonial context. Timothy Mitchell has demonstrated how the cadastral survey became an instrument of territorial control for the central government and for the British who oversaw it.[29] The mapping project was under the supervision of the Debt Commission, which conceived this tool as a means to increase the Egyptian state's income. The Land Registry directly served the interests of Egypt's European creditors. Surveyors' efforts precisely delineated the contours of state-owned land. Many farmers who had survived thanks to the cultivation of plots that had been abandoned by the state decades earlier suddenly lost this source of income. The new maps also ensured a more systematic and heavy taxation of the Egyptian peasantry. British mapping for the cadastral survey implied a fundamental transformation of traditional land management practices. Private property as understood in the West, based on systematic measurements and trigonometric accuracy, was apparently imposed when the work was completed in 1908.

The degree of imperial control exerted by British maps of the region was far from ideal by late 19th-century standards. In 1874, an expedition set off to correct the longitudes of a preestablished set of stations in the British Empire. The simultaneous observation of the transit of Venus made it possible thanks to telegraphic communications with the Greenwich observatory. This expedition determined the coordinates of the Helwân observatory (Cairo).[30] It would be the only point of fixed longitude in the country until the late 1900s. The surveyors relied on this point of reference, but the triangulation of the 1890s and 1900s was nevertheless too hasty to be accurate. The unexplored areas were filled by traverse surveys which were intrinsically rudimentary. A triangulation of a high order was eventually undertaken in 1907; the Helwân observatory being the fundamental point of a geodetic datum based on an up-to-date ellipsoid (the 1906 Helmert ellipsoid).[31] It was only then that their computations and the resulting maps managed to reduce the margin of error.[32] The map of Egypt before 1907 could be used for tax recovery and land expropriation, but its accuracy was questionable regardless of the technical accomplishments of the Survey Department. Like most British and European surveys in the Middle East before World War I, the mapping of Egypt cannot be called a true geodetic survey until the 1910s. This level of accuracy would have been irrelevant considering the Egyptian financial context. The average price of the surveyed acre had to remain low, a shilling per acre according to T. H. Holdich, to meet the requirements of the Public Debt Commission.[33] As a consequence, the number of complaints addressed to the government questioning the cadastral measurements rose considerably in the early 20th century.[34]

The Survey Department published 1:2,500-scale maps for various state agencies and sold them to the general public. They contained a wealth of information for private investors and Anglo-Egyptian funds. British officials working on the cadastral project were often requested to participate in financial and entrepreneurial projects that would clearly benefit from new topographic knowledge. Auckland Colvin, who was the first director of the Egyptian cadastral surveys, sat on the board of the Egyptian Delta Light Railway, a rail company that managed multiple lines in the delta. He also had interests in the Delta Land & Investment Company. The latter was a precursor to modern-day joint ventures. It was managed conjointly with Sir Elwin Palmer (1852–1906), financial adviser to the khedive and the first governor of the National Bank of Egypt in 1898. Two Cairene families, one of which was that of Yacoub Cattaui, were also partners.[35] The purpose of the Anglo-Egyptian venture was real estate development on the outskirts of Maadi, south of Cairo. British agents in Egypt understood how valuable maps could be in that regard. This was a nonnegligible advantage to such enterprises. Some actors knew how to use the Survey Department's cadastral register for effective property management. Precarious titles were easily contested. Maps facilitated expropriation.

Surveys were also implemented to other ends. The Geological Survey was founded in 1896; another enterprise headed by H. G. Lyons, who would be succeeded by William Fraser Hume. The geology department was one of the first in the world to be established as part of a state institution. Once again, technocrats used Egypt as a testing ground. These officials were so intent on modernization that some of their administrative creations far exceeded what was done in the majority of European countries at the time. Geological explorations intensified in a more rational and systematic manner as the century neared its end. Hume, while still an inspector for the Geological Survey, led a pioneering expedition into the Eastern Desert.[36] The results were published in 1902 and a geological map published in color accompanied the book.[37] Much attention was paid to the issue of raw materials such as sulfur and oil. Hume's route surveys participated in the compilation of a map of the little known region. Surveyors began looking past the Nile Valley in search of other exploitable resources. A series of geological maps compiling all the data collected by the Geological Survey was published in 1910. Blanks on the map were already shrinking as geologists and surveyors began venturing south of Egypt.[38]

Hume highlighted the presence of iron ore in the Eastern Desert.[39] To mine it would be operationally difficult, but Hume's publications attracted the attention of investors and mining specialists. It was no coincidence that the rationalization of geological knowledge about Egypt led to the creation of several companies hoping to exploit these resources. In 1901, the Egypt and Sudan Mining Syndicate was established in London to operate a concession in the region explored by Hume. That same year, the statutes of the Egyptian Mines Exploration Company were filed in London with the same goals. The Geological Survey's efforts clearly paved the way for geologists and surveyors who directly served private interests. They made their own surveys and maps for British investors. Following the publication of the Geological Survey, C. J. Alford conducted one of the first private explorations in 1899 and 1900 at the behest of English investors, it being financed by the London-based Victoria Investment Corporation.[40] For a while, the myth of the Egyptian and Nubian gold mines encouraged exploratory fever. Arthur Llewellyn's research is another example of this type of symbiosis between mapping and investing. He was employed by the London firm John Taylor and Sons, specialized in public works and mining exploration. Llewellyn was sent to Egypt and the Sudan to locate any exploitable mines.[41] On the eve of the First World War, it turned out that very few veins were truly viable and profitable. There is no doubt, however, about the relationship between the work of the Geological Survey and the mining speculation which followed it, and it underscores the powerful role which maps played in staking out resources for their eventual exploitation.

Both in Egypt and the Sudan, the impression of control given by the abundance of large- and small-scale maps and the self-serving publicity of surveyors must not hide the discrepancy between technocratic ideals of order and the intricacies of the Nile regions. If limited in its panoptic attempts, British mapmaking in this region nevertheless consolidated a network of experts that was institutionalized by the Western reformers' cartography. The affirmation of Egypt as a nodal point within the empire was partly a consequence of the Anglo-Indian mapping of the country. This led to the reinforcing of the authority of administrators. Mapmaking and other attempts at transforming Egypt and the Sudan materialized a new layer of British influence and its attendant supervised territory. One man in particular embodied these evolutions.

KITCHENER'S MIDDLE EAST AND THE ANGLO-EGYPTIAN NEXUS

Kitchener's 1898 victory at Omdurman turned him into one of the influential voices in Eastern affairs. His early career and later triumph had supposedly made him an expert on the region. As we have seen earlier, Kitchener participated in a few of the most significant surveys in the East. He was a cartographer in Palestine with the PEF and then in Cyprus. He then became sirdar in Sudan and commander in chief in India in 1902. During his time in the Raj, he turned his attention in particular to the North-West Frontier and Persia, which was threatened by Russia. To his great disappointment, he was not elected viceroy in 1909. He spent the next two years in the dominions and eventually refused the position of commander in chief of the Mediterranean: he regarded this administrative division as an obsolete entity in light of the contemporaneous strategic reorganization which gave birth to the very notion of "Middle East." The government offered to turn the position into a sort of command over the transitional East.[42] In the end, however, he declined the offer and eventually succeeded the late Eldon Gorst as British agent and consul general in Egypt and Sudan. Not only did Kitchener's stations encompass the contours of the emerging British Middle East, but his networks also mapped out what the region would become.

Throughout his career in the area, Kitchener met and often selected most of those who would shape British policies in the Middle East in the early 20th century.[43] Gilbert Clayton (1875–1929) served under his command as an artillery officer in Sudan in 1898. Kitchener met Ronald Storrs (1881–1955) in Egypt. Storrs, the Oriental secretary of the British resident in 1909, thus served directly under Kitchener and would later become governor of Judea, from 1920 to 1926.[44] In 1913, Kitchener anticipated an Ottoman invasion of the Sinai. He sent Leonard Woolley and T. E. Lawrence to complete topographic and archaeological surveys in the Zin desert under cover of one of the Palestine Exploration Fund's operations. They explored the region for six weeks between January and February 1914.[45] Archaeology masked the monitoring of the Hejaz line and Ottoman positions.[46] Kitchener also built up his network thanks to his eminent status among British Freemasons. He belonged to fifteen lodges and chapters. He was also district grand master for Punjab, Egypt, and Sudan. His successive appointments actually form a Masonic map of the Middle East. Kitchener and Reginald Wingate, who became the head of the Sudanese lodge, were both Freemasons. Another key figure of British policies in the East, Henry McMahon, had Kitchener for his Masonic grand master in India.[47] Most of the members of this disparate group gravitating around Kitchener were involved in the creation of the Arab Bureau in 1916.[48]

Kitchener's networks, H. G. Lyons's Egyptian surveys, and the Sudan Survey Department are direct reflections of the emergence of a new subimperial sphere in the Middle East. Egypt and Sudan became testing grounds for modernized colonial governance. Surveys, statistical, cadastral, or geological, were an integral part of the experiment. As a consequence, Anglo-Egyptian views gained considerable weight in the process that gave birth to the British Middle East after the First World War.

6.2 THE ANGLO-INDIAN OUTLOOK

FRASER HUNTER'S MAP OF ARABIA

India, which had progressively developed its own subempire in the Persian Gulf, was keenly interested in Arabia. Its strategic importance on the Suez route and growing concerns regarding German advances in the region in the early 20th century explain why Anglo-Indian agents played a significant part in surveying and mapping Arabia at the time. Modernity and shifting geopolitics called into question the region's former isolation. Two railways projects were to frame the Arabian Peninsula and tip the balance there toward an intensifying British influence. In 1903, the Ottomans granted an extension of the Berlin-Baghdad line to a German-owned company. This was a direct threat to the Indian subempire in the Persian Gulf. The Hejaz railway between Damascus and Medina began operating in 1908. The prospect of Ottoman troops moving en masse toward the shores of the Red Sea was another worrying prospect for Britain and India. Arabia was more than ever a territory in contention. A more systematic approach to data gathering and mapmaking in the region was urgently needed.

It is hardly surprising, therefore, that the reference map for Arabia in the early 20th century was compiled by a Survey of India officer (fig. 6.2). The

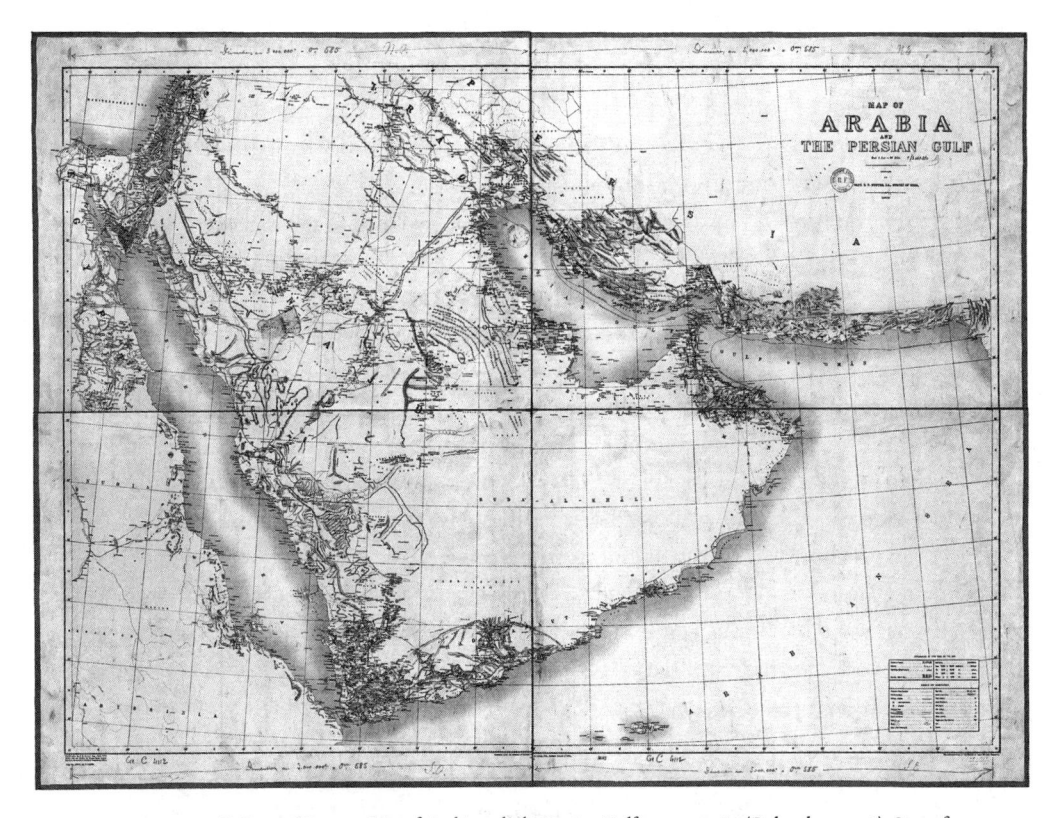

FIGURE 6.2. F. Fraser Hunter, *Map of Arabia and The Persian Gulf*, 1: 3,041,280 (Dehradun, 1908). Size of the original: 85 × 60 cm Photograph courtesy of the BNF (Cartes et Plans GE C-4112 (1–2)).

document is a testimony to the increasing Anglo-Indian focus on the area's strategic value. It was compiled in Simla at the request of the viceroy, Lord Curzon, by F. Fraser Hunter with John Lorimer's assistance. In 1905–6, they both worked ceaselessly to put it together from the most accurate sources available to them. Much of the information used by Fraser Hunter came from local sources, Anglo-Indian agents, and pundits. He had little time for most of the data gathered by legendary explorers such as Richard Burton, or the more controversial William Palgrave. By contrast, Fraser Hunter did not discount local sources such as Al-Hamdani's *Sifat Jazirat al-'Arab*, a 10th-century geography of Arabia. Fraser Hunter explained later how he realized what use could be made of this ancient knowledge. While in Mesopotamia, his colleague John Lorimer saw a "venerable Arab" sightseeing in Baghdad and consulting Al-Hamdani's book in "the same way as a modern tourist would the Baedeker."[49] This interest in local and traditional intelligence was not specific to Fraser Hunter. Lorimer's documentation is replete with information drawn from locals. In the Iraq article from the second volume of his *Gazetteer of the Persian Gulf, Oman and Central Arabia*, part of the information came from non-British informants and officials such as the Baghdad consulate's dragoman at the time, Yacoub Thaddeus, and a broker named Thomas Khalil, and from investigations entrusted to Naoum Abbo Effendi.[50]

The Anglo-Indian agents' trust in this body of local knowledge was sometimes overly optimistic. Local agents and pundits could sometimes manipulate or hide sensitive information for their own purposes. In 1908, Fraser Hunter sent an Indian plane tabler, Abdur Rahim, to Mecca, with a miniature of the first version of his map in his pocket. He gave him a small sextant and a questionnaire, the answers to which would improve his topographical knowledge of the Forbidden City. Abdur Rahim, however, proved too garrulous, and other pundits became aware of Fraser Hunter's project. Someone talked, and the Turkish consulate got wind of the whole matter. On his arrival in Arabia, Abdur Rahim was searched and his topographical instruments seized, but he managed to keep the map, which was mistaken for a handkerchief. Local surveyors trained in India, some of whom were Muslims, were thus able to protect sensitive locations from British gaze on occasions. This was far from being the only illustration of resistance to Western exploration in Arabia. A surveyor's life on the peninsula was far from easy. Not all locals had a positive appreciation of Western topographical rituals. Major O'C. Tandy, a political agent in the Gulf, was shot at on a daily basis by local tribesmen while surveying Oman in the 1900s. With characteristically British phlegm, he eventually responded: "If I was as bad a surveyor as you are marksmen, I ought to be killed. Now why don't you go home and learn how to shoot before you attempt to fire at a man you can't hit."[51] The conversation apparently put a stop to their making a nuisance of themselves. Such tales of exploration and mapmaking are not mere anecdotes. The cartography of Arabia was an intricate process in which counteraction, negotiation, and potential manipulation by local inhabitants participated in the shaping of Western knowledge.

The map was at first much criticized both by the WO and by the Survey of India for its supposed inaccuracies.[52] The map was redrawn after its system of transliteration was questioned. This was more than just a scholarly discussion on Arabic translations. Bureaucratic wars over place-names reflected wider administrative and geostrategic redefinitions. Fraser Hunter's map itself inaugurated a new perspective on Arabia. A few years later, Frederick Fraser Hunter chose to apply a more up-to-date projection than the one he used in the first version of the cartographic document. Fraser Hunter's map materialized the emergence of a new area: Arabia in its broadest conception, including within a single entity the Egyptian Red Sea, the emirates of the Arab-Persian Gulf, and the Shatt al Arab. The projection gave substance to this redefined entity. Traditional projections rarely did justice to the size of the Arabian Peninsula. A pseudo-Mercator projection, one commonly used in 19th-century British atlases, made Arabia appear smaller than Greenland despite being approximately the same size. On this later version of his map, Hunter applied George C. Gore's modified secant conical projection, which improved the relative proportions of the peninsula.[53] The centering of the map objectified a new geographical construct with an Anglo-Indian taint. Arabia was not *Felix* or *Petraea* anymore.[54] Rather, it was a strategic asset with a potential that was henceforth delineated and denominated. Fraser Hunter's map was not, however, the only work to update British Arabia. It was accompanied by the *Gazetteer of the Persian Gulf*, edited by John Lorimer, which appeared in 1908. This comprehensive handbook was to provide all the relevant data for decision making in a region that had by then become a hub for the Raj. Both the map and the *Gazetteer* were used as the basis on which India's Arabian policies would be planned in the 1910s.

DEEPER INTO ARABIA: ANGLO-INDIAN INITIATIVES

A new generation of Anglo-Indian surveyors broke with the old ways of engagement in Arabia, and that with one overall objective, which was to overcome the limitations which had hampered their predecessors. Speaking fluent Arabic or dressing like an Arab had not been the essential prerequisites for the Victorian explorer. Such skills, however, now became indispensable as demand for more accurate knowledge of the Peninsula increased. Two characters dominated this changing approach: Gerard Leachman (1880–1920) and William Shakespear (1878–1915). Leachman began his career in the Royal Sussex Regiment. He served in South Africa and India. In 1909, Sir Percy Cox (1864–1937), then resident in the Arab-Persian Gulf and orchestrator of the Anglo-Indian and British influence in the region, sent him to Ha'il to establish diplomatic relations with the ruling family, the House of Rashid. Leachman was not, in fact, officially mandated there; he was simply sent there on administrative leave. It was there that he met Percy Cox and William Shakespear, also on administrative leave with the unofficial support of India and the WO.[55] The Indian staff regularly used such subterfuges in order to hide informally organized intelligence activities from the Ottomans as well as the FO. After Blunt, few British had ventured as far into inner Arabia as Leachman. Leachman left Baghdad and joined the Bedouins of the Shammar tribe. He then left them to follow the Amarat tribe.[56] He always wore Bedouin garb, and because of his fluent Arabic, he blended with the landscape. It provided him with a privileged vantage point from which he unraveled the tangle of tribal relations. His map localized the major tribes with unparalleled precision, enabling the British to offer protection for one tribe against another more accurately. He was one of the first to report the weakening of the Rashidi dynasty against the House of Sa'ud, whose rear base in Kuwait was protected by the British. The Al Rashid family was, at the time, proving increasing loyal to the Ottoman government. From November to December 1912, Leachman undertook a second trip to Riyadh to meet with Abdulaziz Al Sa'ud, who had taken back the capital of Sa'ud in 1902.[57] The latter made no secret of his hostility toward the sultan's sovereignty. Leachman's goal, ultimately abandoned when Abdulaziz Al Sa'ud refused his offer, had also been to explore the Rub al Khali. Leachman's mission was a way to demonstrate British India's interest in Al Sa'ud. In 1906, Al Sa'ud, who was under British protection in Kuwait, attempted to seek recognition from British authorities. At first, the resident in the Gulf paid little attention to him. That situation, however, changed in the 1910s. Leachman underlined this shift in his accounts, and it played a part in shaping British policy in inner Arabia for years to come.[58] Surveys and maps continued to play a central role in that evolution.

The career and activities of one of Leachman's superiors, William Shakespear, illustrates that centrality. Shakespear, born in India, first enrolled in a regiment of Bengal Lancers, and he quickly became an interpreter. In addition to Punjabi, which he spoke from birth, he also learned Urdu, Pashto, Farsi, and Arabic. At the age of only twenty-five, this outstanding young officer was spotted by the Indian Political Department and sent to the Gulf to act as consul under Percy Cox, the political resident. From Cox, he learned to master topographical instruments and improved his knowledge of photography. In 1908, he drove to England from Karachi across the European continent when he was on leave. Upon his return, he was appointed political agent in Kuwait in 1909. He carried out several expeditions into the hinterland, riding camels or driving around in his automobile, surveying the region with unprecedented accuracy. He conscientiously filled several field notebooks with his observations, in keeping with the Indian surveys' tradition.[59] Furthermore, Shakespear swiftly adopted photography to facilitate the depiction of topographical relief.

Shakespear met Abdulaziz Al Sa'ud in Kuwait in 1910, and again the following year while on an expedition to inner Arabia. It was customary for Indian officers to collect as much topographical and geographical data as possible.

Shakespear's explorations justified his frequent visits to Al Sa'ud. Eliminating the blanks from maps allowed him to develop a friendship with Abdulaziz. Shakespear used his surveys to substantiate the memorandum he sent in 1911 to the FO and the government of India.[60] From that moment on, Shakespear decided on his own account to promote the idea of backing Al Sa'ud in order to control central Arabia and expel the Ottomans from the peninsula. His expeditions continued until the First World War without the FO ever allowing Shakespear to formalize the support he had nurtured on the ground. Leachman and Shakespear's construction of Arabia on paper was soon to contribute to forming the contours of the Middle East. A document addressed to the CID in 1915 contained a sketch that illustrated the geographical and geopolitical space that stretched from Kurdistan to the north of Arabia.[61] It was compiled from data collected by Leachman, Shakespear, and other British explorers and cartographers. Their capacity to improve and overcome linguistic and cultural rifts gave substance to the legitimacy which they claimed for the knowledge that they had collected and changed the perception of the territory itself.

A few British agents went still further. G. Wyman Bury was one of them. He landed in 1896 in Aden, where he was eventually recruited by the British residency. He served as a soldier during the suppression of Aden's tribal uprisings in 1903. He then traveled the hinterland for seven years without any assumed official function. He had gone "native."[62] Bury lived in the Yemeni hinterland, and the local population apparently named him Abdullah Mansur. Like Leachman in central Arabia, he dressed as a local (fig. 6.3). He wrote a handful of reports on the tribes for the WO before his involvement in a corruption scandal in 1905 tarnished his reputation. Thereafter he lost all official support until the First World War. He nonetheless dedicated the book he published in 1911 "to my chief and all those who supported a strong policy on the hinterland."[63] This dedication leaves little doubt as to the nature of Bury's activities. His explorations, which aimed to determine tribal networks and sovereignty in the hinterland, was encouraged by a group of supporters, including the RGS and Anglo-Indian officials. He revealingly thanked John Scott Keltie, the secretary of the RGS, for proofreading his work. His local appearance, though creating a sense of distance between him and the respectable British agents in the area, never hindered him from being an unofficial agent on the field. The ambiguous nature of Bury's status in Arabia reflected tensions between the central government, India, and the RGS, the latter sometimes supporting exploratory projects without Whitehall's approval. Bury's unofficial mission demonstrated the existence of multifaceted British policies in Arabia as well as the existence of a network that could support these explorations outside the traditional administrative framework in order to avoid attracting the attention of the FO, which was officially committed to preserving Ottoman sovereignty.

When the agent was too visibly Western to survive his explorations, the pundit could completely take over.[64] Such was the case for Theodore and Mabel Bent's survey of Hadramaut in 1894. They were a couple of unaffiliated explorers who initially landed at Aden in 1893. They were told there "that a survey of the Hadhramaut by an independent traveler would be useful to the Government."[65] This suggests that the Bents' expeditions were indirectly supported by the British resident in Aden. At the time, Ottoman authorities systematically refused to give authorization to British explorers interested in the Arabian Peninsula. The granting of such firmans allowed them to play on the rivalry between European powers. To give the Bents an unofficial mission to explore the Hadramaut was one way to try to get around the obstacle. However, the Western couple could not viably survey the hinterland. They relied on Imam Sharif, a senior Indian surveyor.[66] He was instrumental to the surveys of the Afghan Boundary Commission (1884–86).[67] Sharif carried out his own surveys alone and under cover. Personnel of this type facilitated the collection of knowledge.[68] The presence of tradesmen from the Indian subcontinent was not exceptional in the area. The long-standing ties between Hadrami merchants and port cities along the Indian Ocean as far as Singapore helped Imam Sharif and his companions travel around unnoticed.

The Bents' adventures in Arabia reveal the growing importance of the pundits in the cartographic unfolding of the expanding Indian subempire. Arthur McMahon's Second Seistan Boundary Commission (1903–5) heavily relied on this corporation of native surveyors as well.[69] They trigonometrically surveyed more thirty-eight square miles between 1902 and 1905. The desert zones of the Dashti Margo were also included in the campaign. The pundits paid a heavy price. One of the most well-known Indian surveyors, Mohi-uddin, died traveling across the desert. One of the few survivors of the expedition, Saidu, brought back the maps drawn up during the tragic odyssey.[70] He wrapped himself up in them to hide the documents and bring them back. The commission was eventually able to complete the work initiated by Goldsmid forty years earlier. Ninety markers were placed along the new border.

How did this apparently disparate mapping of local areas influence the making of the "Middle East"? In an area where imperial supervision was at best erratic, the normative power of the Anglo-Indian gaze was instrumental in shaping views on the area. It was characterized by a combination of the acquisition of knowledge and political negotiation. Surveying overlapped with direct

THE AUTHOR IN NATIVE DRESS.

FIGURE 6.3. George Wyman Bury, *Portrait of the Author*, in Gerard Wyman Bury, *The Land of Uz* (London: Macmillan, 1911), frontispiece. Photograph courtesy of the MISHA Library—Strasbourg University (FP/Man A).

intervention. The mediation between local agents and those British agents who had "gone native" was another crucial feature of Anglo-Indian practices in the field. In the search for collaborators and better access to information networks, explorers connected to the Raj were the vanguard of British involvement in the East. Their maps were therefore not just reflections of an increasingly accurate topographical and geographical image of Arabia, Persia, or Mesopotamia. They conceived, promoted, and substantiated a distinct a distinct set of assumptions on what the East meant to the British Empire and to India in particular. This aggregate of data was mechanically recycled in later maps compiled in London. A large portion of what the WO knew about Eastern place-names, tribal networks, local sovereignties, and geological resources by the 1910s stemmed from Anglo-Indian intelligence.

6.3 EDEN REDUX IN MESOPOTAMIA: THE SPECULATIVE MAP

Mesopotamia was another Middle Eastern territory of critical importance to British India. The obsession with locating and understanding its ancient irrigation systems was one of the constants of the British mapping of Mesopotamia during the 19th century. Layard's accounts, James Felix Jones's surveys, and biblical stories had fostered the idea that the region could once again become the fertile paradise of antiquity. Such a vision was an integral part of British views of Mesopotamia in the early 20th century. Its main promoter, William Willcocks, had already participated in large-scale irrigation projects on the Nile. H. G. Lyons encouraged Willcocks to survey Mesopotamia. He would go on to demonstrate the viability of major waterworks to restore the area's glorious agricultural past. An apostle of civil engineering, he first published a pamphlet on the subject in 1903. At one point in it, he enunciated with Saint-Simonian undertones that "Egypt will always remain the queen of irrigated countries; but next to Egypt may certainly be placed the wonderful land irrigated in ancient days by the Tigris. The resurrection of this old world irrigation is near at hand. The Baghdad railway is steadily advancing from the west, and European civilization is scattering before it the Arabian locusts which have so long held possession of these plains."[71]

Willcocks used his surveying skills to persuade his readership of the soundness of his views. His 1903 publication included nine maps. For years, he continued to promote a project that attracted the interest of the promoters of British influence in the region. He offered an alternative to counteract German advances toward the Gulf. Willcocks gained the support of the Ottoman government in 1909 and began a survey campaign that lasted for over two years. Moved by an unshakable faith, his King James Bible in hand, Willcocks took on the task of re-creating Eden, which he situated in southern Mesopotamia.[72] He was convinced that with limited investments to renovate the ancient infrastructures, the plains could once again be irrigated. For, despite his eccentricities, Willcocks was a highly competent water engineer as well as a publicist. He flooded members of the RGS with arguments in favor of his project. In 1910, his paper, "Mesopotamia: Past, Present, and Future," was discussed by eminent British agents in the Middle East such as Francis R. Maunsell, Gertrude Bell, and Colin Scott-Moncrieff.[73] He designed several maps based on his surveys, some of which were published by the RGS in 1912.[74] They were designed according the most advanced techniques (Willcocks had taken full advantage of his trips to India and Egypt) and illustrate his fascination for the resurrection of the scriptural garden.[75]

Willcocks's scheme was a peculiar combination of two major strains of British cultural representations of the East: biblicism and the constructive imperialist ideal of development. His travel guide was the Bible; his faith led him to believe that irrigation could make the four rivers of Eden flow again. Since the Euphrates and Tigris still flowed out to the Gulf, Willcocks wanted to re-open two canals (Sakhlawia and Hindia) to re-create the divine landscape. In addition to this Miltonian Orient, Willcocks constructed a picture the region as a zone for reform, where the progress of civilization and technology could make Al Iraq green again.

How much trust did British officials actually place in Willcocks? The indirect support from Egypt's Survey Department was evident: most maps were compiled in Cairo, and H. G. Lyons wrote a rather flattering article about the project.[76] In London, any form of constructive interference in the region that could be used to slow German progress was examined carefully, yet Willcocks's extreme religiosity prevented him from garnering widespread support.[77] However, these projects were consistent with long-standing Anglo-Indian dreams of restoring Mesopotamia. The eccentric Willcocks and his maps were not an isolated case. Others promoted the development of Mesopotamia.

Gertrude Bell, a key figure in the construction of the Middle East, was one of them. She joined the inner circle of William Mitchell Ramsay in 1900. As we have seen earlier, Ramsay became one of Charles Wilson's friends when the latter was consul in Anatolia. Ramsay was Oxford's first professor of classical archaeology and an expert in Middle East topography, who gathered around him an informal network of talented students. Gertrude Bell joined the group along with David George Hogarth (1862–1927), whom she met in 1899. She quickly became a familiar figure of the British community in the East. Her voluminous correspondence testifies to her contacts with John Gordon Lorimer, the Blunts, T. E. Lawrence, and William Willcocks. Her letters reveal a labyrinth of interrelations with some of the individuals who eventually drew up some of the Middle East's contours. Bell worked within a nebula of informants who were linked to the government in a more or less official way. She was "in the employ (without pay of course) of the Intelligence Division of the Admiralty."[78] Gertrude Bell helped develop the topographical knowledge of Mesopotamia, alongside other colleagues such as Mark Sykes.[79] This informal surveying enterprise was not coordinated by a central authority; yet it managed to monitor the advance of the Bagdadbahn, a German-supported project of a railway line connecting Berlin to Baghdad, one of the British government's concerns in the region. From 1903, various agents were dispatched to the region by the WO and the Admiralty.[80] The collected data was compiled in a memorandum written in 1907 on the strategy to be adopted in Mesopotamia.[81] From 1907, the WO's Geographical Section of the General Staff (GSGS) printed various maps to document the region's railways to visualize the effects of the new infrastructures on the Middle East's equilibrium. The risk hanging over the Persian glacis and the British-supervised Gulf was an incentive for the British to collect as much information as possible on the Arab tribes, their potential loyalty, their leaders, or the location of towns and villages. This intelligence might one day help counter German influence in the area. Bell's map echoed Whitehall's concern over this new *Drang nach Osten*.[82] Bell's, Sykes's, and Lorimer's observations ultimately confirmed that the railway was advancing exceedingly slowly and the danger was not imminent.[83] A secret Anglo-German agreement was signed in 1914, shortly before the war, to end disputes over Asiatic Turkey.[84]

The purpose of Bell's explorations, aside from her scientific and human interest in the region and London's request for updated information, can be found in the fact that she shared long-standing hopes of reforming the region's economy. One passage in her very popular book *Amurath to Amurath*, where she offers her opinion of William Willcocks's projects, illustrates that belief: "The one ray of hope for the future sprang from the labours of the irrigation survey whose leader was lying imprisoned in midstream at Amarah. 'He who holds the irrigation canals, holds the country,' is a maxim which can be applied as well to Mesopotamia as it was to Egypt."[85] Like many other residents in the region, Gertrude Bell was deeply interested in water resource management and its reform. In addition to adopting Willcocks's views on irrigation, she believed that controlling the rivers was a means to control the population. The success of the Young Turks' movement led Britain to consider the possibility of a more effective development of Ottoman dependencies. This renewed interest in the future of Mesopotamia might account for the new nomenclature Bell adopted in *Amurath to Amurath*.[86] She began to use occasionally the name of the province, Iraq, to designate the populations from Baghdad to Basra.[87] Once again, the place-name was more than just a fashionable neologism. She saw a potential new territorial entity in Mesopotamia. Her work manifests a strong belief in the British regional mission to develop the economy and to promote the emergence of nations. Other British agents, like the Anglo-Indian Arnold Wilson, went even further, straightforwardly expressing their desire "to see it announced that Mesopotamia was to be annexed by India as a colony for India and Indians."[88]

Believers in the hydraulic redemption of Mesopotamia were also to be found in Britain. Arnold Toynbee himself saluted Willocks's scheme in his 1917 book *Turkey: A Past and a Future*. From his standpoint, Iraq could be a new frontier if constructive policies were implemented: "If you cross the Euphrates by the bridge that carries the Bagdad Railway, you enter a vast landscape of steppes as virgin to the eye as any prairie across the Mississippi."[89] Willcocks, Bell, and many others helped shape Mesopotamia into something more than the devastated remembrance of fallen biblical empires. Their colorful maps and no less colorful accounts substantiated these new claims. The developmentalist hopes of the constructive imperialists found an outlet; the most expansionist Anglo-Indian found an opportunity to gather strength; and an array of British agents discovered an unlikely destiny in redesigning the region after 1918.

6.4 PERSPECTIVES ON PERSIA: THE MAP AS AN ARGUMENT

Promoters of a forward policy in Persia became more outspoken as Russian advances threatened the northern border of Persia in the late 19th century. The construction of the Trans-Caspian railway by Russia (1879–1901) made it possible to move armies to the Persian border in a matter of weeks, thus threatening India itself. Many advocated a more activist policy to counter what seemed to be a considerable Russian threat. Maps became instrumental in the debates in British imperial circles about how to represent, and respond to, this peril—a fact not lost on the ambitious Curzon. The reference map of Persia in the late 19th century and early 20th century was the work of this major figure of British policy making in the Middle and Far East. He had journeyed through the region in 1889–90 and was convinced that geography was central to the elaboration of a geopolitically minded approach to foreign policy.[90] Revealingly, he became the president of the Indian Survey Committee in 1905 to reform the institution. He was also president of the RGS from 1911.[91]

Under George Curzon's rule, the Gulf became the center of gravity of the emerging Middle East, and Persia a key territory on the map of the British Empire. In 1903, in order to publicize the significance of the region, Curzon went on a tour of the Gulf which consecrated the concept of the Middle East. Curzon's hagiographer, Lovat Fraser, described the position of this region in the viceroy's thinking as follows: "The Persian Gulf has a place in the written history of mankind immeasurably older than that of any other inland sea. I believe it will one day be demonstrated that the first dim glimmerings of civilization dawned upon the mind of primeval man within its landlocked waters. It was the scene of great events, which determined the course of progress of the human race, while the Mediterranean was probably still unfurrowed by the keels of ships. Take a map, and see how the Gulf lies at the very centre of the Old World."[92]

Curzon first published his map of Persia in 1892 and used simplified versions of it in his books. He supervised its compilation from a wide range of sources, including Russian and German documents.[93] One of Curzon's early published works, *Persia and the Persian Question* (1892) lent substance to his emerging geostrategic grasp.[94] Underlying the latter were his cartographic accomplishments, which the work illustrated in a map at the end of the first volume, which became its defining element. This publication marked Curzon as an expert in Persian affairs, a reputation which considerably aided his entry into politics. The modernity of the map and the comprehensiveness of his book bolstered his influence on the Persian Question. The map was also an attempt to delineate the border between India and Persia (fig. 6.4). Curzon's Persia is a good illustration of the legitimizing effect of such a cartographic document. The book sold more than a thousand copies and was widely read in diplomatic and scholarly circles.[95] While these sales figures may seem modest, his book was widely read in diplomatic and scholarly circles. The map itself was used in various official memoranda.[96] In January 1905, Ritchmond Ritchie, permanent undersecretary of state for India from 1902 to 1909, wrote a report on British interests in the Persian Gulf for the cabinet with reference to Curzon's work.[97] Curzon's case testifies to the converging relationship between cartography and statecraft. Not only did he publish several maps with far-reaching consequences for British strategic views; he also amassed a personal collection of maps for his own use.[98]

Maps were also used by lower-ranking individuals to promote new per-

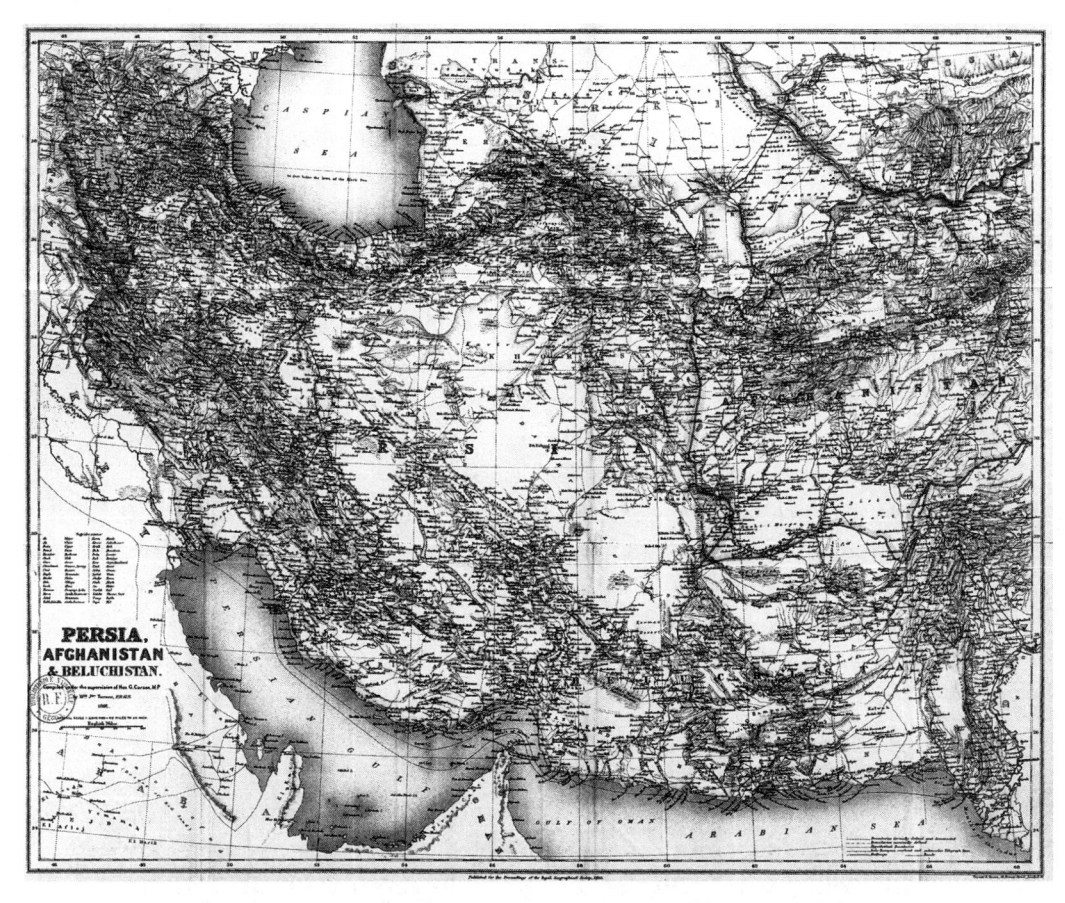

FIGURE 6.4. George Curzon and William Turner (engraver), *Persia, Afghanistan and Beluchistan*, 1:3,810,000 (London, 1892), in *Proceedings of the Royal Geographical Society and Monthly Record of Geography* vol.14/1, 1892, 141. Size of the original: 61 × 50 cm. Photograph courtesy of the BNF (Cartes et Plans GE FF-110).

spectives on the Persian Question. Herbert Sykes's field notebooks offer an insight into the practices of a British agent in a region in that regard. They document his topographical surveys in southeastern Persia as part of a broader scheme to build a telegraph line for the Indo-European Telegraph.[99] Sykes was looking for the best route to build a line that would connect Meshed, where his cousin Percy Molesworth Sykes was monitoring Russian activities, to the Gulf coast. He also formulated suggestions as to the development a potential railroad. His explorations were under the supervision of the Indian Political Service, which needed to update its knowledge of a region explored by Frederic Goldsmid more than thirty years earlier. Herbert Sykes left London on November 2, 1902, and landed in Bombay twenty days later. He explored southeastern Persia, from the Gulf to Meshed region, for almost a year. The official motive for his expedition was that he wished to visit his cousin.[100]

His archives reflect the systematic nature that his surveys had acquired over the course of several months. Various routes were carefully reconnoitered: one from Chah-Rigan to Bandar Abbas, another from Yazd to Torbat, and the last one from Meshed to Seistan. Sykes rigorously noted his calculations of gradients, his barometric readings, the distances he traveled, and the geology of the land. One section of his notebook shows corrections to measurements of longitude and latitude on the left page, while the right page has comments on the relief along with a sketch of what he observed (fig. 6.5). Like many engineers and officers trained in route surveying, Sykes was able to complete fairly accurate sketches in a very short time. Route surveying had become standardized, as demonstrated by the printed notebooks he used.

On his return to Britain, Herbert Sykes communicated the results of his expedition in many forms.[101] In 1903, he also wrote articles for the *Times of India* to publicize his journeys in Persia in a series entitled "Roads to the Gulf."[102] His surveys and publications were one of many reactions to the consolidation of Russia's position and to the threat it represented for British interests in southern Persia. Curzon's administration had done everything to strengthen Anglo-Indian domination in the Persian Gulf, from the creation of Kuwait to the development of commercial networks, notably on the Karun River. However, the fear of a Russian invasion of Persia, intensified by the deep political turmoil of the 1905 Constitutional Revolution in Persia, remained ever present. Most official documents on the region insisted on Britain's inability to organize a counterattack against Russian troops taking the trans-Caspian route. Herbert Sykes's travels were part of a larger movement to gather intelligence and to accumulate preparatory surveys in case of conflict.

These two approaches to Persia demonstrate the new position of surveys and maps in the British geopolitical debate in the late 19th and early 20th centuries. Cartographic visualization had acquired an autonomous authority. The promotion of a new strategic vision could hardly ignore cartography's mastery over space in all its forms through its disciplining capacity to represent mental constructs of that space.[103] The times when the cabinet would frown at Palmerston's display of a map of Asia Minor were long gone.[104]

Both the high-placed Curzon and the lower-ranking Sykes illustrate how a new regime of cognitive tools increasingly fashioned views on the region. Spatial visualization and cartographic proselytizing furthered a new approach to international affairs. Maps became instrumental to geopolitical hermeneu-

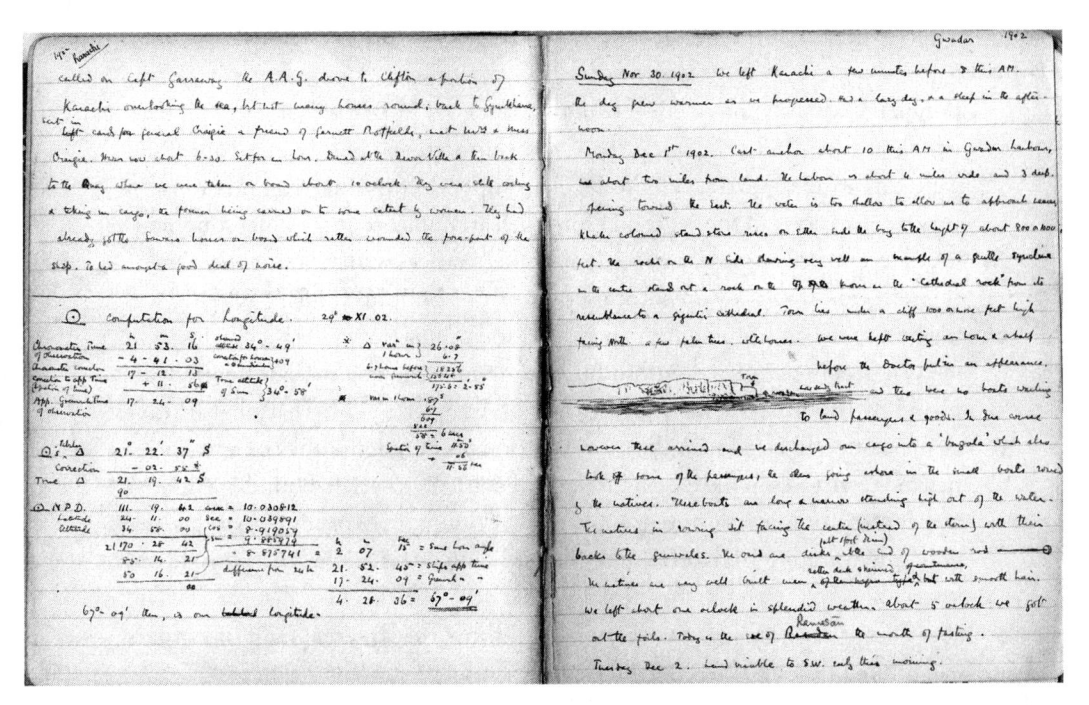

FIGURE 6.5. Herbert Rushton Sykes, *Pages from a Field Book, 1902–1904*, GB165–0274 Herbert Rushton Sykes Collection. MEC Archive, St. Antony's College, Oxford.

tics in a late 19th-century world characterized by global rivalries and a growing need for multilateral agreements. Decision makers in London and propagandists of British expansion walked in the footsteps of Indian officials: they started to think cartographically and geographically more systematically than ever before. Significantly, some key diplomatic notions of the high imperial age, such as the concept of zone, or sphere, of influence, were coterminous with the spatialization of international affairs through the cartographic prism. The 1907 Anglo-Russian Convention, which put an end to the Great Game, and its subsequent translation into maps of foreign spheres of influence in Persia, is a characteristic example of this process.[105] Cartography and geography had reached a point where the delimitation of modern zones of influence had become a viable solution to solve conflicts. They provided standardized data that were considered accurate enough, sometimes preposterously so, to substantiate universally accepted agreements. The trust placed in maps and geographical information to engineer spheres of influence would become blatant a few years later when a single map became the synecdoche of the Sykes-Picot Agreement (1916).

6.5 BRITISH DEMARCATORS: EXPERTS AT DRAWING LINES

Boundary commissions were yet another category of British surveys and maps which participated in shaping the Orient into an apparently controlled and well-categorized territory. The FO had supervised border commissions on various occasions in the region before the early 20th century. Such opportunities to delimitate states and to demarcate frontiers multiplied at the turn of the 19th century. They are a focal point of this book. Boundary making was integral to the development of an international legal framework, which was in its turn deeply enmeshed within European imperial projects. The central part played by Britain in establishing borders, its expertise in international arbitration, positioned London at the heart of a process of redefinitions of communities and territories in the East. This unprecedented growth in boundary surveys and demarcations participated, among other things, in expanding markets, negotiating commercial treaties, and standardizing international relations. It was interconnected with the political ascendancy of British liberalism in the late 19th century. The following sections will show that the mapping of borders had momentous consequences in the fabrication of the Middle East.

ADEN AND YEMEN

During the 1870s and the 1880s, the metropole and the government of India had opposing views on how best to protect the strategic node that was Aden. The increased attention encouraged the mapping of Aden's hinterland and the region of Yemen. Campaigns to explore further inland were conducted by the residency. The territory of Aden was officially under the jurisdiction of the government of India, who conceived this base as a means to secure a portion of the Arabian Peninsula and access to the Suez Canal from the 1870s. This logic still dominated debates under Curzon's viceroyalty. Aden was an enclave in a territory theoretically under Ottoman sovereignty from which the British could build networks to protect the Hadramaut tribes. In the late 19th century, several Yemeni emirs had signed treaties with Great Britain. 'Abdali, sultan of Hawshabi, signed a treaty with Great Britain in August 1895, and he was just one of several emirs to do so in the closing years of the 19th century. Consequently, tensions between the Ottomans and the British in the Gulf were palpable. The sultanate attempted to thwart the development of the alliance system by sending troops and spreading pan-Arab ideas hostile to European domination. India promoted a forward policy against the advice of the

FO, which supported a less interventionist approach. These tensions resulted in a major crisis in July 1901. A tower was built in Ad Dareja by a tribe with close ties to the Ottomans. The British asked for the building to be destroyed before sending troops to blow it up. Rising tensions favored the interventionist stance. A Turkish-British border commission was set up to demarcate the limits of British influence in southern Arabia.

General Pelham James Maitland (1845–1937), the British resident in Aden, who handled the negotiation with Ottoman authorities, was an expert in border delimitation. He had previously worked under Thomas Holdich. Holdich himself had participated in the commission to establish Afghanistan's borders between 1884 and 1886. Colonel Robert A. Wahab, another technician from the Afghan Boundary Commission, joined the delegation in Aden. The Ottomans were thus up against two specialists in border negotiations, whose level of expertise was a force in its own right. The Ottoman authorities were compelled to allow the British to act as judge and jury. All the experience acquired in Baluchistan, on the Turkish-Persian border, and in Afghanistan combined to separate Aden from the neighboring Ottoman administrative unit, which was at the time called Yemen. The network of Anglo-Indian surveyors specializing in border delimitation immediately set out to work. Maitland first defined the territory that was to be mapped; then, along with the Ottoman commissioner Remzi Pasha, they determined on which side of the border to place each tribe. Finally, the cairns materializing the border were installed.[106] The FO followed the process fairly closely. The drafts regularly sent by the two officers provided visual proof of the progress of negotiations.[107] Yet, because of the distance between London and Aden, Whitehall was mostly a spectator. An agreement was signed in June 1904, accompanied by a reference map.[108] The final agreement acknowledged British authority over Aden's hinterland. That was precisely what the Indian government was looking for. The data collected by Wahab and Maitland also helped elaborate the first accurate ethnographic maps of the region. Britain could now assess which tribes supported the Ottoman sultanate and which would welcome London's protection.

The fabrication of these maps of southwestern Arabia illustrates how a seemingly neutral border demarcation process was in fact a military intervention and how it was a warning for the Ottomans. The Anglo-Indian surveyors were accompanied by a detachment of a hundred men, many of whom were armed soldiers. Several hundred people and horses served the commission's logistic needs. Tensions between British soldiers and their Turkish colleagues were not uncommon. The survey also provoked angry reactions from tribesmen, who viewed the theodolites as harbingers of new taxes. The cartographic development of the border was itself an important military operation. It was perceived as such by a segment of the local populations. Spatial dominion was not achieved simply by the needle of a prismatic compass, for the barrel of a gun was never far off. Another indication of the changing nature of the British involvement in the region was the gradual disappearance of former placenames. Wahab, like Leachman and Shakespear, recorded local names with the greatest possible precision. The last traces of the Greek and Latin names still used into the late 19th century slowly disappeared in favor of an updated toponymy, one that could be useful to the negotiator and the traveler. Terms such as "Nejd" or "Yemen" began to be heard at Westminster.[109] The ancient "Arabia Felix" began to be confined to historical atlases.

The same sequence of events led to the delimitation of a border in Sinai that was very protective of British interests.[110] British authorities first triggered a crisis that was then taken as a pretext for a demonstration of military power and a heightening of diplomatic pressure. Then, once the principle of a border commission was accepted by the Ottoman Porte, British surveyors ensured that the resulting frontier reflected largely British interests. Finally, under various pretexts, new topographical missions were sent into Ottoman territory. The diplomats of the Levant Consular Service and the Indian Consular Service updated their geospatial intelligence, sending the results to the Raj, to London, or to Cairo. Shortly before the war, Gilbert Clayton was thus able to develop a network of Bedouin informants in the Sinai Peninsula. They were ready to report to Cairo on any movement of armed Ottoman troops in the region.[111]

OIL AND SOIL

The long-running disputes over where the Turco-Persian frontier lay afforded a further opportunity to consolidate British interests in the Middle East. The mid-19th-century attempts to solve the issue had borne little fruit. Yet that was set to change in the early years of the 20th century. The Tehran Protocol of December 1911 provided for the creation of an international commission in order to determine the border once and for all. By that time, however, what was at stake for Britain in the demarcation process was changing as the information from geological surveying revealed more about the resource portfolio of the region in question.

In the early 20th century, the Survey of India printed a series of geological maps that reflected this changing perspective on the region. G. E. Pilgrim's

1907 *Geological map of part of southern Persia* was one of the reflections of this fast-evolving dimension of the British conception of the Middle East.[112] Pilgrim traveled the Gulf from 1904 to 1905 and published an account of his survey in addition to the map.[113] This was the first comprehensive geological study of the region. Pilgrim's work was indicative of a wider evolution of the Western perception of the East. Persia and Mesopotamia were no longer backward and unproductive spaces. Geological exploration gradually revealed the sources of wealth that lay underground. Bitumen that surfaced in the Mosul area, an exotic attraction until then, drew attention shortly before World War I.[114] The first use of oil as fuel by the Royal Navy dates back to the 1880s. The 1900s witnessed an increasing awareness within the Admiralty of the potential of hydrocarbon as a fuel for its vessels.[115] Oil made its entry onto the Middle Eastern stage. The British had seen their advantage reduced by the concession granted by the Ottomans to the Germans as part of the construction of the Bagdadbahn. German investors were authorized to exploit mineral resources on a strip several kilometers long around the line. In reaction, Whitehall's diplomatic efforts secured concessions for British companies as well.[116] Ernest Cassel (1852–1921), a director of the Bank of England with a successful record of investments in the East, founded the Turkish National Bank in 1909 with a view to gaining access to Mesopotamian resources. But it was in Persia that the British had to secure their first significant oil production. Although Baron de Reuters's investment in Persia in the 1870s had miserably failed, another figure from the City, William Knox D'Arcy (1849–1917), would achieve a return on his investments in the Persian economy in an unexpected way. In 1901, he managed to obtain a concession from the Shah, Mozaffar ad-Din. D'Arcy hired an engineer named George B. Reynolds to start prospecting. The cost of operations led D'Arcy to the brink of ruin, despite his 1904 association with Burmah Oil. It was not until May 1908, when the company was nearly bankrupt, that oil was discovered in Masjed-Soleyman. The Anglo-Persian Oil Company (APOC) was set to thrive.

One of the men who took part in the discovery of this first well was Arnold Wilson (1884–1940). The young officer, like many of those who were to play a significant role in the Middle East during the First World War, was a former student of Clifton College in Bristol. Wilson was in Persia with a detachment of Bengal Lancers to protect the operations headed by George Reynolds. Once the oil gushed out, Wilson set out to complement his topographical knowledge of the surrounding territories, in particular the Zagros Mountains and neighboring Luristan. He traveled throughout these two regions for months in order to compile a reliable map. His diary reflected his motives: "The Indian I.B. has made me responsible by the CID for this part of the world but not for Mesopotamia and Arabia which come under the W.O. With small funds at its disposal it must do its best. Major H. H. Austin, who was at Clifton in your day, has sent me a draft of the ½ inch sheets for this part of Persia—mere outlines with next to no details. Great areas shown as 'unexplored': it is an immense incentive. I have extended my triangulation outwards from the fields with vast labor, for I have to climb four or five thousand feet every day."[117] Percy Cox, the political resident in the Persian Gulf, soon recognized Wilson's talents.[118] Like most of his colleagues in the Indian Political Service, Wilson understood very well the authoritative power of maps. Wilson wrote a long report that included a map designed by "the best engraver" under Major Austin's command.[119]

The intelligence officer's expertise proved very helpful to British interests as tensions between Persia and the Ottoman government began to threaten the long-term stability of a region where British-operated oil drills started to be productive, forcing the various players into fixing the Turkish-Persian border. Arnold Wilson played a key role in ending this decades-old dispute. He joined the Turco-Persian Frontier Commission with C. H. D. Ryder (1868–1945), the head of the British Survey Party. The fixing of the border was facilitated by telegraphic communications, through which the exact time and longitude could be determined. Stakeholders gradually came to demarcate a border on which markers were placed. The issue of the Shatt al Arab, which would long remain to poison relations between Middle Eastern states, was resolved in the short term in favor of the Ottomans, yet part of the delta was officially assigned to Persia for the first time. Overall, Wilson and the British delegation had achieved their goal. They had successfully managed to promote a settlement that protected British oil concessions in the long term.

Interestingly, the border delimitation process was of direct concern to the Admiralty, a department that was not historically involved in the definition of British policy toward Mesopotamia or Persia's hinterland. This shift was related to the Royal Navy's conversion to the policy of oil-fueled ships under Lord John Fisher's command, starting halfway through the first decade of the century.[120] Fisher, nicknamed the "oil maniac," wanted to secure Persian resources to that end. He understood how maps could be instrumental in legitimizing the oil concessions. As a consequence, the Admiralty saw to it that the APOC's rights were part of the Constantinople Protocol of November 1913, which eventually settled disputes regarding the Turco-Persian boundary. Admiral Edmond Slade's 1914 report, which eventually convinced the British government to take a 51% stake in the APOC despite opposition from the IO, included a map that was circulated around the lobbies in Westminster in order to promote the navy's perspective on the region.[121] The cartography contained in Slade's report was a speculative projection, however, since the systematic surveying and mapping by the international commission was still underway and the official survey had yet to appear.[122]

British expertise was instrumental in designing the last border to be delimited before the post-1918 restructuring of the Middle East. Boundary making was not only a way to enjoy a comparative advantage in terms of knowledge acquisition, but it also gave the expert leverage in the bargaining process. In Aden, Sinai, and Mesopotamia, Britain took great care to define territories that protected its interests. More important perhaps, British agents involved in boundary commissions gained a legitimacy which would help them play a key role in the creation of the new Middle Eastern states after the First World War.

6.6 THE PROSPECTIVE MAP: THE "SICK MAN" IN AND OUT OF EUROPE

The Ottoman Empire had, as we have already seen, long been regarded as the "sick man of Europe." Events around the turn of the century, however, conspired further to alienate the Ottomans and the British Empire. Internal Ottoman difficulties, culminating in the Young Turk Revolution of 1908, moved the Ottoman Empire away from the reach of traditional British influence in the East. The fact that the Young Turk regime sided with Germany and the newly found entente between Britain and Russia, the former enemy, accelerated the process. The remnants of pro-Turkish attitudes in Britain in the 1870s had almost completely evaporated from the British cultural landscape by 1900. This paradigm shift in British policy is illustrated by the evolutions in the cartography of both Turkey in Asia and Turkey in Europe.

The British map of the East had to take account of the fast-moving events that were reshaping the Balkans. Charles Eliot (1862–1931), governor of Kenya between 1900 and 1904, had served as a diplomat in Turkey in the 1890s. In his book *Turkey in Europe,* he summarized the alterations in interregional relations as observed by a diplomatic agent: "The Balkan States in the last ten years [. . .] have ceased to be the pivot of the Eastern question and the centre of gravity has moved from the Near to the Far East."[123] One of the most revealing consequences of the region's mutations was the near-final institutionalization of a new place-name to describe it. The expression "Turkey in Europe" was still in use, especially for the Adrianople region that remained under Turkish control on the eve of the First World War, but "Balkans" and "Balkanic States" had become preponderant. On the eve of the war, the WO maps systematically used the expression "Balkanic states" or "Balkan Peninsula."[124]

The map of the Balkans became more elaborate as regional changes began pushing Turkey out of Europe. Until then, precise knowledge of populations and their locations was mostly compiled from maps produced outside of Great Britain or by individuals who were not officially employed by the government. The FO and the WO sought more accurate information on the political and ethnographic characteristics of the Near East. In 1903, the Mürzsteg Reform Plan provided for British officers to be dispatched to Macedonia to reorganize the local police.[125] They promptly began to collect topographical data and to compile maps of the region. William Fairholme, former military attaché at the Embassy in Vienna, was appointed political representative in the area in 1908. He wrote various reports with a specific emphasis on the human geography of the Balkans, as shown in one of the maps on "races and religions" he compiled for the FO.[126] Their maps pushed the Balkans just a little closer to Europe.

Ottoman sovereignty on the Middle East was weakening in the early 20th century. Prospects of a complete overhaul of the region's geostrategic balance called for a more methodical mapping of what was known as "Turkey in Asia" and its dependencies. The piecemeal cartography of the Victorian era could not serve the purpose of modern prospective military planning. The main architect of this systematic survey and cartography was Francis R. Maunsell (1861–1929). Maunsell first held the position of military vice-consul for Sivas, Trabzon, and Van from 1897 to 1899 before becoming a military attaché at the embassy of United Kingdom in Constantinople from 1901 to 1905. He coordinated the efforts of several attachés in order to collect knowledge on different parts of Asiatic Turkey. Mark Sykes, Louis Molyneux Seel, and Aubrey Herbert all worked under his supervision, and his cartographic legacy is monumental. It coincided with a period during which the construction of the map of the East became truly structured and systematic. Great Britain, whose influence over the Ottoman Empire was declining rapidly, was faced with the spread of German intelligence and diplomacy in the region. The comparison between the Prussian topographical services during the Franco-Prussian War of 1870–71 and its British equivalent had already forced the WO to develop a dedicated Intelligence Department. The context of the 1900s led to the same results, and Maunsell was commissioned to perform a comprehensive examination of the Ottoman patient, using topography.

Maunsell was from a new generation of British officers: the product of London's new ability to train surveyors and cartographers who were as able as their Anglo-Indian counterparts were. One of his first appointments took him to Kurdistan, where he traveled from Erzurum to the Tigris in 1893.[127] Maunsell was actually engaged in cover surveying, however, and to do so, he traveled with a set of easily transportable instruments that had been designed by the Indian Survey Department. His kit included a six-inch sextant, an artificial horizon, a stopwatch, an aneroid barometer, and a prismatic compass. He established fifty-nine trig points in order to establish a reference grid. He corrected latitude errors during his stops in several large cities where he could check the precision of his chronometer. Maunsell then compiled the first accurate map of Kurdistan. Europeans had already traveled across the zone, but Kurdistan itself had never been fully mapped and delineated. The map was published as a supplement to an article published in 1894 in the *Geographical Journal*.[128] In his article, Maunsell referred, quite logically, to Kurdistan as the "country of the Kurds."[129] Ethnographic criteria thus seemed to define the country he was introducing to his scholarly audience. Yet a glance at the map shows that it was physical geography, and more specifically the hydrographic system, that was in fact used as a frame of reference. The Euphrates constituted one segment of the border to the north of the 38th parallel and the Tigris to the West. Maunsell was well aware that Kurdish populations lived well beyond these boundaries, but the seductive rationality of hydrographic and physical boundaries got the better of him. This would be a consistent feature of the Western outlook on the region's border definitions in the long term.

Maunsell's map of Kurdistan was at the intersection of a variety of British interests. It first illustrated a more prospective approach to the Middle East from a Whitehall standpoint. Second, Maunsell's covert surveying was supported by the RGS. Many of the society's members wanted to plan the topographical unveiling of Asia Minor in a coordinated fashion. In 1905, Clements Markham's annual address to the society underlined the necessity of mapping Kurdistan, the Taurus, and the rest of Asiatic Turkey.[130] The RGS consistently supported explorations in Kurdistan, even after Maunsell had completed his surveys. Vice-Consul Percy H. H. Massy (1857–1939), who was stationed in Cilicia at the time, was encouraged by the RGS to continue his colleague's undertaking. The RGS lent him all the surveying equipment and, as shown by the illustrations of the article he published in the *Geographical Journal*, and the society also encouraged him to take photographs.[131] This new interest in Kurdistan was reflecting the priorities of a third institution: the government of India. Kurdistan's nodal character, at the intersection of Mesopotamia, Persia, and Ottoman Turkey, made its protection a priority for the Raj. It partially explains why the British were among the first to put Kurdistan on the map.

Maunsell's explorations in Kurdistan were a small part of his contributions to the British map of the Middle East. His most impressive achievement was his monumental map *Asiatic Turkey*, a series of forty sheets drawn to a scale of 1:250,000. It was the first truly systematic map of the region. It took Maunsell over a decade, from 1901 to 1915, to complete this work. He used all the latest sources, including his own work, and systematically listed them on each sheet.[132] Maunsell only referred to very recent data, never going back further than the 1870s.[133] A mathematically precise grid was established using Mount Ararat, a visible landmark, as a reference point.[134] A forty-mile-long geodetic base was measured from the British consulate in Van.[135] The era of route surveying was over. Modern surveying techniques replaced them to figure Anatolia and the Taurus accurately. When data were lacking for some areas in the grid set by Maunsell in 1901, British agents were sent to fill in the blanks, such as Mark Sykes in northern Mesopotamia.[136]

The map itself benefited from the latest developments in printing. The Intelligence Department of the War Office printed it in color. Maunsell applied colored contour lines and hypsometric tints. The document is illustrative of the progress of cartographic standardization and of the changing nature of intelligence collection in the Edwardian era. The cartographer, who only represented topography, relief, and political data at first, gradually included more varied information. The sheets showing Mesopotamia, which were drawn at the start of the First World War, displayed ethnographic (Muntafik Arabs) and geological (bitumen wells) data, which were missing from the first drafts.[137]

Renewed maps and modernized names helped in conceiving a different East. Turkey in Europe and Turkey in Asia were not working categories anymore. Many observers foresaw in the Ottoman Empire's agony the first signs of a complete transformation of the geopolitical framework. Maunsell's cartographic project is one of the many traces of this shifting British vision. Not only was statecraft evolving, with an increasing demand for prospection, but maps were also being updated in order to grasp a fast-evolving area where the old equilibriums seemed to be put into question one after the other. Maunsell's mapping and mapmaking also illustrate how the changing scale of many early 20th-century maps of the Middle East profoundly modified people's perspectives of the region. They figured an increasingly fragmented area and visualized, with greater detail than before, distinct ethnographic and economic patterns. This zooming-in operation reflected the increasing degree of precision of mapping in the Middle East. Large-scale maps were also promises of better planning, be it military, diplomatic, or even economic.

CONCLUSION

By the late 19th century, British mapping and mapmaking materialized a transitional East that was more and more visibly an integral part of the broader imperial structure. Geological, ethnographic, and prospective maps began to project a very different picture from the traditionally received one of a stagnant region, unreceptive to modernity. The new global balance of power as well as late Victorian anxieties as to the sustainability of British hegemony called for an overhaul of geographical categorizations. Curzon's division of Asia into the Far East, the Central East, and the Near East testifies to this process.[138] Many in India or Egypt underlined how significant this hitherto ill-defined "middle region" was to the security of the British world system as a whole. Russia's expansionism appeared to threaten Persia and the Ottoman Empire. Maps and books such as Curzon's *Persian Question* not only identified a threat, but also visualized it, and in the process of doing so, heightened the sense of alarm at its apparently sinister existence. Some also started to pinpoint the interconnections between the different parts of the "Muslim World," a phrase of common currency by the turn of the century, as well as the relevance of the holy sites of Islam in Arabia to Indian stability. By 1898, Rafiuddin Ahmed (1865–1954), a barrister and a member of the Muslim Patriotic League, emphasized how Britain had become the "greatest Mussulman Power the world has ever seen" after the victory of Omdurman.[139] From Egypt to Persia, British conceptions of the area on the "Highway to India" began to crystallize into a single category. The apparent completeness of British cartographic and geographical knowledge undeniably participated in the reification of the regional entity. Experts professed a panoptic gaze. Texts and maps had preordained the area. The cartographic codification of the Middle East, even if far for complete in reality, was a manifestation of power in itself. The ever-growing assemblage of maps, statistics, and data substantiated a nascent geopolitical gaze on the region.

From that perspective, mapping and mapmaking contributed to the promotion and legitimation of the various forms of British expansionism in the area. Anglo-Egyptian surveys are a test case in that regard. Geological or cadastral maps demonstrated the merits of British interventionism. Mapping and mapmaking played an instrumental role in the rhetoric of modernization then central to the late Victorian outlook on the East. British geographies of the East became ever more tied to cartography. Maps and geographical expertise became instruments of persuasion at a time when laissez-faire advocates and Imperial Federation enthusiasts brashly discussed the future of the empire. As a consequence, the surveying and mapping of the East followed stricter scientific imperatives than before. Yet mixed views from India, Britain, or even Egypt, as to which function and definition to assign to the area, did not disappear. The power relationships behind the British mapping of the area were a complex of contradicting influences. It is also noteworthy that the men behind these new visualizations constituted professional and personal networks that were to become very influential after the First World War. Kitchener's and Curzon's protégés were all entangled in the post-1918 redefinitions of the Middle East. Their presuppositions cannot be fully comprehended outside the late Victorian and Edwardian context in which they were nurtured.

As classical and biblical topographies seemed to recede, the surveys and maps studied in this chapter participated to a greater or lesser extent in the territorialization of the East. Legal uses of cartography, boundary making, new map projections, and emerging geographical categories such as the Balkans all participated in an informal consolidation the Eastern state system.[140] The shift from maps of exploration to maps of construction and modernization should also be underlined. The East was not a romantic fringe anymore. It was soon to be replaced by a "Middle East" on whose imagined integrity the empire relied.

CHAPTER SEVEN

ENTER MIDDLE EAST

It is often assumed that the Middle East did not find its legitimacy as a geopolitical concept until the interwar years, decades after its first occurrences, when ministerial departments and army experts officially started to use the denomination.[1] While it is no doubt exceptional and premature, Bacon's 1913 map (fig. 7.1) offers an interesting rejoinder to this widely accepted affirmation. This exceptional example of an early use of the term was intended for schools and for candidates to the civil service examinations. The publisher sold it as *New Contour Map of the Near and Middle East (the Land of the Five Seas)*. It delimited a large transitional East from Bizerte in North Africa to Hyderabad near the Indian frontier and from the Crimea in the north to Berbera in Somaliland. The map projection, which displayed a territory bordered by the Mediterranean, Black, Red, and Caspian Seas and the Persian Gulf, drew from the most recent analyses. The expression "Land of the Five Seas" was borrowed from H. J. Mackinder's regionalizing constructs developed in his groundbreaking 1904 article "The Geographical Pivot."[2] Bacon's efforts to depict the interrelationship of the various states, histories, and environments mirrored Mackinder's perspective.

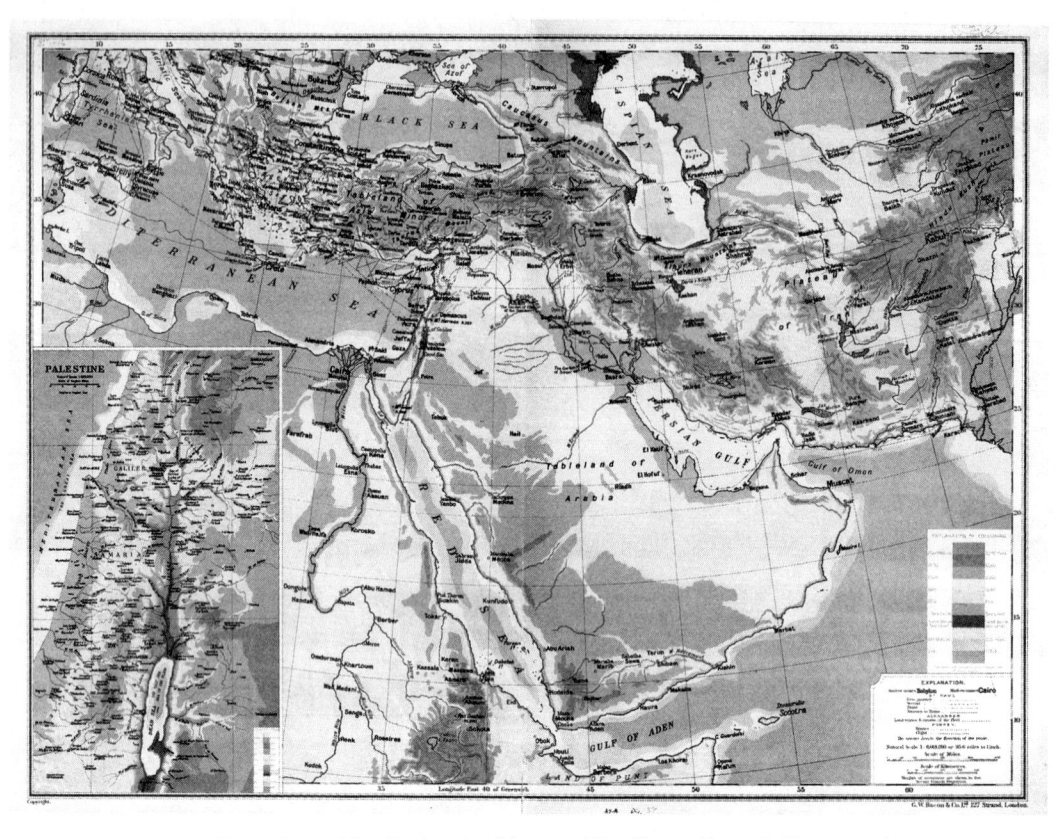

FIGURE 7.1. George Bacon, *New Contour Map of the Near and Middle East (the Land of the Five Seas)*, 1:6,019,200 (London, 1913). Size of the original: 71 × 96 cm. Photograph courtesy of the RGS (Map Room mr Asia Div.38).

Bacon's map was still caught between the old and the new. Both the title of the map and the indications found in the document testify to its hybrid nature. Place-names were a combination of the antique and the modern. While the Suez Canal was shown, along with updated borders, the key focused on the journeys of Paul, Alexander, and Pompey. Willcocks's location of the Garden of Eden and an inset orographical map of Palestine completed the picture of an East which was still interpreted in terms of the legacies of a Greek, Roman, and biblical past. This feature was praised in the various reviews of the map.[3] Yet it was more than just another of the many instances of the mid and late Victorian scriptural geographies. With its outward signs of presentational novelty, such as the lavish ten hypsometric tints and the centering of the map, the document took on a different shade. As shown by Huseying Yilmaz, this combination of a historically charged "Near East" and a contemporary-minded

"Middle East" was to become an enduring characteristic of the West's outlook on the area.[4]

The map's targeted audience was composed of secondary school pupils and students. It was not the first its kind. W. and A. K. Johnston published a test map of the "Middle East" as early as 1908. It was advertised in the *Geographical Teacher*.[5] This, in itself, is evidence that the denomination "Middle East" had begun to surface outside the limited circles of experts and scholars who first coined the concept. School maps show that the category's gestation period was coming to an end. They are illustrative of the first phase in the invention and emergence of the "Middle East" as a frame of reference in the British cultural landscape, which the present chapter wishes to address. I shall therefore analyze the networks by which the new category was promoted and the various layers which supported this process of regionalization. Throughout the following sections, specific attention will be paid to the intellectuals of statecraft and their presumptions. To this end, I shall also address the manner in which constructive imperialists, along with other proponents of an overhaul of the British imperial structure, confronted what amounted to a more academic grasp of the East and the way metropolitan strategists competed with men on the spot over the definition of what could be the last addition to the British Empire and its structuring axis.

7.1 NEARER EAST VERSUS MIDDLE EAST

GENEALOGIES OF A PHRASE

Several expressions were coined in the course of the 19th century to delineate an intermediate East stretching between the fringes of Europe and India. Phrases such as "Nahen Orient" or "Mittlern Orient" are attested from the first decades of the century, in Goethe's *West-Oestlicher Divan* and the works of Carl Friedrich von Rumohr.[6] French writers used variations on the same structures, such as "Orient Moyen," from the mid-19th century. Jules Michelet, in his *Histoire de France*, referred to the "Orient occidental" as early as 1856.[7] The first occurrence of the term in an English-speaking publication is often wrongfully attributed to an Anglo-Indian officer, Thomas Edward Gordon. He used the phrase in a 1900 article, before Alfred Mahan, an American strategist of considerable influence in certain British circles in the 1900s and alleged inventor of the "Middle East," began to spread the neologism.[8] A forgotten occurrence in an article written by Seth Low (1850–1916) in 1899 seems to indicate that the expression was already circulating informally in American diplomatic spheres, before Gordon used his own version of the notion.[9] Low, at the time president of Columbia University, was a member of the American delegation to the Hague Conference in 1899. As Alfred Mahan was one of the other delegates of the United States, it would not be unreasonable to assume that the concept was already in use inside Western diplomatic spheres.[10] Enrico Catellani (1856–1945), professor of international law and diplomacy at the Bocconi University in the early 20th century, tackled the "question of the Middle East" in his 1902 syllabus.[11] There are earlier occurrences of the phrase. An anonymous piece published in 1898 in the London-based *Outlook* is titled "The Problem of the Middle East."[12] Another article, which appeared in *Blackwood's Magazine*, mentions the "Middle East" in 1895.[13] It is attributed to Alexander Michie (1833–1902) by *The Wellesley Index to Victorian Periodicals*.[14] Michie was established in China as a banker and a correspondent for the *London Times* during the Sino-Japanese War of 1895. The earliest occurrence of the phrase is an isolated mention in a review of John P. Newman's *The Thrones and Palaces of Babylon and Nineveh from Sea to Sea* in an 1876 issue of *Zion's Herald*.[15]

In Europe and America, the lack of plasticity of antiquating notions such as "the Eastern Question" called for new geopolitical subdivisions to grasp Asia, well before Alfred Mahan purportedly popularized the concept of "Middle East." Whoever actually coined the phrase responded to the need to change scales in order to comprehend Asian affairs in the late 19th century. The multiple authorship that characterized the concept's birth is responsible to a degree for the vast range of definitions of what the "Middle East" was from the start. Mahan himself did not give a very clear definition when he first used the terms in 1902: "South Persia is in fact the logical next step beyond Egypt; [. . .] the middle East, if I may adopt a term which I have not seen, will someday need its Malta, as well as its Gibraltar; it does not follow that either will be in the Gulf."[16] He acknowledged the expansion of the Indian subempire in the area. It called for a distinction between the Farther East and the East stretching from the Mediterranean to India. The Suez-India route was the key to the new geopolitical construct. Mahan's articles were published by the *National Review*. Leopold Maxse (1864–1932), ardent imperialist and editor of the conservative publication, was central to the enterprise. The two conducted a friendly correspondence for years. Maxse was convinced of the necessity of putting more pressure on the cabinet for a stronger British stance on the protection of the Gulf.[17] A few months later, it was Valentine Chirol (1842–1929), director of the foreign department of the *Times*, who publicized the term by writing several articles about the "Middle Eastern" issue.[18] This series of articles published

FIGURE 7.2. Valentine Chirol, *Map of the Middle East*, on the cover of Valentine Chirol, *The Middle Eastern Question; or, Some Political Problems of Indian Defence* (London: J. Murray, 1903), book cover. Photograph courtesy of the Bibliothèque Interuniversitaire des Langues Orientales (BIULO GEN. III.1619).

between 1902 and 1903 was the logical conclusion to an overview of the geostrategic situation of Britain in Asia, which he had started in 1896 with a book entitled *The Far Eastern Question*.[19]

Chirol's definition of the Middle East was more specific than Mahan's broad outlines. On the cover of his book, a golden map showed the contours of the region (fig. 7.2). Chirol was less versed in naval affairs than the American strategist. His primary interest was in the defense of India as a whole. His delimitation of the Middle East was centered on the Arabian Sea. It included Tibet and Afghanistan and most of Southeast Asia. This configuration echoed the conceptions conveyed by his social and professional network.[20] Chirol met Curzon while traveling in Asia. Both his eastern experience and his encounter with the viceroy were crucial to his decision to write on the Middle East. He was a member of George Curzon's entourage when the latter embarked on a momentous tour of the Persian Gulf in 1903. Among the British officials traveling on the SS *Hardinge* was Winston Churchill, who would eventually develop the concept of the Middle East after the First World War into a fully fledged administrative entity. Chirol thus became the propagandist of Curzonian strategic views.

Mahan's analysis and Chirol's articles participated in the larger geopolitical debate which developed in Edwardian Britain in the aftermath of the Boer War. The CID founded by Balfour in 1902 carried hopes of an unprecedented coordination. It represented yet another sign of Whitehall's firm grip on the management of Indian or Egyptian affairs. As new administrative divisions emerged, potential careers waned or prospered. The carving out of unofficial kingdoms, such as the Bushire resident's little-known prominence in the Gulf, was to become more difficult. At the same time, the disparity between metropolitan and Anglo-Indian designs on the Middle East persisted. Balfour's views often contradicted Curzon's own assumptions. Even if the terms of the dispute were not simplistic or clear cut, the divergence was acknowledged by Balfour himself, who wrote in 1905 that India was becoming "an independent and not always friendly power."[21] His defensive policies and his attempts to slow down the expansion of India's sphere of influence, which was continually pushing beyond the official borders of the Raj, were at odds with the intentions of the government of India. Some stakeholders brandished their geographical and cartographic weapons in order to foster their views. Curzon, who consistently understood maps as an adjuvant to statesmanship, was not the least active in this regard. Relevant geographical categories and classifications were fought over. While Balfour's CID did not use the phrase as such, several of its memoranda brought forward its own aggregates of interrelated states, such as the "Persian Gulf" or the Nile basin. After a few years, a subcommittee on the Persian Gulf thus saw the light of day.[22]

The concept of "Middle East" was soon taken up by various individuals, most of whom were related to the Anglo-Indian world. The term was repeatedly used in the House of Commons by MPs who shared such a background. One of the term's main promoters was John David Rees (1854–1922), a conservative from the Indian civil service. His 1908 *Real India* was very Curzonian in tone.[23] Rees recurrently used the new terminology in his speeches at Westminster.[24] Donald Mackenzie Smeaton (1848–1910), who had served in the Bengal civil service, also made references to the Middle East as an MP.[25] Angus Hamilton, a correspondent of the *Times* and the *Pall Mall Gazette*, wrote a book on the region entitled *Problems of the Middle East*.[26] Hamilton included three maps in his work. They were not original documents, being reprints from Stanford's stock, presumably to save the cost involved in the production of a new map. The addition of the three illustrations nonetheless defined a Middle East that resembled that of Chirol, extending from the Red Sea to the Ganges delta.

One of the most active officials to spread the phrase was Percy Molesworth Sykes.[27] Sykes, a relentless advocate of a forward policy in the East, became something of an expert on Persia. He used the new nomenclature in many of his numerous articles and books on the region's geography and ethnography.[28] In 1903, his cousin Herbert Sykes wrote an article entitled the "Middle East" in the *Manchester Courier*.[29] Another example of an Anglo-Indian promotion of the expression can be found A. C. Yate's writings. Yate was an officer from the Bombay staff corps who served in the Afghan Boundary Commission.[30] He published a detailed article on Persian railways in the *Scottish Geographical Magazine* in 1912, in which he found the "Middle East" to be a useful category.[31] Anglo-Indian writings were not the only ones to embrace a regional construction which lent support to their particular concerns with territorial substance and geographical legitimacy. The larger circle of those who were committed to the expansion and strengthening of the British Empire swiftly incorporated the neologism within their rhetoric. Such was the case of Arminius Vámbéry (1832–1913), a Hungarian national whose marked Russophobia and no less fervid Anglophilia pervaded the many writings he published over decades, with the FO's support. In 1906, he published a book entitled *Western Culture in Eastern Lands; a Comparison of the Methods Adopted by England and Russia in the Middle East*.[32] H. G. Lyons, head of the Egyptian cadastral survey department from 1901, used the notion in a conference for teachers dealing with the basin of the Nile.[33] Arthur Wilberforce Jose (1863–1934), whose writings echoed John Robert Seeley's works on the empire, published a map entitled "Near and Middle East showing spheres of influence," extending from Baluchistan to Egypt, in his revised edition of his *Growth of Empire* (1909).[34]

Constructive imperialists, Greater Britain sycophants, and Curzonian expansionists were thus among the first to come to grips with the "Middle East." The new phrase acted as an echo chamber for other lobbies involved in Eastern affairs. In 1911, H. F. B. Lynch (1861–1913), the descendant of a line of Irish entrepreneurs in the East, wrote an article entitled "Railways in the Middle East."[35] Lynch, who had traveled in Armenia in the early 1890s, was one of the leading voices in Britain denouncing Ottoman exactions against Armenians.[36] His firm ran shipping lines in Mesopotamia and traded with Persia. Lynch the entrepreneur was also a liberal imperialist, who expressed his dissent from London's Iranian policies in the 1900s. He became the vocal leader of a network which promoted a different policy in the Middle East. He was on the front line in the criticism of the government's handling of the region's affairs following the 1907 Anglo-Russian agreement, which divided Persia into two spheres of influence, and in the context of the Constitutional Revolution

(1905–11). Lynch was the chairman of the Persia Committee (1908–14), a pressure group which took a strong stance against the policies of Edward Grey. It defended Persian independence and territorial integrity. The committee counted E. G. Browne among its members. Its resolutions were supported by Ramsay MacDonald. His appropriation of the new geographical construct was symptomatic of a maturing and shift in the debates on Eastern affairs.

Even if the adoption of the neologism by the wider decision-making elite was still uncertain before the First World War, the "Middle East" was relevant enough to the interests of a number of different groups and interests to be successfully adopted in British public debates. Chirol's, Mahan's, Jose's, and Lynch's delineations had one thing in common: they carved out a region that was not described in terms of stagnation, ruins, and biblical remnants. Their Orient was caught in powerful cross-currents, from geopolitical tensions and imperial competition to the various shades of a modernization dynamic. Their Middle East was on the move.

A BATTLE OF WORDS? HOGARTH'S NEARER EAST

Another sphere which was involved in the exploration and comprehension of the Middle East was simultaneously trying to impose its own particular regionalization. From universities such as Oxford and Cambridge, learned societies, and specialist weeklies to the London School of Economics, presided over by Mackinder from 1903 to 1908, there were the outlines of another informal assemblage, which tended to propagate what amounted to a different breakdown of the Orient. This network, which was predominantly academic, including figures such as A. J. Herbertson, William Ramsay, and L. Wooley, was at first considered with aloofness by Whitehall. These archaeologists, historians, and explorers favored other phrases, such as "Near East," "Nearer East," and the later "Fertile Crescent," first coined by the American Henry Fowler.[37] These phrases involved what was more an affinity than any coordinated attempt to impose a terminology. David Hogarth, for instance, published a book entitled *The Nearer East* in 1902, part of a series entitled *Regions of the World*, edited by Mackinder. For Asia, the series included the volume *The Farther East* and another on India and the Russian Empire, which integrated Central Asia. Hogarth's book met the standards of Mackinder's new geography, which would bring him to formulate his theory of the geographical pivot in May 1904. Mackinder believed in a geography that focused on how territories, resources, infrastructures, physical constraints, and history influenced the development of societies. This led him to devise what was one of the first systematic geopolitical theories. In his view, the pivot of the world was the heartland, the central part of the Eurasian continent from the Volga to the Yangtze River, command of which depended on Eastern Europe. Whoever controlled this hinterland dominated the world.[38]

Hogarth consistently conformed to Mackinder's geographical divisions.[39] The many maps illustrating his work delineate a Nearer East centered on the Nejd and stretching from the Balkans to eastern Persia (fig. 7.3). They substantiated Hogarth's claims in *The Nearer East* mirroring the tenets of the new geography. An exceptional array of economic, demographic, relief, ethnographic, and historical cartography was marshaled in support of the apparent novelty of his work and to enhance the scientific character of an academic geography which was still in its infancy. It would seem that despite the modernity of his outlook, Hogarth, who was a specialist on the ancient East, still viewed the region through the prism of Herodotus's *Geographies* and Alexander's conquests. Like many of his contemporaries, he had a classical education that peopled his Orient with memories of Greek colonies and Macedonian feats of arms. This was not, however, the sole explanation for his reluctance to place the Balkans outside the East:

> The Nearer East is a term of current fashion for a region which our grandfathers were content to call simply The East. Its area is generally understood to coincide with those classic lands, historically the most interesting on the surface of the globe, which lie about the eastern basin of the Mediterranean Sea; but few probably could say off-hand where should be the limits and why. The land-marks towards the West are somewhat doubtful and apt to be removed. Fifty years ago the author of Eothen saw the portal of the East in the walls of Belgrade. Today the western visitor, though conscious that the character of the life about him has undergone some subtle change since his train steamed over the Danube bridge, would not expect to find himself in the "East" until he should sight the minarets of Adrianople, or, at earliest, the three mastoid hills of Philippopolis.[40]

Hogarth's and Mackinder's attempts at renovating the "Near East" were not completely unsuccessful, but their "Nearer East" did not enjoy the posterity of Mahan's "Middle East."[41] Mackinder, Hogarth, and Herbertson even used the phrase "Middle East" on occasions. The fact that the North-West Frontier of India was left out of their regional construct did not favor their views, at a time when the Raj was trying to convince everyone of the need to

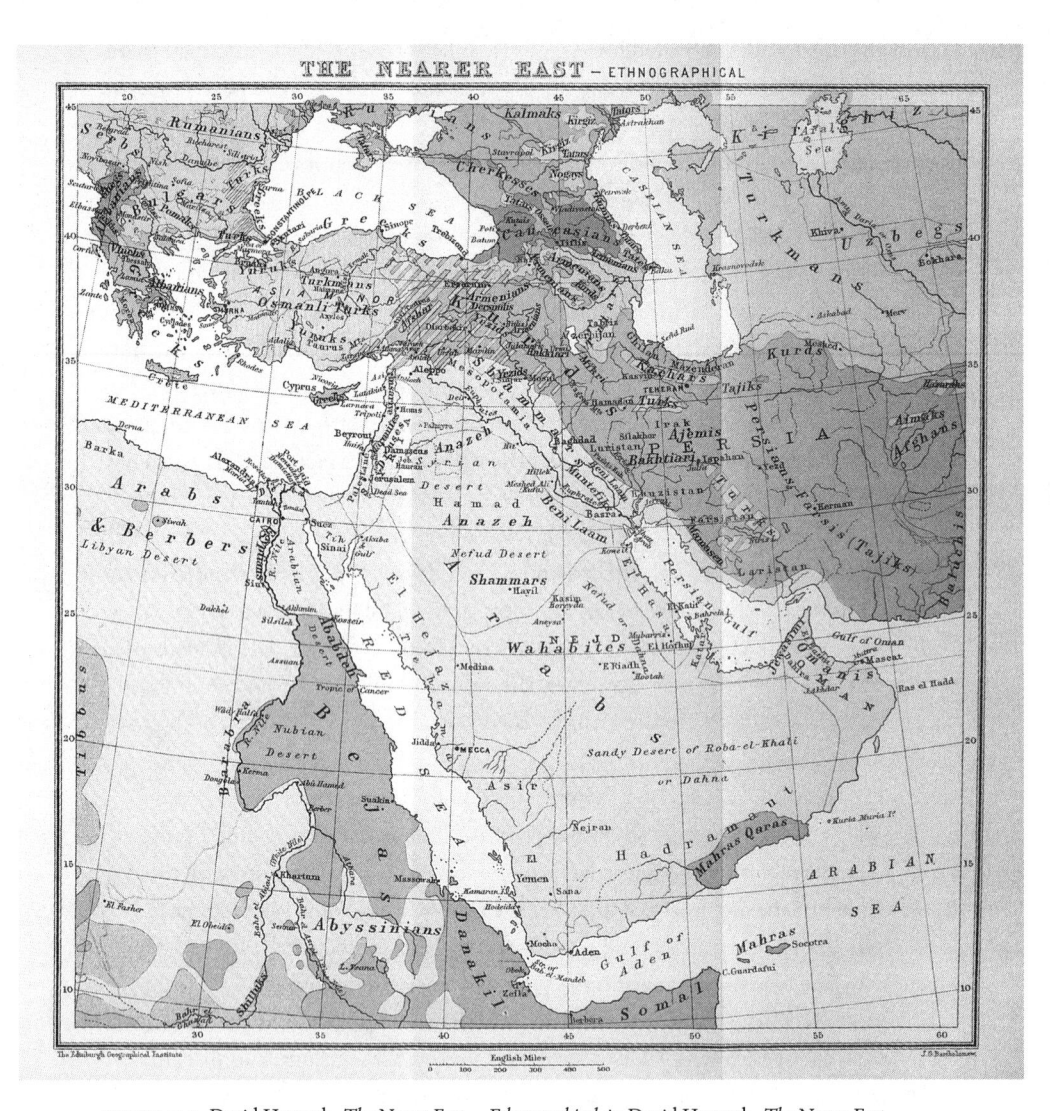

FIGURE 7.3. David Hogarth, *The Nearer East—Ethnographical*, in David Hogarth, *The Nearer East* (London W. Heinemann, 1902), facing p. 176. Photograph courtesy of Bibliothèque Interuniversitaire des Langues Orientales (BIULO AP.IV.250).

protect the transitional area that separated India from the metropolis. Before the First World War, then, the links between up-and-coming Orientalists and Whitehall were at best remote. The contribution made by this line of thought to the geography of the East was by no means negligible. Hogarth's "Nearer East" emphasized the weight of the Arabs in the region. His ethnographic map conveyed the message. This feature of the British outlook on the Middle East would prove decisive a few years later, when Hogarth became the acting director of the Arab Bureau during the First World War, with the result that part of these assumptions would shape the destinies of the region.

The emergence of these geopolitical neologisms was intrinsically linked with the news coverage of world affairs. The articulation of Chirol's articles on the Middle East, the echoes of the new concept in the regional press (in the *Western Times*, the *Derby Daily Telegraph*, and the *Western Daily Press*), and Curzon's 1903 tour in the Persian Gulf is an illustration of these interconnections.[42] The study of the journalistic map of the Middle East is thus crucial to the understanding of the invention of the region.

7.2 A "GOOD GRASP" OF THE WORLD

In December 1910, the *ILN*'s subscription advertisement promised the reader an unrivaled access to the news. It showed an Indian-centered globe surrounded by searching hands, a metaphor of a supposedly knowledge-hungry readership with the world at its fingertips. The geography-minded advertisement proudly proclaimed that the ILN gave the prospective subscriber "a good grasp of the world." Such a commercial use of cartographic imagery, one among many other contemporary examples, is a sign of the popularization of maps in early 20th-century news publishing. Journalistic cartography, already well developed in the late Victorian era, became increasingly popular in Edwardian times. The narratives of imperial expansion in Africa, the troubling difficulties of the British army in South Africa during the Boer War, and the growing agitation in the Ottoman Empire leading to the first Balkan War (1912–13), were all opportunities to provide a copious coverage through mass-produced cartographic imagery.

Once again, technological advances were crucial to the process. The first of these involved a series of new methods for the reproduction and printing of illustrations. By the early 20th century, the intermediary of the mapmaking business, that is, the engraver, had virtually disappeared. Photoengraving processes had replaced him. As this stage fell into disuse, the distance between the designer of the map and the printer vanished, leaving less room for reinterpretations and transformations of the cartographer's initial work. The perfecting of the halftone process in the late 19th century, which made it possible to reproduce any illustration as an image composed of small dots, further diminished the printing costs. By the 1910s wood engraving had become a thing of the past for the press. Not only were reproduction and printing methods transforming and simplifying the illustration of the news, but the very rhythms of journalism were also changing. The telegraphic and telephonic interconnection of the world accelerated the flow of information with an unprecedented efficiency. As seen earlier, the Middle East was not in a dead angle from that point of view. Communication lines and networks of correspondents were well established in the region.

A look at several examples of journalistic maps from the early 20th century highlights the consequences of these evolutions. Whereas it had taken W. T. Stead's *Pall Mall Gazette* a fortnight to report on Gordon's tragic fate, some thirteen years later the delay in reporting Kitchener's expedition and the crushing victory over the Madhists in 1898 was reduced to two or three days. The *Morning Post* even published a map of the Madhist positions on September 2, 1898, to illustrate a telegraph dated September 1.[43] The battle started in the early hours of September 2. The *London Daily News* had its own map of the battle ready for September 5.[44] Modern telecommunications and war correspondents supplied the papers with daily reports on Kitchener's campaign in what was one of the remotest areas of the Middle East. While the pioneering *ILN* only printed its first map of the battle on September 10, the weekly could still count on Caton Woodwille's lavish illustrations to confront competitors.

Another striking feature of the Edwardian journalistic map is its generalization to all types of newspapers, coupled with the cartographic spectacularization of information. From the cheapest dailies to the *Times*, thanks to a range of technical solutions, the addition of a map became quite commonplace. The *Penny Illustrated Paper*, to take an example, covered the deaths of several British soldiers on the border of Baluchistan in January 1898 by way of a very sketchy map of the area. The depiction of such a specific and unexpected terrain in one of the cheapest illustrated newspapers is in itself a testimony to the popularization of cartography as a journalistic adjuvant.[45] An elite newspaper such as the *Times*, with its own cartographic department, saw its daily average number of maps peak in the 1910s and 1920s.[46] A large proportion of these maps illustrated articles on geopolitical matters, for which the Middle East was a consistent provider. In 1907, a map of Persia adorned an article on the Anglo-Russian convention.[47] The deteriorating situation in the Balkans was documented by a war map in 1912.[48] In May 1913, the respected paper published its own map of the "Near and Middle East."[49] The *Times*'s cartographic illustrations did not incorporate the spectacular features of the illustrated papers, or the very rudimentary information of the lesser dailies. The paper's authoritative documents actually testified to the new enhanced standing of the journalistic map.

The study of the journalistic map of the Middle East gives us an opportunity address the vexed question of British culture and its imperial dimension. Did the mass media of the time bristle with colonial preoccupations? It must first be noted that these maps were not the unbiased illustrations of the news coming from every part of the world. The decades of the late 19th and early 20th centuries saw the emergence of what Simon Potter described as an imperial press.[50] The Edwardian era witnessed the development of informal journalistic networks which were liable to produce information that was supportive of the agenda of empire. Reuters's monopoly, for instance, prevented any real pluralism in the presentation of information coming from overseas and the Middle East in particular. Moreover, most correspondents for the major metropolitan newspapers in the region occupied official positions for the Raj, the British administration of Egypt, or the FO. In fact, the sources of information on the East were generally related to the Anglo-Egyptian and Anglo-Indian world. Therefore, the use of cartography in the mainstream press generally fostered the views held by the various proempire lobbies in Britain. It was an effective instrument for those who wanted to defend and illustrate empire building. The cartographic illustration of the fight against the Madhists was, for instance, an integral part of the imperial narrative. Maps thus undoubtedly participated in the consolidation of the popular imperialism of a British domestic readership.

Whether for purposes of mockery or denunciation, maps were also used by a range of papers as a means to underline the most imperialistic attempts at cartographic propaganda. The renowned *Punch* was obviously on the side of the laughers. The failures and tribulations of British cartography was a running joke for *Charivari* in the early 20th century. In 1901 the satirical paper invented a mock "Civil Service Examination Paper (Inspired by South Africa)." Question 12, "draw a map of the railway from the Cape to Cairo, not forgetting to insert the dominant Rhodes," was an undisguised attempt to scoff at the megalomania of one of the British Empire's most famous schemes.[51] The same year, an inventive glossary of war terms defined a map as "a chart upon which names are sprinkled without any special significance as to exact locality": a cruel reminder of the defective uses of maps during the Boer War.[52] *Punch*'s rather innocuous derision was not the only symptom of a divergent rhetoric. If instances of anti-imperial maps in the radical press were still to be found, there were various denunciations of the intense propaganda mapping that developed in the Edwardian era. J. A. Hobson, for instance, warned his readers against the seductions of the imperialist map of Africa: "Those who see immense areas of land marked on a map, and are informed that the population for the Colonies and Republics is less than one and a half persons per square mile, and that the climate and soil are favourable to white colonists, may entertain the notion of some great scheme of settling British agricultural labourers or small farmers in South Africa. But the least reflection and investigation of the country will serve to check such wild optimism."[53] Radical stances in the press added to the polyphony, as a first generation of anti-imperialists such as Marmaduke Pickthall, E. G. Browne, and Benjamin Horniman started to criticize openly the very nature of British influence in the Middle East.[54]

Maps of the East could also substantiate other anxieties. The rising late Victorian concern for Eastern Christians and the subsequent calls for humanitarian intervention continued to prosper. Missionary networks and philanthropic societies adopted cartography in order to call attention to the urgency of some of the situations they dealt with. Such concerns sometimes found a way into the commercial press. In 1912, for instance, the *ILN* published an ethnographic map of the Balkans.[55] It showed the distribution of the various religious denominations. The purpose of the map was to highlight the fact that most of Turkey in Europe was inhabited by Christians. This was not new to the British public; the Bulgarian atrocities of the late 1870s had educated a generation in this respect. In the context of the First Balkan War of 1912, the message was clearer than ever: Ottoman Turkey's Muslim rule was not to extend beyond Asia Minor. The *ILN*'s map was one of the many materializations of the Balkans as a distinct and European entity by the early 20th century. The illustration indirectly promoted the full emancipation of Eastern Christians.

Visualizations of the Middle East in the press show that the cartographic discourse was not a totalizing one. It reflected broader debates as well as the multiplicity of imperial and emerging anti-imperial approaches of the Edwardian era. These multiplying maps testify to the fact that there never was a monolithic imperial gaze on the East in the early 20th century. The imprint of cartography as such, regardless of its underlying assumptions, deeply shaped modern consciousness. As we will see in the next section, the fact that more and more people started to envision the world cartographically is essential to the understanding of British cultural projections on the Middle East.

READERSHIP AND MENTAL MAPS

Berlin-Bagdad, Berlin-Herat, Berlin-Pekin—not heard as mere words, but visualized on the mental relief map—involve for most Anglo-Saxons a new mode of thought, lately and imperfectly introduced among us by the rough maps of the newspapers.[56]

H. J. MACKINDER

Mackinder's evocation of the "mental relief map" of the general public reveals how familiar cartography had become in the first two decades of the 20th century. Michael Heffernan's quantitative analysis of the newspaper maps confirms the impressive increase of cartographic illustrations from the first decade of the 20th century.[57] This mass-produced cartographic imagery had visible consequences. Data tend to indicate that a desire to improve one's knowledge of the world was shared by increasingly diverse social groups. Paul Thompson's interviews, conducted by the University of Essex in 1970, of people who lived through the Edwardian era offer an insight into the cultural practices of late 19th- and early 20th-century Britons.[58] Several interviews emphasize the seductions and attractions of the world's geography for a growing segment of the general public. M. W. Charles, a mail carrier born in 1894, and Mr. Griffith a blacksmith born in 1877, explained how they eagerly attended evening classes in geography after their day's work. Denis Cosgrove notes that the level of geographical culture was surprisingly high in North America in the same period, because of the addition of a more substantial geographically based content to the curriculum, the mass diffusion of maps, and the unusual interest in the world that was cultivated by the contemporary press. The same could very well apply to Britain. As indirect evidence of it we can cite the fact that it was not uncommon to see the map and geography highlighted in the advertisements of the first decades of the century.[59]

Another measure of the popularity of geography and maps emerges from the study of the membership of the local geographical societies. John Mackenzie has observed a peak in membership around 1910.[60] Charles Withers found similar results for Scotland. In both cases, the new members were from more socially diverse backgrounds than previously. A statistical study of journal subscriptions and a profiling of the readers would usefully complement these analyses, yet they are one of the many manifestations of a crucial fact: that the attention the general public paid to the world beyond Britain considerably increased as cartographic illustrations spread in the press.

7.3 EVERYMAN'S LIBRARY

In 1907, W. and A. K. Johnston published an *M.P. Atlas*, which gave the firm's old maps a new lease on life.[61] The reviews were not good.[62] Some of the plates from the *Royal Atlas* had been recycled, and the new ones were often rough. But the title itself was revealing. The expensive, twenty-five-shilling book was intended to offer its readers the same vantage point on the world's affairs that the professional politician had. Johnston's skillful marketing was an indirect indication that the received wisdom in the Edwardian age was that MPs were well versed in geography, a somewhat optimistic assumption. It is also evidence of the increasing role played by geographical information in the democratic debate. British citizens needed to understand what was at stake for Britain in the global competition.

Numerous publishers took notice of this new demand for modernized atlases. British mapmakers faced stiff competition from their Continental counterparts. Adolf Stieler's *Atlas*, a German work translated into English in 1907, was a reference in the early 20th century.[63] Mapmakers resorted to traditionally successful marketing methods such as serialization, as exemplified by Johnston's *Royal Atlas of Modern Geography*, which sold in twenty-eight installments from 1895. The choice of the right title could also boost sales, such as Ernest Rhys's *Literary and Historical Atlas of Asia*, which was part of series entitled Everyman's Library.[64] But the most characteristic feature of the Edwardian era's map market was certainly the growing involvement of prominent press titles in the publication of reference atlases. The *Times* started publishing a version of Cassell's *Universal Atlas* from 1895.[65] The maps were reprints of Richard Andree's *Handatlas*.[66] Ten thousand copies of the thirty-five-shilling 1900 edition were sold.[67] Alfred Harmsworth (1865–1922), editor of the *Daily Telegraph* and competitor of the *Times*, decided in his turn to publish an innovative *Atlas and Gazetteer* from the middle of the century's first decade.[68] It was sold in fortnightly installments. Each part cost sevenpence. The lavishly colored maps introduced the reader to an interconnected world. A map of the Near East placed the region within a system of transportation, communication, and trade networks. While the *Times* insisted on the German origin of its maps, a synonym for good cartography, Harmsworth's *Atlas* was advertised as an industrial and commercial atlas useful to the economics-minded British gentleman.

J. G. Bartholomew's *Twentieth Century Citizen's Atlas* was also released in fortnightly installments by George Newnes from 1902.[69] Bartholomew's *Citizen's Atlas* is one of the best examples of the efforts by mapmakers to broaden their customer base. On its first edition, in 1898, the book was lauded as the advent of a new genre of popular atlas. It presented the buyer with 120 colored maps, and it followed with more than 150 in its 1900 and its 1910 editions. Notwithstanding its high price, the first edition was sold for two guineas (about eighty pounds in 2014); the atlas was a bestseller. The first edition sold more than forty thousand copies. In 1903, the new edition was sold for one guinea (about forty pounds in 2014). The success of the first edition caused the publisher to lower the price of his atlas even further. The documents were regularly updated. This is especially true of the sheets showing the Middle East. The map entitled *Turkey in Asia, Arabia, Persia* correctly located Ha'il. This city of central Arabia was located precisely on the British charts following the journey of Charles Huber (1883–84). Wilfred Scawen Blunt's and Charles Doughty's maps presented a significant error in longitude, which was corrected at the end of the 19th century. The *Citizen's Atlas* map must therefore have been compiled from recent sources: Huber's travel notes were published posthumously in 1891. This analysis is borne out by the design of several other maps, which seem to be largely influenced by the Anglo-Indian outlook on the region. The very centering of a map presenting the routes to India, showing maritime roads, railways, and telegraph lines, echoed the ongoing redefinition of the British perspective on the Middle East. The atlas also included a map of Central Asia, which included the 1906 McMahon commission's border delimitation. A significant detail also demonstrates the interconnections between the most outspoken experts in Eastern affairs and the drawing up of maps which were designed for the general public: the Lynch Brothers' steamship lines on the Tigris and Euphrates were shown on the map, despite their very low tonnage. All in all, the *Citizen's Atlas* was evidence of the increase in both the quality and the quantity of geographical information made available to the British public in the Edwardian period. It was part of a larger cultural trend which insisted on the necessity of educating a population confronted with "a new age of wider world interests, world empires, and world commerce."[70] The Middle East was caught up in the general movement. Though some, like Mark Sykes, favored an Orient with "a past" rather than "a future," early 20th-century British maps tended more and more to depict the region in terms of its utility.[71]

THE HUMAN SIDE OF THE MIDDLE EAST

Philip and Sons' commercial map of Southwest Asia (1907) typifies what was a very widespread cartographic genre in the late 19th and early 20th centuries (fig. 7.4). Philip's *Chamber of Commerce Atlas,* published from 1912, which sold

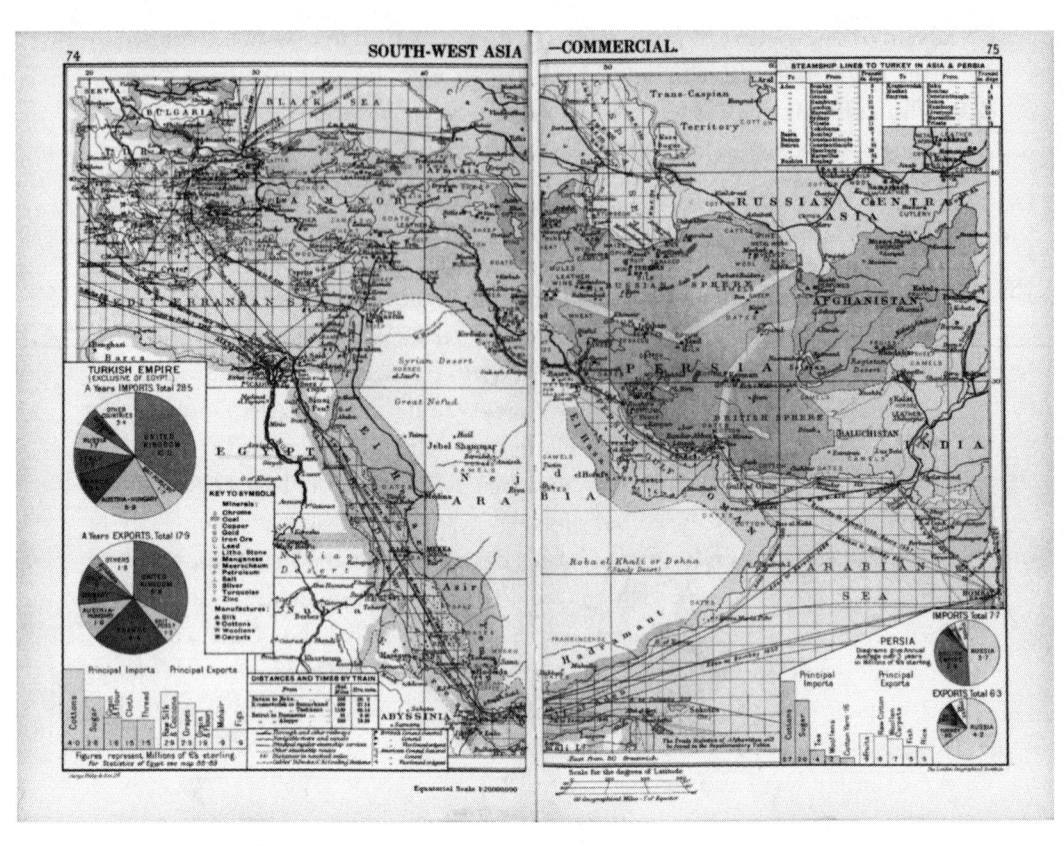

FIGURE 7.4. G. Philip, *South-West Asia—Commercial*, in *Philip's Chamber of Commerce Atlas* (London: G. Philip, 1912), 74–75. Size of the original: approx. 26 × 52 cm. Photograph courtesy of the BNF (Cartes et Plans GE FF-12292).

for six shillings, is a perfect example of the new representations of the world which were fostered in the Edwardian age: commercial atlases became a distinctive feature of the British map and atlas market. Maps charting the flow of products and commercial geographies challenged the traditional depictions of the region. Most major establishments published their variation on the theme. Bartholomew and Newnes's *Atlas of the World's Commerce* was published in 1907, while *Pitman's Commercial Atlas* was first released in 1914.[72]

Such documents relocated the East in the present, and even into the future. Biblical ruins and stagnant despotisms were of little relevance to their intentions. Philip's projection revealingly mirrored the Middle East of Mahan and Chirol. Indications of the Russian and British spheres of influence in Persia projected the region into the overall mapping of contemporary geopolitical issues. Mention was made of oil resources, the exploitation of which had only begun in the region two years earlier. Pie charts based on the Board of Trade surveys displayed the paramount importance of the Turkish Empire for British trade. Railways, telegraph lines, and sea routes transformed the region into a geopolitical axis. The Middle East was exploitable. Its economic qualities revealed.

GEOGRAPHIA SACRA AND MISSIONARY ATLASES: THE CHRISTIAN OUTLOOK

Biblical atlases were still a very popular approach to the geography of the Orient in the late 19th and early 20th centuries. The long tradition of sacred geography managed to survive the development of the new human geography. The gradual de-Christianization of British society had not as yet caused the *geographia sacra* market to disappear; quite the contrary. Masterpieces of the genre were still to be published. Eminent biblical scholars were able to use masterfully modern cartographic rhetoric to publish works which were veritable monuments to the geography of the Holy Land. The most celebrated reference work on the subject was George Adam Smith's *Historical Geography of the Holy Land*, which passed through many editions.[73] Lloyd George made it one of his handbooks during the negotiations at the San Remo Conference in 1920. General Allenby had high regard for Smith's book as well.[74]

The religious perspective remained a key feature of the British mental map of the Middle East. Smith was a minister of the Free Church. His *Historical Geography* was one of the many pointers to the vitality of nonconformist denominations in the Edwardian period.[75] They thrived on the opposition of a part of the population to the reforms of the educational system, the Education Act of 1902 in particular, which implied a decline of the churches' influence on the school curriculum. The last map in Smith's atlas provides another explanation. It located the main missions in the Middle East, and it connected biblical history in a sacred past to actual missionary work in the present. It is a sign of the unprecedented intensity of missionary activity in the late 19th and early 20th centuries. The number of single women enrolled in missions increased by as much as 168% between 1880 and 1914, according to Jeffrey Cox.[76] Biblical atlases were thus part of a larger geographical and cartographic documentation, which substantiated this impressive burst of Evangelical enthusiasm in the age of high imperialism.

Missionary atlases and maps which prolonged scriptural geographies participated in the shaping of a specific image of the Middle East. Missionary cartographies, which contributed to British scientific maps of the East in the mid-Victorian period, had, by the turn of the century, become a distinct ensemble. Figure such as G. P. Badger were unlikely to emerge by that time, given the growing mistrust shown by the established surveying and mapping institutions toward missionaries. As shown by Michelle Tusan, the churches' Middle East was often depicted as one of the most difficult terrains for the proselyte: a Muslim world which confronted the missionary with astonishing obstacles.[77] The missionary outlook on the area had its own rationale: the Christian was confronted with a geographical and religious entity stretching from North Africa to India.[78] In his presentation before the First Missionary Conference on Behalf of the Mohammedan World held in Cairo in 1906, S. M. Zwemer displayed a map of the ensemble entitled *The Evangelization of the Muslim World in this Generation*.[79] For the most fundamentalist of the nonconformists this was not just any other part of the world. The millenarian comprehension of the biblical lands was still very much alive in some circles and also gave rise to a considerable number of vocations in the period before the First Word War.[80] A multiplicity of often unsuccessful missions developed in the region from the 1880s, such as the Palestine and Lebanon Nurses Mission to the Druses, which reached a peak in donations and membership in the decade after 1900.[81] The Christian outlook on the Middle East continued to shape a significant facet of British representations of the region.

7.4 NEW GEOGRAPHIES

> From East to West—Page from a Business-Man's Diary.
> Friday.
> Popped into the Board Meeting of the Undiscovered Island Development Company. All going well according to the experts; but I admit, as I said to a companion director, that geography and minerals were never among my strong points at school. By the way, at University never did any geography. Nobody ever did. Quite forgotten if we had any maps.
>
> *Punch*, February 27, 1901, 172

Punch's philistine businessman pointed to what was a gap in the curriculum of British universities at the time. There were few departments of geography at the turn of the century. Those which did exist were recent creations, whereas in Germany the first full chair of geography had been established in Leipzig as early as 1871.[82] Yet the fictitious board meeting of the Undiscovered Island

Development Company was in itself revealing of a greater awareness of the importance of "maps," "geography," and "minerals" in the interconnected world of 1901. Those vulnerable to the charge of being geographically illiterate were more vulnerable than ever to the jokes of the satirist. In this context of shifting cultural frames, the British education system underwent profound changes between the 1890s and the First World War. The gradual increase in the minimum school leaving age to thirteen in 1899, along with the proliferation of secondary schools following the 1902 Balfour Act, increased both the supply and demand for education. From 1907, the Liberal government decided that a quarter of secondary school places were to be reserved for students from the most modest backgrounds. The traditional obstacles dividing primary from secondary education were beginning to be lifted. While inequalities still characterized the British school system, the government was making its first attempts to improve social mobility. As a rule, the supervision exercised by the state over education was strengthened. Geography benefited from these evolutions in every respect. The struggle waged by the RGS for its recognition as a separate subject eventually proved successful. As we will see, these evolutions significantly overhauled the representation of the Middle East in the classroom.

FROM UNIVERSITY TO SCHOOLS

If geography gained its hold in the intellectual landscape, it was in part a consequence of the wider awareness of Britain's shortcomings, in a context of growing international competition. The findings of a survey conducted by Reginald Esher on the mismanagement of the Boer War pointed, for instance, to the level of geographical ignorance prevalent among British soldiers. The apparent British inability to read or design maps was regarded as a matter for concern, given the German expertise in this regard. Significantly, Erskine Childers, in his best-selling *Riddle of the Sands* (one of the first espionage novels), mocked the "the prehistoric rottenness of the English charts" and the fact that since they were "relatively useless," the spy and hero of his book could use them as "shining proofs of [his] innocence" if he were to be caught.[83] The most active propagandists of empire, led by the British Empire League (1896), the League of the Empire (1901), and the Empire Day Movement (1903), also pointed to the Continental emphasis on maps and geography as a means to strengthen international influence. Learned societies also insisted on the issue. A 1906 article from the *Geographical Journal* insisted that "Germany even then had professors of geography in nearly all its universities, and a number of thoroughly trained and earnest students who devoted themselves to the investigation of the subject in all directions."[84]

As stated earlier, the academic institutionalization of geography in Britain began in the 1880s, under the auspices of the RGS. The process was still an ongoing one in the early 20th century. In 1900, a Department of Geography was inaugurated at Oxford. Mackinder saw to it that geography took on a significant position in the curriculum of the London School of Economics.[85] Teaching departments in geography were established in 1908 and 1909 at the Universities of Edinburgh and Glasgow. Geography was taught as a special subject from 1904–5 at Oxford, where a diploma was given to geography students who attended Mackinder's classes.[86] A sixty-pound grant was also put in place in the mid-1910s in order to stimulate vocations. Specific courses of studies which incorporated geography classes were also opened in the 1910s to provide for the training of colonial officials. Once again, this was a belated reaction to Continental educational projects such as Germany's Colonial Institute (Hamburg), established in 1908, and France's Ecole Pratique Coloniale in Le Havre.[87]

The birth of the Geographical Association (GA) in 1893 was the signal for improved connections between the world of academic geography and primary as well as secondary education. Membership of the GA peaked in 1910 at a figure of nearly two thousand members. William Maclean Carey welcomed these advances in 1910, underlining that "the teaching of geography has made great strides within the last ten years."[88] This success installed the new geography in the educational landscape. In 1914, H. J. Findlay listed the objectives of the new geography at school as follows:

1 · That our object in teaching geography is not primarily the acquisition of unrelated place-names, but a knowledge of the physical conditions which unite man into communities and govern his activities therein.
2 · That in our selection of geographical facts their interest to the life of man is of more importance than their mere relations in space.
3 · That the great importance of climate in determining man's activities should be emphasized, and the home climate, as the only available standard, should be intensively studied.
4 · That owing to their importance, climatic considerations should at first be the basis of our world divisions, not political considerations as at present.
5 · That our ideal of an ordered world is to be reached inductively, not deductively and, in view of the immensity of the subject, discrimination in the choice of topics is imperative, only those being chosen which have a direct influence on the life and work of man.[89]

SYLLABUSES AND CURRICULA

In 1912, the inspectors of the Board of Education noted that "probably no subject in the Elementary School curriculum has changed more during the recent years than Geography."[90] It was not an idle remark. These changes were brought about as a result of the combined efforts of the university boards, which had responsibility for the setting of standards, with their local examinations, the GA, the RGS, and the Board of Education, which started issuing its annual *Regulations for Secondary Schools* in 1904 with a strong emphasis on geographical education.

From the late 19th century, the most influential British Universities had their system of local examinations. The Universities of London, Oxford, and Cambridge, as well as Manchester, issued syllabuses which shaped the curriculum in many secondary schools. These competing injunctions laid a growing emphasis on geography, as civil service or army examinations gave an increasing importance to the subject. The London Chamber of Commerce tests and the Institute of Bankers competitive examinations also included questions on geography, and even cartography. The most geographically conscious institution was the University of London, where "physical and general geography" was made an independent subject from 1902.[91] In the first decades of the new century, Oxford and Cambridge local examinations were also requiring an ever-greater geographical proficiency on the part of their potential students. The geography of the Middle East was not overlooked. In 1907, the Mediterranean was one of the subjects to be studied for junior examinations. In a 1902 revision test paper for the senior Oxford local examination, candidates had to satisfy the following requirements: "(1) On an outline map of the Balkan Peninsula draw carefully the Lower Danube and the Maritza river basins [. . .]. Mark the boundaries of the various states."[92] The same was true in Scotland. The revision test paper for Scotch leaving certificate examinations for 1902 included the following questions: "(2) Give an account of one of the following as regards climate, products, means of communication, people and government: Japan, Persia, Mexico, Morocco, Argentina, Venezuela. (3) Draw a sketch map of Norway or Spain, or Egypt or Brazil or New Zealand [. . .]. (5) Explain carefully [. . .] the connection between geography and politics on the North-West Frontier of India."[93]

This was the culmination of the sustained efforts of the RGS and other associations such as the GA. The RGS released its "Syllabuses of Instruction in Geography" in 1903.[94] They were to orchestrate a progressive enlargement of pupils' horizons, starting out from their own county and extending to the remotest parts of the earth. This first frame of reference was thus discussed and propagated by the GA. The honorary secretary of the association, A. J. Herbertson (Oxford School of Geography), was particularly persuasive in that regard. He played a prominent role in the redefinition of the recommendations issued by the Board of Education. His approach had a lasting influence in the definition of syllabuses by local education authorities. Herbertson elaborated a typology of six regional ensembles, divided into subregions. This regionalist approach was adopted in one form or another by many institutions. The regulations issued by the Board of Education in the late 1900s reflected Herbertson's views. His influence was also felt through his *Senior Geography* and *Junior Geography*, which quickly became authoritative books for the teaching of geography. He and his wife also organized summer schools and conferences for teachers to spread the gospel of the new geography. The couple sold an estimated 1.4 million textbooks books from 1902.[95]

Where was the Middle East in Herbertson's school geography? He did not ignore the new projections, which were developed at the turn of the 19th century. His *Junior Geography* included maps that were reproduced from Hogarth's *Nearer East,* as early as 1906. An even clearer illustration of the author's ability to popularize modernized geographical regionalizations is to be found in his chapter on Asia. He designed an exceptional political map of Southwest Asia centered on the Gulf, mirroring the most recent definitions advanced by Mahan.[96]

There is little doubt that these new school geographies were in the service of imperial ideas.[97] Herbertson himself would publish his own *Geography of the British Empire* a few years later.[98] It is worth noting that the most ardent promoters of geographical education believed the subject to an effective means by which to convince the country's youth of the centrality of the empire in British identity. W. P. Welpiox, a lecturer on education at the University of Leeds, summarized this position in 1914: "We seek to give them some orderly account of the growth of national life and inspire them with a sturdy national sentiment [. . .] because they are the future citizens of a great Empire that has world-wide commercial and political relations they need to be given a 'world' point of view, to be led to an interest in and understanding of problems and movements of international and world importance."[99] The propagandizing dimension of geography was a key feature of the new outlook on the teaching of the subject. Significantly, the GA maintained close relations with the Royal Colonial Institute. Some of its most active members, such as C. P. Lucas from the CO, who presided over the association after the First World War, were among the most influential scholars to use geography and history to foster the imperial cause.[100] Lucas was the renowned author of a series entitled *A His-*

torical Geography of the British Colonies. These monumental narratives of British imperial expansion provided teachers with a rationale through which to defend and illustrate imperial unity. The position of the Middle East in this architecture was described in one of Lucas's volumes, which focused on Mediterranean and Southwest Asian dependencies, as the crucial "high road to the East."[101]

A PEAK IN GEOGRAPHICAL LITERACY?

Various indications found in the *Reports of the Board of Education* testify to the systematization of the teaching of geography. There were obvious regional and local differences. The specificity of Scotland, where the insistence on geographical education was greater than in the rest of Britain, must be emphasized. The figures for England show that geography was taught in every primary school by 1902.[102] Overall, the inequalities in terms of educational opportunities were significantly narrowed compared to the Victorian era. Even evening schools offered more geography than before. Out of 3,465 such institutions, 1,817 provided courses in geography in 1903.

According to a 1907 report from the Board of Education, the position of geography had improved considerably, even in the remotest rural areas. In one grammar school, an inspector was glad to find that "the work in Geography is a specially good feature of this school [...]. There is an admirable and most skillfully made collection of maps and diagrams, some of the best of them being the work of members of the staff."[103] Not all institutions offered the same environment. In the new municipal secondary schools, where funds were often lacking to pay for a teacher correctly trained in the topic, the picture was somewhat different: "This subject is taken by five teachers, none of whom possesses any special knowledge of it [...] and the custom is for the teacher to endeavor to compensate their meagerness by dictating lengthy notes."[104]

Nuances aside, the picture offered is of a marked contrast with the mid- and late Victorian situation. Rachel McDougall's quantitative assessment of Bartholomew's production of school atlases from 1880 to 1922 unambiguously shows a sharp rise after the Second Boer War, then a first peak in 1914.[105] The joint forces of the propagandists of the empire, a printing market producing cheaper cartographic apparatus, and the heralds of the new geography converged in the first two decades of the 20th century to lecture a generation of pupils about an East which was less marked by sacred geography and by the Victorian era. Knowledge gained within the confines of the school was not out of step with the major cultural changes which were underway. The images and words of an Edwardian youth echoed what was to be observed in the press. The eagerness to understand the world had probably never before reached such a degree of acuity in Britain. What the teaching of the geography of the Middle East reveals in particular is the accelerated popularization of complex geographical and geopolitical conceptions.

CONCLUSION

The Edwardian fear of an eclipse of the British Empire triggered a number of wide-ranging reforms, some of which had far-reaching consequences for the geographical literacy of the general public. The mobilization of the educational system, the development of cartographic propaganda, and more map-minded public discourses favored a visible improvement of the geographical understanding of the global interconnections of the empire. Britain, much like the United States at the same time, had become a "map immersed society."[106] This resulted in a modernization of the geographical imaginations of the Middle East, which was more and more understood in terms of commercial and geopolitical interests. This phenomenon was not specific to the East. As a result of its particular location in British traditions, the cultural fabric of the area came to be profoundly modified. The established frames of reference had previously placed the Orient in a singled-out position on the British mental map of the world in the 19th century. It was an area embedded in the layers of its biblical and classical past, characterized by legacies and promises of stagnation more than by its potential development and its increasingly visible interrelation with the future of the British imperial structure. Along with war and trade, tourism brought it closer to Britain. In the decision-making circles, imperial anxieties also led to an overhaul of the traditional grids. The end of the Great Game from 1907 along with the revolutions in Persia and in the Ottoman Empire condemned such long-standing categories and the "Eastern Question" to obsolescence. New problems required new conceptualizations, new phrases, new categories. The emergence of the "Middle East" was therefore much more than a footnote in specialized articles. It was evidence of the reinvention of the region as a key component in the global European imperial structures by powers who would play a crucial role in shaping it after the First World War.

CHAPTER EIGHT

FALLING INTO PLACES

On March 12, 1919, the Middle Eastern Political Section of the British delegation in Paris met at the Astoria Hotel to discuss the "territorial arrangements in Mesopotamia, Syria and Palestine, in the event of Syria being assigned to France."[1] The discussion focused on Clemenceau's memorandum of February 5, 1919, which recognized the transfer of Mosul to Mesopotamia in exchange for the establishment of a French mandate over a large, unified Syria. The meeting brought together some of the most renowned figures involved in British Middle East politics. Louis Mallet (1864–1936), the head of the British legation's Turkish section, had a long experience with the East. He had been assistant undersecretary and head of the Eastern Department at the FO from 1907 to 1913. He was ambassador to Constantinople when the war broke out. Author of several memoranda on the future of the Middle East, Arthur Hirtzel (1870–1937), a senior official, represented the IO. T. E. Lawrence, basking in the glory of his bold military campaigns in Arabia, also attended the meeting, as did Gertrude Bell. Arnold Toynbee, Mallet's assistant, was an expert on Eastern affairs at the FO. Richard Meinertzhagen was Allenby's chief political officer during the Palestine campaign. He had firsthand knowledge of the terrain.

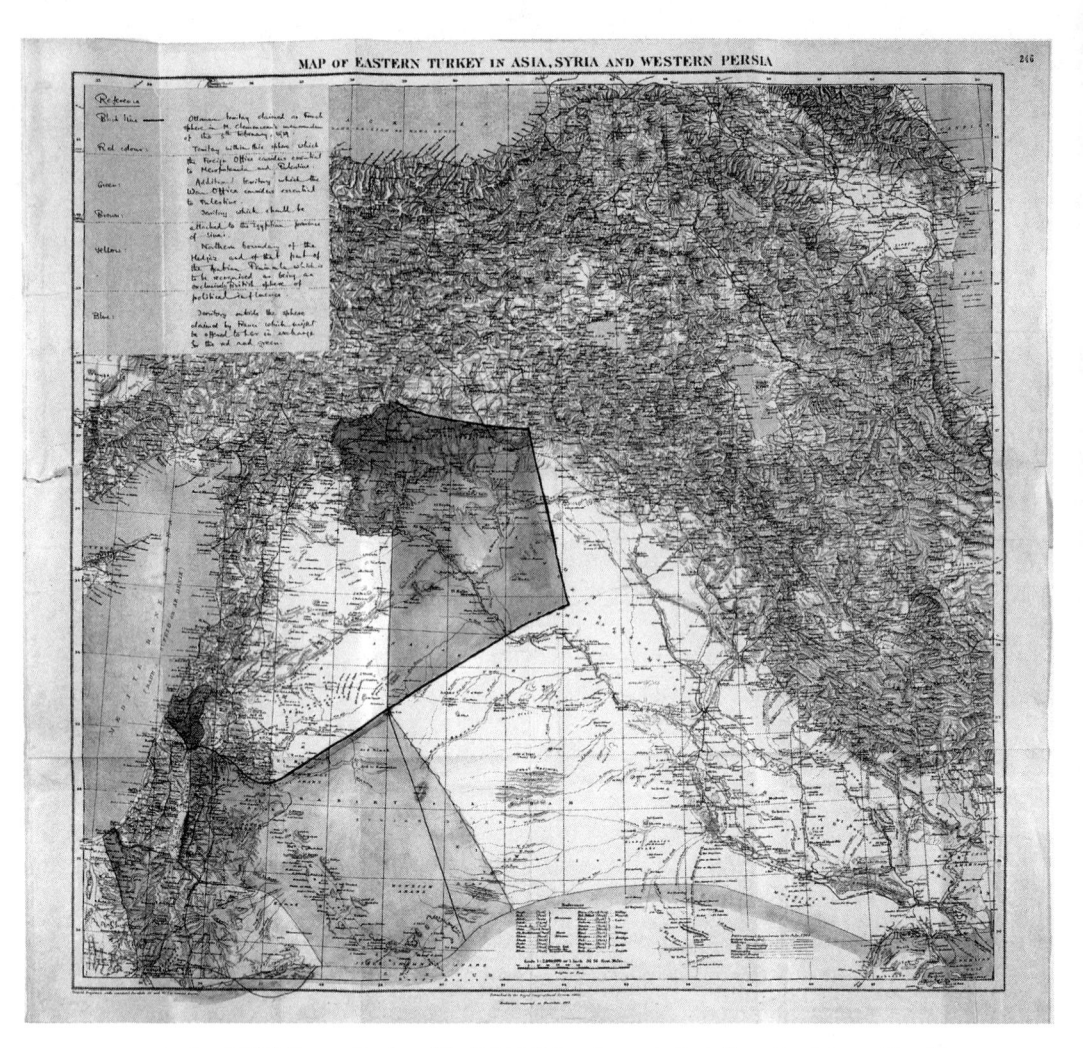

FIGURE 8.1. Middle Eastern Section of the British Delegation to the Paris Peace Conference, *Map of the Middle East with additions*, scale not given (London, March 12, 1919). National Archives of the UK (ref. Fo 608/83/3, 246).

The memorandum and the attached map reveal the complex decision-making process leading to the elaboration of a new East to replace the Ottoman Empire (fig. 8.1). In the first months of 1919, Mallet's Middle Eastern Political Section was absorbed by unwieldy discussions on the future of the region. Like that of the peace conference in general, the functioning of the British delegation was characterized by its disorganization. The delegation, in particular, was confronted with an overwhelming flow of paper representing the interests of numerous individuals, groups, and institutions.[2] The British legation staff was obliged to process hundreds of pleas and petitions from an array of associations and individuals. During the few days before the meeting, Louis Mallet had to read a letter and an article by William Mitchell Ramsay, an expert scholar on the East, recommending the creation of an American mandate on Anatolia, a memorandum on the creation of a new Palestine by the Council of the Anglo-Jewish Association, and a letter from A. Papadopoulos advocating annexation of the Turkish Vilayet of Aidin to Greece.

The task of compiling and summarizing this mass of information was not the least of the difficulties facing the British delegation. The conflicting desiderata of the various departments resulted in confusion that is visible in the archives. The map is a perfect illustration of this phenomenon. The red color represented the views of the FO: the upper course of the Euphrates could not be left to the French. Willcocks's grand projects of irrigation were still very in vogue: a coherent control over the hydraulic resources of the region was imperative. Unopposed access to the oil fields of the Mosul area was also an objective. The demands of the British administration in Egypt appear under the color brown. They focused on Palestine. In contrast to many officials in the FO for whom the contours of Palestine were those defined by the Bible, from Dan to Beersheba, Milan Cheetham, the high commissioner in Egypt, wanted to establish a more northerly border in the Sinai without reference to biblical tradition.[3] The WO's desiderata in Syria, displayed in green, were a reflection of the memoranda issued by Zionist organizations a few days earlier.[4] For reasons of its own, the WO was in agreement with the underlying principles put forward by Aaron Aaronsohn (1876–1919), a close relation of Chaim Weizmann, the influential defender of the Zionist cause. The northern extension of Palestine that the WO suggested had the double benefit of limiting French influence in the Levant and providing for a better management of Palestine's hydraulic resources by the full inclusion of the Litani area within the British-controlled territory. The desert-sand yellow painted on the map to figure "the exclusive sphere of British influence in the Hejaz" was perfectly compatible with T. E. Lawrence's designs. The terrains of his grand-styled Arab Revolt were thus delimited. The participants at the meeting acknowledged that all this coloring would certainly not suit the French. A blue area delineated the territory outside the sphere claimed by the French which might be offered to them in exchange for the red and green. The set of colors was in itself revealing of the map culture of those attending the meeting. The FO's red was an indirect reminder of the traditional British atlas representation of the empire; blue called up memories of the violet shades showing the French colonies on the cartographers' figurations of the world.

The proposals of the Middle Eastern Political Section were far from definitive. Discussions were ongoing, and the Middle East was not the central focus of a legation which was primarily concerned with Europe. The meeting says more about the nature of British decision making than about what the Middle East was eventually going to be. In order to understand this, we have to read the handwritten comments added by the various experts who were to give their opinion on the map. The file circulated among the various members of the British delegation, and each jotted down his remarks, thereby presenting the differing viewpoints that were an integral component of British plans for the redrawing of the Middle East. On March 18, 1919, Meinertzhagen, who attended the meeting, explained his point of view: "I dissented entirely from the whole proposal. I could not do so during the sitting of the Committee as it would have disclosed Lord Milner's proposal of his French zones and French Arab mandate to Col. Lawrence and Miss Bell, which I thought undesirable." This communication breakdown reveals how distrustful some were of the two heroic mavericks. General W. Thwaites (1868–1947), the director of military intelligence, had his own views on the question and bluntly asked Louis Mallet: "Is this to be considered a serious proposal?"[5]

The British legation experts were still in the early stages of formulating a coherent approach to the question. Divergent considerations were successively taken into account. For some, such as Lawrence, tribal questions were not to be overlooked; others, like Mallet, thought that it would be useful "to have the views of the military section on the relative merits of a lateral and longitudinal division." Thwaites, for his part, authoritatively took the liberty of "making erasures" to reduce the blue area on the March 12 map. Two new voices were added on March 29: David Hogarth from the Arab Bureau in Cairo and Arnold Wilson, from the IPS. Wilson quickly made it known that his administration in Baghdad "could assume responsibility for the administration" of an extended Mesopotamia. Hogarth believed that it was possible to follow the Zionist's maximum definition of Palestine, therefore contradicting the views of other fellow Anglo-Egyptian actors such as Cheetham. Finally, the American legation's proposal for the southern boundary of Armenia, which provided for a very large Mesopotamia, was favorably received by some members of the Middle Eastern Section. Very little blue was left on a new map designed on March 29 (fig. 8.2).

A form of closing argument was voiced by a weary Lord Curzon, head of the Inter-Departmental Conference on Middle Eastern Affairs in London, in *Present Position of Middle Eastern Question* (April 18, 1919). The revealing introduction to his argument accounts for yet another set of discordances in the

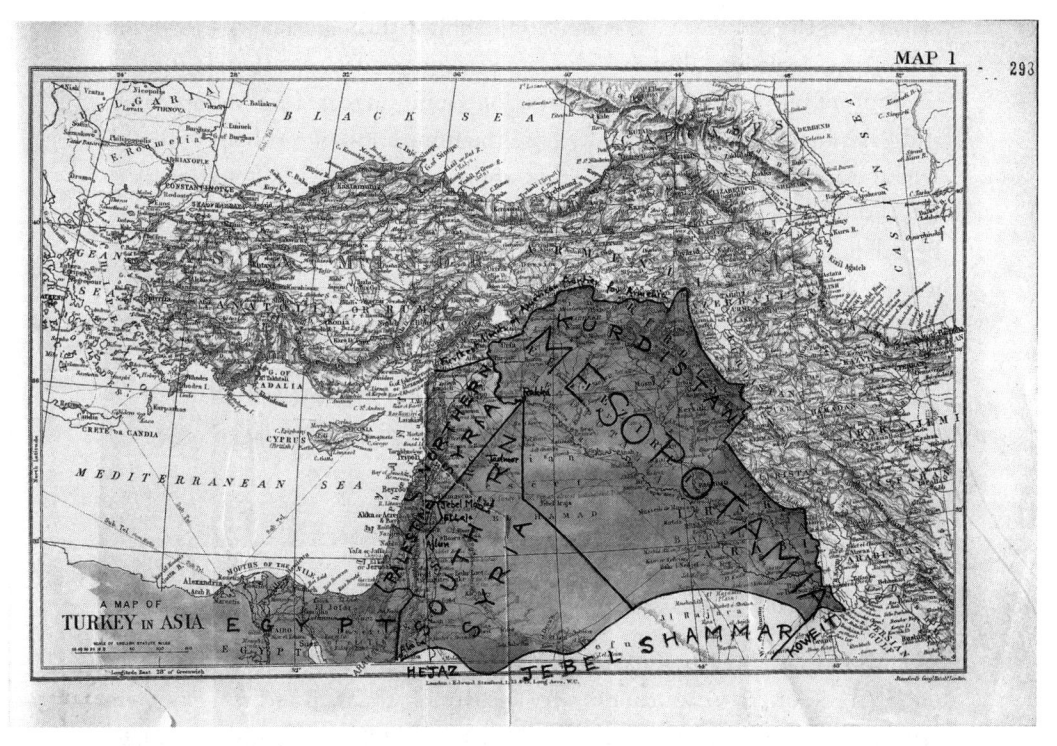

FIGURE 8.2. Middle Eastern Section of the British Delegation to the Paris Peace Conference, *Map of the Middle East with additions*, scale not given (London, March 22, 1919). National Archives of the UK (ref. Fo 608/83/3, 293).

governmental machinery: "I am reluctant to criticise in London a policy which is being pursued in Paris, as to the phases of which we are not fully informed [. . .]. At the same time [. . .] the policy said to be favoured by the peace conference is likely to produce widespread disturbance, if not disaster."[6]

The reader might easily be lost in these intricate negotiations, which Elie Kedourie justly called a "labyrinth."[7] In presenting these examples, it is not our intention to provide a detailed narration of a story which has already been well researched.[8] This very imaginative cartography merely enables us to underline some of the key evolutions of the British map of the Middle East. First, the war afforded an unparalleled opportunity to map out the Middle East, with far-reaching consequences.[9] The effects of the militarization of the region's cartography will be one of our prime concerns in this closing chapter. Second, the very existence in 1919 of a Middle Eastern Political Section in Paris and of an Inter-Departmental Conference on Middle Eastern Affairs in London calls into question the traditional assumption that the "Middle East" only became an effective category during the Second World War. Many of the individuals involved in the postwar talks endorsed the phrase through their own varying definitions. Anecdotal, but nonetheless revealing, evidence of the adoption of the concept is to be found in the delegation's minutes regarding Mesopotamia and Palestine. They systematically bore the impression of a rubber stamp on which was engraved "Political Turkey and Middle East." The question as to when this frame of reference was institutionalized after 1914 is crucial. Third, British postwar maps make audible a multiplicity of voices, each of them singing in a different key its particular idea of what the Middle East was to become.

8.1 MAPS GALORE: BRITISH CARTOGRAPHY OF THE MIDDLE EAST DURING THE FIRST WORLD WAR

Like crawling ants whole armies are
That strive across a coloured map.

FROM CROSBIE GARSTIN, "THE FLYING MAN," *Vagabond verses*, 1917[10]

The Middle East became a stake in the war on November 2, 1914, when Russia declared war on the Ottoman Empire. The latter had signed a secret agreement with Germany in August earlier in the same year. While the region was initially a sideshow, the British army fought several campaigns in the area. The largest offensive was launched in the Dardanelles for the campaign of Gallipoli (1915–16). Fierce debates separated two strategic schools in Britain: the so-called easterners and westerners. Russia had asked for the opening of a new front. The capture of Istanbul and the Dardanelles was to be a severe blow to the Triple Alliance. Though many believed any large operation against the Ottomans to be a waste of energy, Churchill, then first lord of the Admiralty, convinced the government that an offensive on the straits was worthwhile. The operation ended in a disaster for the Allies, who lost more than 180,000 soldiers. Churchill resigned in November 1915 and enlisted to fight in the trenches. In January 1915, an unsuccessful Ottoman attack on the Suez Canal was the starting point for the Palestine campaign, which resulted in General Edmund Allenby's capture of Jerusalem in 1917. The Mesopotamian Campaign was launched to protect British interests in the Gulf and key oil fields. The British surrender to the Ottomans on April 29, 1916, after the appalling siege of Kut-El-Amara was a tremendous blow to British prestige, but General Frederick Stanley Maude's counteroffensive in 1917 and the ensuing occupation of Baghdad restored confidence. The armistice with the Ottomans was eventually signed on October 30, 1918. Minor campaigns were fought in the nominally neutral Persia, where a very effective German spy, Wilhelm Wassmuss, was able to incite revolt against British interests. Percy Molesworth Sykes recruited and trained a local force, the South Persia Rifles, to counter German attempts at destabilizing the country. In Arabia, another spy named Alois Musil tried to turn Ibn Rashid's Shammar tribes against British-supported Wahhabi tribes.

Despite Britain's relatively limited military involvement in the Middle East, the deployment of thousands of British troops and of modern equipment enabled the British to map the area much more accurately than had previously been undertaken. This consequently revealed that pre-1914 British topographical enterprises, which present-day historians comprehend in terms of power and control, were inaccurate and, thus, unreliable for the purpose of military planning. The British suffered heavy losses because of their widely held contempt of the Ottoman's topographical and cartographic capacities and their overconfidence in their own grasp of the geography of the area. Most maps of the Middle East were lacking in the characteristics which might have given them actual power over the territories they were supposed to represent. Accurate coordinates were scarce. Triangulation nets of the first order, the only systemic and truly reliable surveying method, were to be found only in Egypt on the eve of World War I. Hachuring, of little use for the artillery, was still the prominent representation of relief. European cartography of the Orient was defective, and as a consequence the maps could be misleading. The longed-for well could be miles away from its location according to the map. The level, smooth landscape drawn on paper could in practice reveal itself to be impassable. Even more telling of the absence of a European mastery of the Orient, despite the traditional discourse of power, was the fact that nothing could be done effectively without local assistance.

MASTER OF ALL THEY SURVEYED?

The study of the British map of the Middle East during the war provides an opportunity to discuss the profound overhaul of the Western geographical and cartographic comprehension of the region after 1914. An exploration of the uses and misuses of cartography during the eastern campaign, from Gallipoli to Gaza, will underline these transformations.

There is a widespread myth about the Dardanelles campaign. Incomplete intelligence and lack of adequate maps are generally blamed for the catastrophic outcome of the battle. Peter Chasseaud and Peter Doyle have demonstrated that there is another side to the story.[11] British command actually had a better grasp of the terrain than the received wisdom acknowledges. Gallipoli was not the Crimea. The actual understanding of the field in 1915 far exceeded what the officers embarked to fight Russia in Malakoff knew.[12] Nevertheless, there may at times be some truth in myths.

When the decision to organize landings near the Dardanelles was made, General Ian Hamilton recounted that he was summoned to Kitchener's office on March 12, 1915. Kitchener, then secretary of state for war, provided him with vague directions and nothing but a "featureless map" upon which to draw a plan.[13] Hamilton was told that "the cross fire from the Fleet lying part in the Aegean and part in the mouth of the Straits must sweep that flat and open stretch of country so as to render it untenable by the enemy. Lord K. demonstrated this cross fire upon the map."[14] Hamilton's diaries are certainly highly unreliable. Postwar explanations for the disaster were needed. Yet the centrality of maps in the narratives of the battle is in itself proof of their complex function in the campaign.

Hamilton quickly discovered the shortcomings of Western mastery of the East: "The Dardanelles and Bosphorus might be in the moon for all the military information I have got to go upon. One text book and one book of travelers' tales don't take long to master and I have not been so free from work or preoccupation since the war started. There is no use trying to make plans unless there is some sort of material, political, naval, military or geographical to work upon."[15] Maps might be lacking; their shortcomings could be used as an excuse. The ability of officers to read them was also an issue. Hamilton had a 1:63,360 map of the peninsula, based on a French survey dating back to the Crimean War. It had been updated in 1915 and could serve the general's intelligence needs well enough.[16] Yet the value of maps in warfare depended on the officers' ability to visualize the battlefield thanks to them. Defective cartography, exemplified by inaccurate contour lines, and the lack of training in reading this type of document explain Hamilton's feelings when confronted with the peninsula's relief: "These cliffs were not in the least like what they had seemed to be through our glasses when we reconnoitred them at a distance of a mile or more from the shore. Still less were they like what I had originally imagined them to be from the map. Their features were tumbled, twisted, scarred—unclimbable."[17] High-ranking officers were not the only ones who were left without a topographical clue during the Gallipoli campaign. Former ANZAC soldiers underline the perplexity provoked by their maps. Most of them had a map case and showed "much interest [. . .] in the issue of maps." They used the long hours on the ships to copy their own handmade sketches of the area. Like many of his comrades, Private Sydney Callaghan from Australia's

Second Field Company Engineers had one in his pocket when he landed on the beaches of Gallipoli in April 1915.[18] As Fred Waite recalled, however, when the attacks started, the fog of war blurred the outlines: "Owing to the darkness, our unfamiliarity with the country in front, and our misleading maps, we were brought to a standstill."[19]

The Allies promptly filled in the gaps. "Aeroplane photography" was used to grasp the terrain. Seaplanes borne on the H.M.S. *Ark Royal* complemented the information, with unsatisfactory results at first, since no fixed point could be established on the front trenches. On the ground, soldiers and officers surveyed the battlefield.[20] Maps were compiled and reproduced at general headquarters, which had its own printing section to continuously update the trench maps on a 1:10,000 scale. Officers made sure the documents would not fall into the wrong hands. Trench diagrams of the Gallipoli area (GSGS no. 3126) all bore a warning that they were "not to be carried into any attacks." Captured Turkish maps soon permitted the British command to perfect their understanding of the terrain.[21] The reforms of the late 19th and early 20th centuries had come to fruition, and the Turkish general staff could rely on effective cartographic resources.[22] Good Turkish maps undoubtedly played a great part in the Allies' defeat in Gallipoli. Undue naïveté as to the capacity of the Ottomans to wage a modern war thanks to the most accurate topographical information was quickly dispelled. Maps not only served tactical and strategic aims; they also became part of larger schemes that were devised to deceive and manipulate. The most significant example is to be found in Hamilton's dispatch of December 11th, 1915.[23] In his report, the general explained how he intended "to order [. . .] a whole set of maps of Asia in Egypt" so that Turkish informants would hear about it and report it to their headquarters. The Operation Fortitude–like plan was intended to give the impression that a major operation was being planned to land near Mitylene on the Syrian coast. The Ottoman army would then send some of the troops stationed near the straits to fight on the new front.

Maps became an integral part of the war in the East. Significantly, Reginald Teague-Jones (1890–1988), an Indian intelligence officer who tried to counter Wassmuss's schemes in Persia, had only one war trophy of which he was especially proud: the German spy's atlas, a 1912 edition of Justus Perthes's reference work.[24] To capture maps, to collect, and to process topographical data were essential aspects of warfare in the Orient, where accurate and precise documentation was lacking. Any piece of intelligence stolen from the enemy could prove to be useful. Edmund Candler, a war correspondent for the *Manchester Guardian* reporting on the Mesopotamian campaign, recalled the prizes found after the capture of Baghdad in 1917: "In one of the offices we found the survey maps;

another contained the trade returns of the notorious Wonkhaus [...]. Hanging on the walls and littering the floor were the maps and plans drawn up by Sir William Willcocks which might have made Iraq as fruitful as Egypt."[25] This worked both ways. A set of captured Turkish maps held in the RGS's library today were copied from Francis Maunsell's monumental work on Turkey and translated into French.[26] Stolen information could not suffice; local intelligence was also crucial. A perusal of the *Military Handbooks* compiled and published for the use of the army shows how critical this knowledge was. As noted in the introduction of the volume *Routes of Arabia*, "a great deal of the information contained in these pages is drawn entirely from local sources."[27] In Cairo, K. Cornwallis from the Arab Bureau tried, with constrasted results, to cross-check an incredible mass of information that he received from other British agents or that he himself collected from travelers and traders passing through Egyptian bazaars.[28]

In many theaters, such as Mesopotamia, army surveyors were confronted with an ungraspable landscape. Candler noted: "Nomenclature in a land of flat mud is difficult for the surveyor. There is nothing salient for him to give a name to [...]. So the sketch maps became dotted with 'mud huts,' 'Arab village,' 'Broken huts,' 'Two mud walls,' until confusion became the mother of invention, and the Old Testament [...] was called in with its none too fanciful associations. We had 'The Walls of Jericho,' 'Eve's Crossing,' 'Sodom and Gomorrah.'"[29] This biblical reflex and its mid-Victorian substrate conditioned the perception of the territory by soldiers who had little more than scriptural geography to apprehend the battlefields. O. Teichman, from the Medical Corps, noted in his diary for December 27, 1917: "On looking at a map of Ancient Palestine that evening, we noticed that we had crossed from Ephraim into Dan."[30]

The straits, Mesopotamia, and Palestine campaigns revealed this twofold feature of the cartographies of the East in the early 20th century: accurate mapping was far from being the monopoly of the West, while prewar British surveys were useless to anyone actually wanting to establish an effective military control over the territories in question. The imperial gaze of the 19th century was a myopic one. This led the British forces fighting in these campaigns to remap the East, with an unprecedented accuracy.

THE AIRMAN' VIEW

Late Victorian bird's-eye views were no longer a chimera after 1914. The Middle East was viewed from above.[31] The First World War replaced the sketchy outlines of the Middle East with what were unprecedentedly accurate maps.[32]

While in 1919 David Hogarth had no option but to underline the extent of the destruction wreaked by modern warfare, he acknowledged that "it is a fact that during the late war science has achieved [. . .] much constructive progress," and specifically as far as the geography of the East was concerned.[33] The Middle East became a field of experimentation for the British army. The pilots met no opposition from the Ottomans, who had no planes. They could, thus, fly in a straight line across an often cloudless sky. The main difficulty in compiling maps from aerial photography consisted in the lack of verified coordinates for fixed points. In Arabia and Mesopotamia, the 19th-century calculations were defective. Positions remained uncertain.

Close cooperation between the air force, the survey headquarters, Allenby's headquarters, and artillery units in Palestine was regarded as an example to follow after the war.[34] The PEF's surveys were lacking in precision. Indirect artillery fire, which requires accurate altimetric data, would have been impossible without improved cartographic information. Modern warfare was dependent on good intelligence and, specifically, on accurate topographical and geographical data. The practice of predicted fire ("map shooting") by field artillery could not prove effective in the absence of relevant topographic and altimetric data. The old direct fire of the Royal Field Artillery and its empirical approach became obsolete. Major General Geoffrey Salmond, commander of the air force in the Middle East, and Captain Thomas, his photographic officer, put in place a highly effective organization aimed at providing Allenby with the best maps possible. Thanks to aerial photography precise relief maps of Palestine were issued, which permitted officers to visualize the battlefield and organize rehearsals of attacks. A view of Gaza from this secret set of maps shows how detailed the representation of the relief in fact was (fig. 8.3). Pictures of the terrain were taken every day. New targets and positions were communicated to the survey headquarters, which promptly updated the blank monochrome maps of Palestine printed by the Egyptian survey. Allenby's headquarters was close to the survey. Wireless communications fostered an effective circulation of the new information and the transmission of orders for a more accurate indirect fire. From early 1918, surveyors could count on better planes, such as the Bristol Fighters, which could fly as high as twenty thousand feet.

The Mesopotamian forces developed their own techniques, drawing upon the Anglo-Indian expertise in the field of aerial photography. Brigadier General Hedley, of the Geographical Section of the WO, described in a report how a new map of Baghdad was compiled by field officers of the Mesopotamian Expeditionary Force after the capture of the city in 1917 (fig. 8.4). Squadron 30 of the RAF supplied photographs of the city, which were pasted

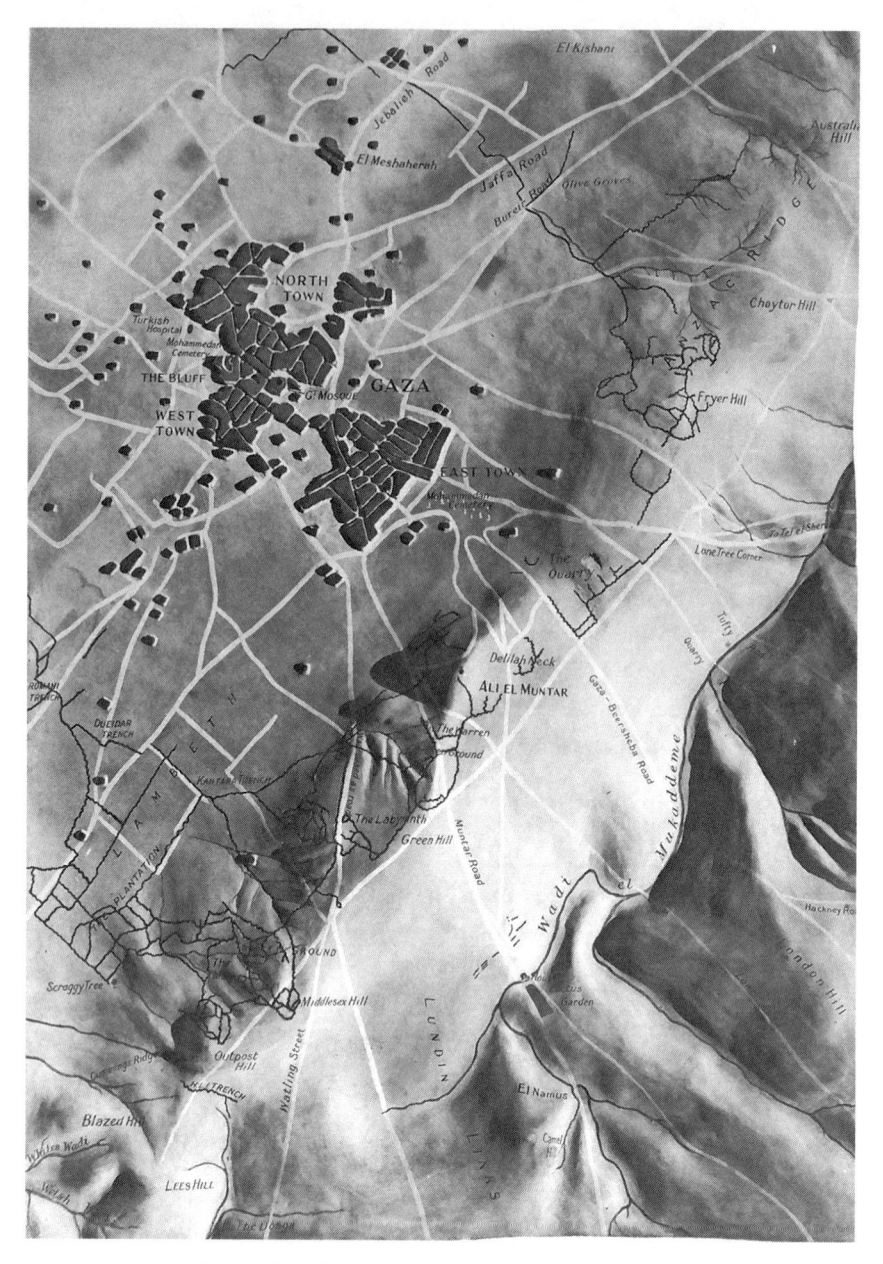

FIGURE 8.3. Survey of Egypt, *Bromide print of Gaza Operations Model, n° 3, Gaza*, 1:12,500 (London, 1917). National Archives of the UK (ref. WO 303.496).

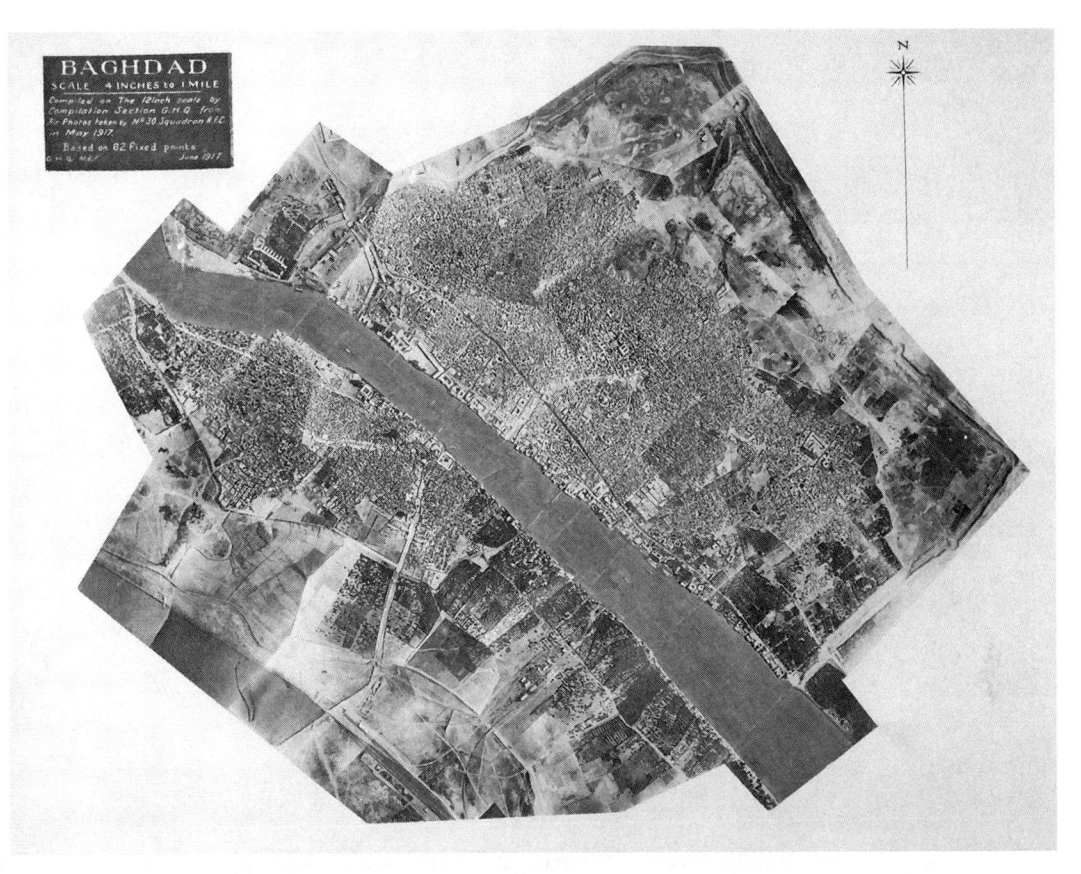

FIGURE 8.4. Map Compilation Section, GHQ, Mesopotamian Expeditionary Force, *Photo-compilation of the Baghdad City Plan from Air Photographs* (Baghdad, 1917). Size of the original: approx. 85 × 105 cm. National Archives of the UK (ref. WO 302/551).

in position on board. Since most of the pictures were taken with a camera that was not truly vertical, an enlarging lantern that could move in all directions was used to correct any distortions until the images coincided with a series of sixty-two points fixed by triangulation. A photographic enlargement of the puzzle was made in order to provide a single coherent figuration of the city from above. Inconspicuous agents, such as Iltifat Hussein of the Survey of India, did the fieldwork to make blueprints of all main buildings and collect place-names. The map was compiled by C. G. Lewis from the Survey of India and J. H. Cole from the Survey of Egypt. This was further testimony to the cartographic expertise of the two subempires. It took a month to make the first prints. A zincographed plate was ready within ten weeks. The first aerial pictures were taken on May 9, 1917. From July 21, 1917, the Survey Department

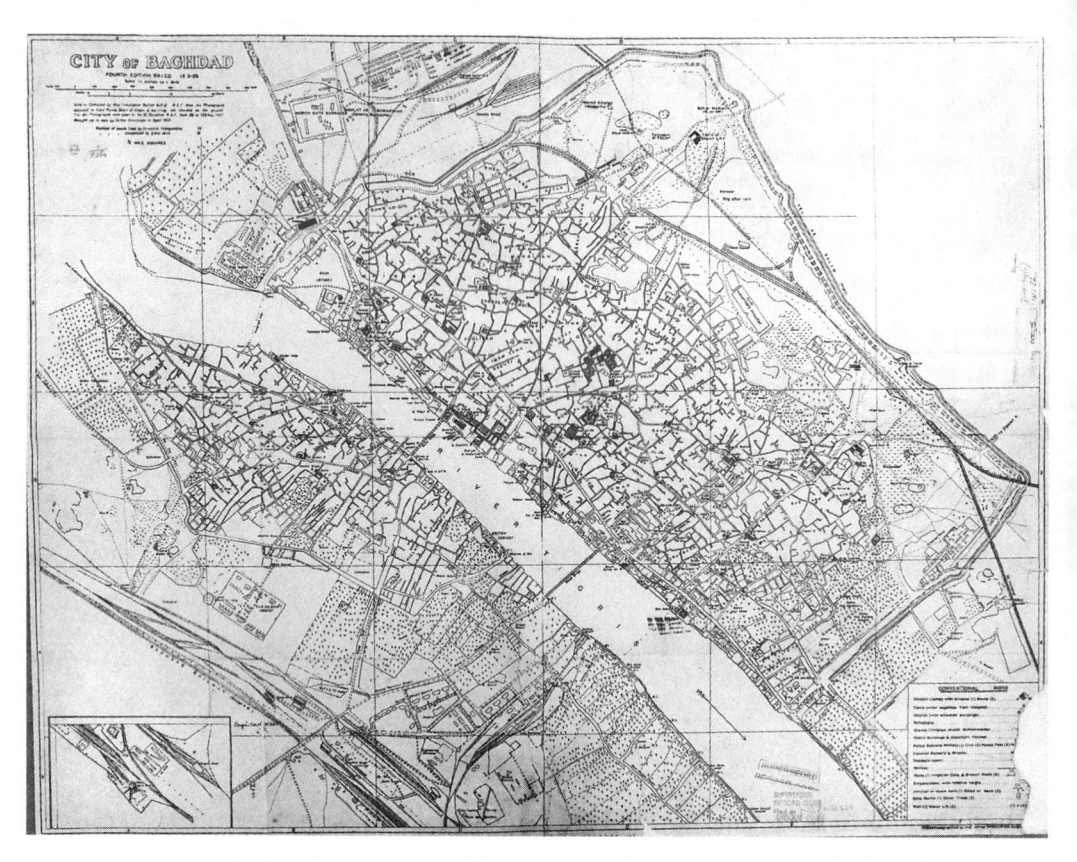

FIGURE 8.5. Map Compilation Section, GHQ, Mesopotamian Expeditionary Force, *Baghdad City Plan*, 1:5280 (Baghdad, 1917). Size of the original: approx. 85 × 105 cm. National Archives of the UK (ref. WO 302/551).

could print forty maps of Baghdad per day. James Felix Jones's hardships of the mid-19th century were a fading memory.[35]

These two examples are far from anecdotal. The RAF took note of the efficiency of the various processes implemented by British officers in the Middle East. Not only had British maps of Mesopotamia, Sinai, Palestine, and Arabia improved to a standard no 19th-century cartographers could have hoped to reach, but also the effective collaboration between the air force and the other branches of the military produced the impression that the East could also be controlled from above. This, at least, was Churchill's approach to the policing of the region at the 1921 Cairo Conference. A year later, the RAF was officially established as the predominant army branch in Iraq.[36]

Air surveys, along with a unique experience of the Middle East during the First World War, gave Britain a vantage point. France unsuccessfully tried to compete by sending troops, such as the Détachement Français de Palestine, but it was a case of too little too late. There was no serious challenge to Britain's self-confidence in its capacity to understand the Middle East in 1918. An apparently superior expertise in Middle Eastern affairs combined with an unrivaled cartographic grasp of the region allowed Britain to claim a prominent position in the designing of the structure that was to replace the falling Ottoman Empire.

After years of unfavorable comparisons with German maps and topographical intelligence, British cartographic competence reached a level comparable to that of the Germans during the war. The RGS became the center of the mass production of accurate maps for military purposes.[37] Both the Geographical Section of the General Staff (WO) and the Naval Intelligence Department had offices in the RGS buildings at Lowther Lodge (1 Kensington Gore) in London. A multitude of cartographers, intelligence officers, academics, and secretaries overcrowded the rooms of the society.[38]

PROPAGANDA, NEWS, SCHOOLS, AND A SURGING GEOGRAPHICAL LITERACY

The war intensified the cartographic focus on the Middle East. As underlined by Jeremy Black, "the First World War and its aftermath greatly increased interest in political maps and mapping [. . .]. Readers of newspapers became ever more familiar with campaign and battle plans."[39] Michael Heffernan also observed the "dramatic increase in map use during WWI."[40] Never before was the British population so inundated with such a quantity of cartographic objects.

In 1914, the FO first established a Propaganda Office at Wellington House. The Department of Information, supervised by John Buchan, was established in 1917 to coordinate efforts. "Wellington House," as British wartime propaganda was named, released masses of pamphlets, posters, maps, and artifacts during the conflict.[41] The East was just a sideshow until 1916, when Lloyd George, acting prime minister from 1916, an "easterner," began to foster the idea that victories in the Middle East and an aggressive propaganda effort against the Ottoman Empire could counterbalance difficulties on the Western Front. A coordinated campaign was therefore organized by Buchan with a single motto: "The Turk must go."[42] Pamphlets, articles, and maps deftly and insistently spread this opinion to the general public. Arnold Toynbee, who was to join the Middle Eastern Section of the British legation in Paris from January 1919, was a productive purveyor of propaganda. He published various pamphlets attacking Turkish misrule and atrocities.[43] A map "displaying the scene of the atrocities" in his work on massacres of Armenians buttressed his Turcophobic stance.[44]

From 1917, the victorious march of British forces both in Mesopotamia and Palestine provided the war cabinet with good reasons to intensify propaganda. The Middle East campaigns could become what trench warfare in France was not: a source of glorious narratives featuring a conquering army on the move. The War Office Cinematograph Committee, in collaboration with the Topical Film Company and Jury's Imperial Pictures, produced a variety of newsreel series documenting the Middle Eastern battles. Films dealing with the years 1917–18, such as the series entitled, tellingly, *With the Crusaders in Mesopotamia*, revived the myth of the crusader.[45] Marching troops, masses of Turkish prisoners, and picturesque views conveniently balanced the terrifying testimonies of the soldiers fighting in the Somme, where movement, clear-cut victory, and exotic scenes were obviously lacking. One of the most successful pieces of propaganda was the newsreel showing Allenby entering Jerusalem.[46] The capture of the city was a momentous occasion for the WO, enabling it to stage a British victory. The myth of the new crusader was once again revived for the home front. While the medieval reference had already been used for Charles Gordon, it was deemed inappropriate as an element of British Orientalism because of its Catholic connotations. Mark Sykes, the great architect of British propaganda for the Middle East, seized the opportunities provided by Allenby's and Maude's victorious advances to multiply variations on the crusader's theme.[47] He believed that "what is wanted is popular reading for the English church and chapel folk [. . .]. Articles should give striking actualities [. . .] rivet the British onto the Holy Land, Bible and New Testament."[48] The underlying biblicism that determined mid- to late Victorian educations and imaginations was still vivid during the First World War. The most surprising cartographic byproduct of this is probably the commemorative handkerchiefs printed with a map of Palestine that were produced after the conquest of Jerusalem in December 1917.[49]

The innovative propagandists were also prompt to use maps in films. From 1918 onward the Topical Film Company produced films showing animated maps of the various fronts.[50] The 1918 film *Deir Al Belah* shows the Survey Department and printing presses of the Palestine forces.[51] A later movie, *Allenby's Campaign,* released in 1919, narrated the crusade with a great wealth of animated maps.[52] There were also films depicting the battles of Kut and Ctesiphon with animated battle plans.[53]

As of the early stages of the war, the demand for cartographic depictions of the conflict rocketed. The circulation of journalistic maps, a product of the official propaganda efforts, was unprecedented. The market of general atlases also boomed. Three hundred thousand copies of the *Times War Atlas* were sold in just four days in September 1914. The traditional maps of the seat of war were also commercially successful. Gallipoli was the first Middle Eastern battle to be extensively covered by journalistic and commercial cartographic endeavors.[54] Despite its apparent unimportance, the area continuously attracted the mapmakers' attention. Philip's 1916 *Strategical Map of Mesopotamia and Asia Minor* and Johnston's *War Map of the Middle East* were two lavish illustrations of the genre.[55] One of the most popular publications in this field was the *Daily Telegraph* colored war maps designed by Alexander Gross of the Geographia Map Company.[56] The series included a *War Map of the Near East* (no. 6, 1916), the *Balkans and Eastern Europe* (no. 10, 1916), a *War map of the Gallipoli Peninsula* (no. 12, 1916), a *Picture map of the Dardanelles* (no. 18, 1916), and a *War Map of Palestine* (no. 23).[57] The increasing circulation of British newspapers during the war did much for the geographical education of the public. The *Daily Herald* sold two million copies daily, for instance.[58] Specialized war magazines such as *The War Illustrated, A Pictorial Record of the Conflict of Nations*, edited by J. A. Hammerton, had wide circulations (750,000 in 1918). They provided their readers with a wealth of maps.

To compensate for the lack of photographic documentation artist cartographers designed a range of more and more spectacular cartographic objects. George F. Morrell was one of the most productive of these skilled illustrators. He mainly worked for the *Graphic*. One of his first works during the war was a world map in two hemispheres titled *Germany's Vanishing Place in the Sun*.[59] He drew several illustrations on the Western Front, notably an impressive map of Verdun.[60] Two of his creations from 1917 are particularly revealing of the propaganda shift engineered by John Buchan at the instigation of Lloyd George.[61] The first was a graphic map entitled *The Restoration of Mesopotamia*, in July 1917.[62] It anticipated the future of the area after the war and revived the old scheme of the Euphrates railway from Alexandretta to the Gulf. Morrell designed an exceptional bird's-eye view of the Middle East for the illustrated paper in November 1917.[63] It depicted the Middle East as seen from the south, with the progressive encirclement of the Ottomans. Both documents mirrored Lloyd George's stance on the obligation to fight the Turkish Empire. Like the newsreels of the Topical Film Company, these maps infused optimism, something that the news from the Western Front could scarcely foster in 1917. They were rooted as well in the decades-old presumptions about the possibility of reviving a supposedly stagnant Orient. British cartographic propaganda presented the possibility of building nations once they had been delivered from the yoke of Oriental despotism, an echo of the enduring views that can be traced back to the second half of the 19th century, as we have seen earlier.

This geography-minded outlook was reinforced by the daily experiences of the British population. The letters sent by soldiers from unknown places, the daily publications of maps on the evolution of the battlefields, and, after the war, the internationalist concerns fueled by the hopes for a definitive peace settlement all contributed to further the knowledge of the world. This was particularly visible in the educational system. As shown by Rosie Kennedy, "the war entered the curriculum."[64] Specific works such as Albert Cock's *Syllabus in War Geography* were designed to connect the conflict to geographical issues for classrooms.[65] The war contributed to an enhanced institutionalization of geographical education from primary schools to the university. The first Honours School of geography was established in 1917 in Liverpool.[66]

The geographical literacy of British youth was also improved through the war-related evolution in children's everyday culture. Battlefield-inspired board games were quite popular in the first years of the conflict, with visible consequences for the players' grasp of the geography of the war.[67] Juvenile literature also helped to foster the pupils' geographical grip on the world.[68] The prolific Frederick S. Brereton (1872–1957) took up G. A. Henty's legacy. His children's novel on Gallipoli, his *On the Road to Baghdad*, an exceptional piece of propaganda for the Indian Political Service, and his *Latest Crusade*, an account of Allenby's campaign, all include sketch maps of the Dardanelles, Palestine, and Mesopotamia to educate the young reader.[69] These novels and maps resorted to motifs and notions that were rooted in late Victorian and Edwardian imaginations regarding the East. Brereton's novel on the Mesopotamian campaigns could resort to worn-out stereotypes going back to the 18th century, such as the cliché of the lazy Turk smoking his narghile, an occupation which "for a Turk that may be good enough, sufficient exercise both for mind and body; but the fresh blood, the keen intellect, the wonderful energy of Anglo-Saxons require more movement." He could also modernize his discourse and use information gathered from the early 20th-century debates on the region. The young reader learned from the hero, and with the help of a sketch map, that "of course, there is the pipe line [. . .]. It cannot fail to be of great importance to Britain. You see, numbers of our battleships now use oil fuel almost exclusively."[70]

The combined effects of the war on children's everyday culture clearly increased the level of geographical literacy. One of the first statistical traces of this shift, in itself a sign of a greater awareness of the importance of the subject in the curriculum, is to be found in the *Scottish Geographical Journal* after the war.[71] Wallace Whitehouse, a lecturer at the University College of Wales and contributor to *Geographical Teacher*, wrote a statistical analysis of the geography papers of 1,205 pupils who had sat for the senior stage of the 1919 Central Welsh Board examination. Some of the questions, such as 3B, which involved drawing "large sketch maps" to indicate the reasons for the development of "three towns of some antiquities on the coast of the Mediterranean," required at least some knowledge of the East. An overall 68.4% of the pupils passed; 41% with distinction, that is, with a more than 70/100 mark: a result J. S. Keltie could only have dreamed of when he wrote his report on geographical education in the early 1880s.

8.2 CARVING KINGDOMS

> When statistics failed, use was made of maps in color. It would take a huge monograph to contain an analysis of all the types of map forgeries that the war and the peace conference called forth. A new instrument was discovered: the map language. A map was as good as a brilliant poster, and just being a map made it respectable, authentic. A perverted map was a life-belt to many a foundering argument.[72]

When Isaiah Bowman, an adviser to Woodrow Wilson in 1919, wrote these somewhat cynical lines after the long negotiations of the post–World War I peace conferences, he was pointing to an innovation in the conduct of foreign affairs. For better or worse, never before had maps been so influential in decision making and in determining the destinies of the world. President Wilson himself could be found on the floor of his French residence analyzing large maps before resuming the afternoon's negotiations. Topographers, cartographers, and mapmakers became the agents of truth. Seldom were members of the participating delegations in the negotiations without a map by which they sought to vindicate their rights and to establish new borders. Maps and geographies of the region played an instrumental role in its post–World War I redefinition. Spatial information, both during the war and in the failed attempts at a global peace that followed, became more crucial than ever. The conflict triggered a series of convulsions which completely reshaped the region. The war hurried the downfall of the Ottoman Empire. Newly born states were to grow on its ruins; new names would appear on the maps as peace treaties and agreements were signed after the armistice. The death of the "sick man" unleashed a host of new powers, as tribes, leaders, Western experts, political movements, religious parties, and a nexus of actors looked for a new geopolitical balance in the East.

"A MAP, A PENCIL, TRACING PAPER"[73]

May 13, 1919

I spread out my big map on the dinner table and they all gather round. LG, AJB, Milner, Henry Wilson, Mallet and myself. LG explains that Orlando and Sonnino are due in a few minutes and he wants to know what he can offer them. I suggest the Adalia Zone with the rest of Asia Minor to France. Milner, Mallet and H. Wilson oppose it: AJB neutral. We are still discussing when the flabby Orlando and the sturdy Sonnino are shown into the dining room. They all sit round the map. The appearance of a pie about to be distributed is thus enhanced. Lloyd-George shows them what he suggests. They ask for Scala Nova as well. "Oh no," says LG, "you can't have that—it's full of Greeks!" He goes on to point out that there are further Greeks at Makri, and a whole wedge of them along the coast towards Alexandretta. "Oh no," I whisper to him, "there are not many Greeks there." "But yes," he answers, "don't you see it's coloured green?" I then realized that he mistakes my map for an ethnological map, and thinks the green means Greek instead of valleys, and the brown means Turks instead of mountains.[74]

Is Harold Nicolson's surprising account a reliable depiction of the uses of the map during the peace negotiations in Paris, or is it just a colorful story to enliven his book? However trivial the treatment of the Italian claims on Asia Minor might seem in his narrative, and however surprising Lloyd George's mistake about the map legend, it is abundantly clear from contemporary records that maps and the guesswork that surrounded their interpretation and their manipulations were a feature of the discussions on the new postwar order. Nicolson's telling story also testifies to the fact that maps, and specifically British maps, had become central to the negotiators' works.[75]

A look at the preparatory phase of the peace conference shows how crucial geography was to government planning in the aftermath of the war. As early as 1917, a set of ad hoc departments and committees had started to collect data and to issue reports in preparation for the forthcoming peace conferences. A department of the FO, the Political Intelligence Department (PID), was established in March 1918 to coordinate and filter the knowledge collected by a variety of governmental institutions. One hundred seventy-four peacebooks, secretly designed monographs sketching the history, geography, demography, and economy of each country relevant to the peace negotiations, were published by experts for the widespread use of British agents in Paris.[76] The PID peace memoranda of October 10, 1918, provided an advisory commission matching each inter-allied committee. Their division followed a territorial logic. The Middle East section was placed under the direction of Sir Louis Mallet. It comprised Turkey, Arabia, parts of Libya, Egypt, Abyssinia, the Caucasus, Persia, and Central Asia.[77] As shown in the introduction, maps and geographical intelligence constituted a key feature of the reports issued by the commissions. Such information was meant to substantiate their expertise and to provide the negotiators with matter through which to make their case. The chaotic organization of the Paris conference itself, which soon proved too large and unmanageable an undertaking, fostered the growth of the advisers' influence, thus crowning a long-term process leading to the ascendancy of the specialist in government planning.

Ethnographic maps, ideally suited to the nation-building creed of the negotiators, were particularly useful. By identifying populations in a seemingly accurate manner, the experts could propose ostensibly logical borders for the successor states to the Ottoman Empire.[78] The FO archives provide an excellent example of a pioneering use of cartography and statistics. This consisted in a map attached to a report of the Intelligence Department of the naval staff dated June 25, 1919, dealing with the population of the Armenian provinces of Transcaucasia. The map displayed the racial majorities by district, to elucidate the distribution of "Mohammedans" and "Armenians" in the area.[79]

One can find in the archives other instances of inventive cartography. These sought, however mistakenly, to display the intricate cultural realities in the Middle East. One of the most interesting is the WO map published by the GSGS under the number 2901 (fig. 8.6).[80] Given its scale, the multicolored document could not render accurately the religious, tribal, and human variety of the area. For lack of a better solution, the cartographers chose to use overlapping rectangular shapes to figure regions where various populations coexisted, thus giving the impression of a well-ordered territory. The confusing legend itself, a mixture of religious categorizations and ethnicities, is revealing of the cartographer's perspective on the racial organization of the region. Russians, Greeks, and Christians were on top of the list, Circassians at the bottom. The map showed the separation between the Arab East and the Levantine Orient. The harmonious blue of the Arab tribes, entirely forgetful of the Shia-Sunni divisions in Mesopotamia, or of the presence of Christian communities in the hinterland, contrasted with the colored mosaic of the Mediterranean coasts. This was an echo of one of the most enduring British ethnographic divisions of the Middle East: between the French- influenced Levantine coastal regions and the British-protected Arabistan. This distinction was given lapidary formulation by T. E. Lawrence in 1917: "Morally the peoples (of Syria) somewhat resemble one another, with a steady gradation from neurotic sensibility, on the coast, to reserve, inland."[81] This somewhat artificial, fabricated depiction of the peoples of the Middle East was in fact referred to during negotiations, as shown by the inclusion of a map entitled *Mesopotamia: racial division* in Erle Richards's memorandum on Mesopotamia, dated January 24, 1919. Its very simplistic reference avoided troublesome details, crudely opposing "Arabs" to an all-encompassing category of "mixed races."[82] This map illustrates how evidentiary documents used in the course of the conference were the expression of preexisting beliefs and assumptions, such as those consistently promoted by many British agents in the East from the late 19th century. Their preference for the supposedly unstained Arabs of the desert shaped their understanding of the area, for instance. Moreover, the technical limitations of these documents, coupled with the predominant use of small-scale maps, served conscious purposes.[83] The GSGS ethnographic chart reflects how simplification was a key feature of the representation of the Middle East during the negotiations. It also demonstrates how pervasive social Darwinism and racial theory had become in early 20th-century cartography. The document is one of the many examples of how thematic mapping of the East reflected a growing concern with the role of race in the shaping of human interactions.[84]

Oil maps are another example of how the ill-informed policy maker learned about the "realities" of the Middle East.[85] Mark Sykes explained the significance of the Middle Eastern oil reserves as early as June 1916 in a memo entitled "The Problem of the Near East."[86] The enclosed map, a surprisingly amateurish document, displayed "oil areas" in the shape of roughly drawn circles.[87] The Naval Intelligence Department focused heavily on the strategic importance of the Mesopotamian oilfields.[88] This was to become a decisive factor in the British determination to include the Mosul area in the British mandate of Iraq. The use of some of these oil maps by British policy makers is well documented: this cartography was not only designed to amuse the experts. In what can be summarized as an illustration of the growing cartographic culture of the decision makers, the members of the imperial war cabinet "consulted a map showing the Mesopotamian oilfields" in August 1918.[89]

The representatives and leaders of the various populations of the Middle East whose destinies were discussed in Paris, Sèvres, and subsequently Lausanne

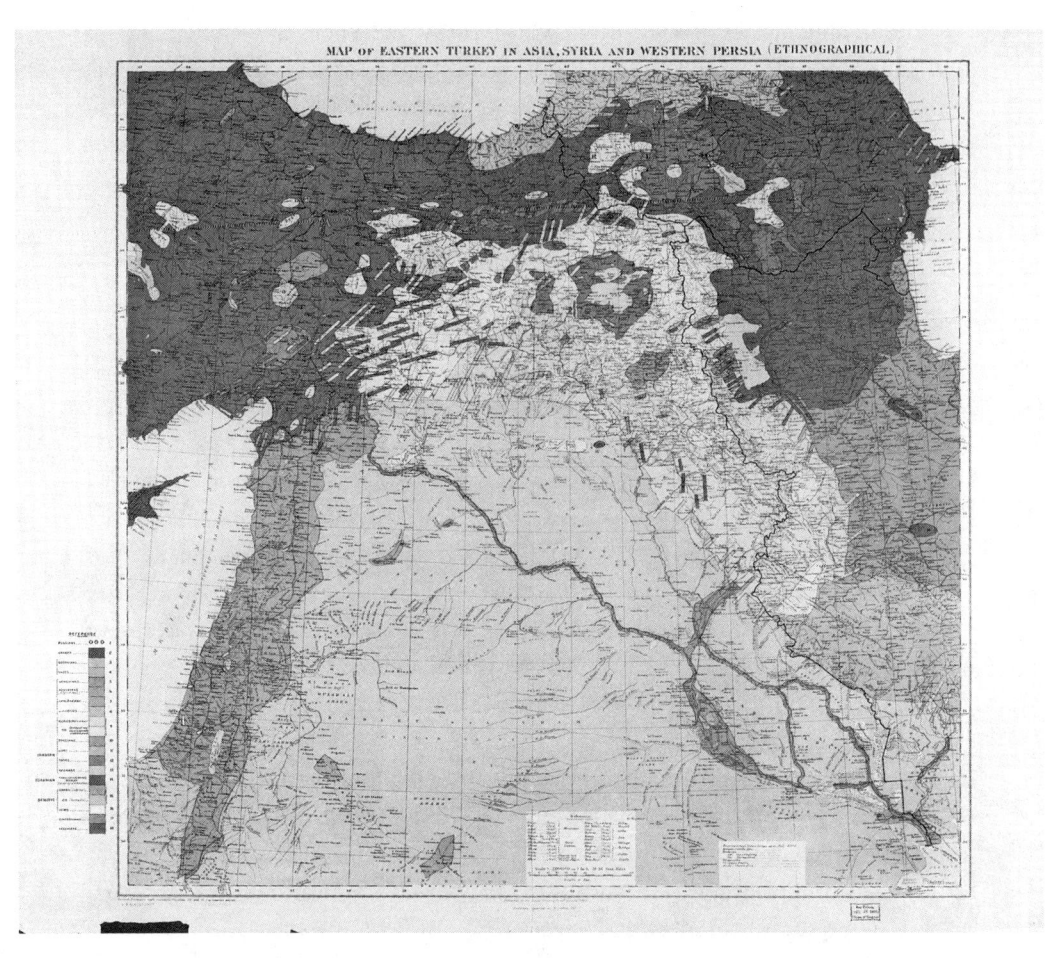

FIGURE 8.6. War Office, *Ethnographical Map of Eastern Turkey-in-Asia, Syria and Western Persia*, 1:2,000,000 (London, 1917; first edition, 1910). Size of the original: 71 × 73 cm. Photograph courtesy of the Library of Congress (catalog number: 2007633929).

adopted the cartographic rhetoric that so visibly pervaded the discussions. They did so with varying degrees of proficiency. The various communities of the region laid their claims before the powers by way of memoranda and attached maps. Some delegations and groups proved to be highly inventive. The most surprising map of all was probably the one sent by the women of the Moslem Trades School in Beirut. They had woven a pan-Arab rug in the shape of a United Syria.[90] For lack of the adequate resources for their own cartographic propaganda others simply used a pen to draw the borders of an imagined state on a British or a French commercial map. The Assyrian delegation, for instance, cut a piece of a French map centered on northern Mesopotamia and roughly traced the outlines of a Chaldean-Christian state.[91] The map complemented a letter in which Lazar Yacouboff, president of the National Assyrian Council, who probably placed too much trust in the Allies' will to protect their co-religionists, wrote: "We want to add that for centuries our nation has trusted that they would be liberated from the yoke of the wild Mussulman tribes [. . .]. This confidence in the reign of Christian peace has been preserved [. . .] and during the war our people were awaiting with impatience the coming of the British, the French and the Americans to deliver them from oppression."

Other delegations, more intent on bringing the participants over to their position, invested much more in maps, with effective results. The Greek delegation, for instance, was particularly active in publishing books and cartographies to support their *Megali Idea*, that is, the establishment of a Greater Greece on the ruins of the Ottoman Empire. A series of maps aiming at showing that Thrace had historically been Greek rather than Turkish or Bulgarian were issued.[92] Persuasive ethnographic maps of the eastern Mediterranean, such as Georgios Soteriades's 1918 *Map Illustrating Hellenism in the Balkan Peninsula and Asia Minor*, were designed to demonstrate the ineluctable facts underlying Greek petitions.[93] Zionist mapping also inundated the participants.[94] A nebula of associations, sometimes backed by the British government, put forward a series of reports, statistics, development schemes, and cartographic justifications.[95] The other powers, such as France, Italy, and the United States, did exactly the same, drawing on far more powerful instruments and experience. British maps, however authoritative they may have appeared, were caught up in a maelstrom of data which made it impossible for any clear picture to emerge.

As a result of this new balance between politics and expertise, postwar diplomacy drew heavily on emerging field of geopolitical science. With the birth of geopolitics, brought about by the work of Mahan and Mackinder, among others, a new geographical and cartographic perspective on the world affairs had developed.[96] Decision makers started to think cartographically on a more systematic basis.[97] The shortcomings of the Congress of Vienna, a catastrophic example for most members of the British legation, were to be avoided through recourse to new tools of government, such as scientific maps or statistics.[98] As noted by Jacques Ancel, the negotiations ended up representing "Geography's revenge on History."[99] The noticeable multiplication of maps in the reports and proceedings of the peace negotiations did not systematically alter the long-standing views of the key decision makers on some issues. Lloyd George never really departed from the belief he had expressed during the war: "Palestine if recaptured, must be one and indivisible to renew its greatness as a living entity."[100] The prime minister had strong Zionist assumptions which were, at least in part, the result of the biblical reminiscences going back to his mid-Victorian education. His position on the matter did not waver, despite the many countermaps and memoranda that were addressed to the British legation.[101] He resisted Curzon's and Cromer's calls for the annulment of Balfour's declaration and an end to Britain's experiments in the area. Cromer, who had held a central position in the Middle East before the war, was an opponent to an extension of Jewish colonies in the region. He willingly evoked a conversation with a well-known British Jew, during which his interlocutor said: "If a Jewish kingdom were to be established in Jerusalem, I should lose no time in applying for the post of Ambassador in London."[102] The innovations in the rhetoric of the negotiations, as well as the apparent influence on the part of the experts, did not always manage to convince leaders whose geographical outlook on the East had been shaped during the Victorian era.[103]

Cartographic justifications had gained momentum during the negotiations held to reorganize the Middle East after the war, those resulting in the treaties of Versailles, of San Remo in April 1920, of Sèvres in August 1920, and of Lausanne in 1923 and the 1921 Cairo conference. Attitudes to this approach varied among British decision makers. Some, like Lloyd George, were well aware of the shortcomings of British intelligence in the Middle East. As shown by David Fromkin, the influential Arab Bureau in Egypt was a defective structure which was often reliant on a single informer. It multiplied maps and reports in order to substantiate an expertise that its agents did not possess. Cairene intelligence was not always reliable.[104] On becoming head of the new Middle East Department in 1921, Churchill tended to ignore reports from the field, privileging the political over the technocratic. Others, like Reginald Wingate, believed on the contrary that the British "geopolitical position and our connection with the Arabian Provinces nearest to us, has given us opportunities for understanding the situation there—and the views of the Moslems of the Holy Places—better than any others."[105] They placed their trust in the flow of paper and maps that materialized British expertise on the region. Curzon studiously read and collected geographical and cartographic documentation with a more critical perspective, as shown by the large set of files which he examined before writing his memoranda.[106] Ultimately, reality would prove to be the true test, as Arnold Toynbee realized when he finally visited the locations which he had hitherto only observed through maps: "On the eve of the peace conference I was given the job of suggesting with a map the bounds of a possible Greek enclave round Smyrna. I carried out these instructions, and learnt, in doing so, that this plan was a geographical absurdity. It was not till I visited, in 1921, the then Greek-occupied area that I realized how small the Greek minority was, even with the area that I had delimited."[107] Even if, as is shown above, defective geographical and cartographic data determined some decisions, these were above all the diverging expressions of presuppositions which had been developed years, and even decades, before the war.

IMPERIUM IN IMPERIO: THE INDIAN EXAMPLE

The mass of maps and data produced from outside British circles presented difficulties of interpretation to the few key individuals responsible for taking decisions. The competition between the various factions to define the new Middle East under British authority is a well-known characteristic of the postwar settlements.[108] Far from unified, "British" reaction to imperial realities was fragmentary and plural. The "British" experience in the Middle East testifies to this. The internal divisions of the empire were obvious to the outside observer. The French envoys in the East repeatedly stressed this fact, no doubt because their own imperial structure suffered from similar defects. General Maurice C. Bailloud (1847–1921) noted in a report (October 4, 1917): "For a number of English civil servants from Egypt—I am not saying English civil servants from England—the Sacred Union does not apply outside the continent."[109] A 1918 telegram from a Middle East–based French translator named Poidebard confirmed the impression: "English officers from France are nice to us, Civil Servants and officers from the Indian army that are in charge of the campaign and administration in Mesopotamia are opposed to us. They put up a political and commercial struggle against us."[110] The various subempires whose existence I have traced in previous chapters through the conflicting protocols for the drawing up of maps were to clash in their attempt to impose their rival view on the Middle East from as early as 1918.

In the end, those who proved to be least effective in doing so were the representatives of the Anglo-Indian center, which was crippled by internal divisions. Despite their strenuous campaign during the previous decades to bring the East closer to British influence, Indian officials were not promoted to positions of decision making in the 1920s. None of them attended Churchill's momentous Cairo Conference in 1921. Nonetheless, the story of how they tried to influence the peace negotiations is in itself symbolic of the debates which eventually gave birth to the British Middle East.

We have already met earlier some of the most prominent officers of the Indian Political Service.[111] Two names stand out: Percy Cox and Arnold T. Wilson. Their private correspondence, which is at the disposal of historians, provides a much more revealing dimension of the archives than the official reports.[112] Percy Cox was the chief political officer in the Indian army which fought the Mesopotamian campaign. He became acting minister in Tehran in the last months of the conflict and was eventually appointed high commissioner for the Iraq Mandate from 1920. Wilson assisted him until he became civil commissioner for Mesopotamia. As a consequence of his mismanagement of the Iraqi revolt of 1920 he was replaced by Cox. Both colonial administrators were friends. Their exchanges reveal in clear terms how they developed their own approach to the Middle Eastern settlement.

They fully understood that their grasp of the terrain would be an asset, once the Ottoman army was defeated. Understanding as early as 1916 that they "shall be called upon to administer this country for military purposes," Cox encouraged Wilson to draw up a list of "names of reliable Civil Service Men." For the following years, the two officers went to great lengths to identify those who could participate in a network of people with whom Cox had worked while in the Persian Gulf area a decade earlier. They were ready to administer Mesopotamia. Their preparation was justified. On March 16, 1917, the war cabinet established a Mesopotamian Administration Committee under Curzon, who declared that occupied Mesopotamia would be administered by London, not India. Whitehall was nevertheless forced to accept Indian personnel, since no one else could do the job. Cox and Wilson built up the colonial administration of Iraq with their men. Eager to protect his newly forged administration, Wilson attended the Paris peace conference as a member of the IO delegation, along with Gertrude Bell. He regularly reported to Cox, his mentor. He described to his colleague his relentless efforts "to put forward Mesopotamian views upon various subjects" in the presence of Lloyd George, Balfour, Hirtzel (assistant undersecretary of state for India), Curzon, Chirol, and the British delegation.[113] On March 28, 1919, he met Chaim Weizmann in order to evoke the "future position of Mesopotamia and the respective spheres of influence of British, French and Italians." His views, as always, were supported by a wealth of data and convincing maps.[114]

Wilson's and Cox's Indian-Mesopotamian perspective came into conflict with Metropolitan and Anglo-Egyptian views. Rejecting with disdain the Arab Bureau's solution in favor of the Hashemite dynasty for Arabia, they supported their own champion, Al Sa'ud.[115] Wilson rejoiced in one his letters to Cox in August 1919: "Bin Sa'ud has behaved very well indeed [. . .]. I have had by no means an easy game to play, as Cairo suspected me, quite rightly, of wishing to use Bin Sa'ud to down the Hejaz."[116] Wilson also constantly denounced the widely backed solution of a local government in order to rule over the new Iraqi state. In January 1920, well aware of his weakened position in the eyes of the metropolitan government, he went so far as to make the following suggestion to Cox: "If you think as one of the old gang my presence might hamper you by making H.M.G. suspicious, I shall willingly disappear."[117] These shadowy maneuvers, which would eventually prove to be ineffective, were not the exclusive preserve of Wilson: the Arab Bureau in Cairo and the legendary T. E. Lawrence kept their fellow British colleagues in the dark.

British agents inherited a vast array of long-established presumptions as to what the East could become and how it should be governed. Wilson believed in the potential of Willcocks's schemes to reclaim the Mesopotamian wasteland. He started to apply his Anglo-Indian methods of administration to Iraq. He was aware, like Cox and Bell, of the importance of the Shiite element in the Middle East, an issue that was overlooked by some of the most adventurous agents sent by Cairo to support the Arab revolt. The British administration in Egypt had produced its own perspective on the region. It replaced the myth of the infallible Indian officer, legitimated by long experience in the field, with a more technocratic approach to governance, developed in the late 19th and early 20th centuries, at a time when the area became a field of experimentation for indirect rule and constructive imperialist methods of development.[118] The birth of the Middle East as an imperial entity was a synthesis of these entangled British outlooks on the region's future.

8.3 AN IMPERIAL PROTOTYPE

The war familiarized British decision makers with the phrase "Middle East." Contrary to what is sometimes said, the expression became quite commonplace in official documents. It became a frame of reference well before the 1920s. In 1916, Mark Sykes reported on the policy in the "Middle East," a term that had become interchangeable with the "Near East."[119] A "Middle East Brigade" was put in place by the RAF in July 1916. As early as March 1918, Balfour believed there was "much to be said for the establishment of a Committee dealing with the Middle East."[120] In October 1918, in a letter to Stephen Pichon, the French minister of foreign affairs, Robert Cecil (FO) planned the "Future government in the Middle East."[121] Arnold Toynbee, from the British delegation in Paris, insisted on the importance of the "Middle Eastern point of view" in a message to Eyre Crowe in 1919.[122] WO officials described the situation in the "Middle East" in 1920.[123] In postwar Britain, most major players in the region's destinies had adopted the phrase.

The peace proceedings in Paris did much to establish the geographical entity called the Middle East in the international diplomatic lexicon. Specialist periodicals such as the *Bulletin de la Société des études coloniales et maritimes* and the *Revue des sciences politiques* mentioned the *Orient Moyen* or the *Moyen Orient* from 1918–19.[124] In February 1920, the French communist newspaper *L'Humanité* surprisingly adopted the English phrase "Middle-East" in order to discuss the postwar settlements, for lack of a better expression to describe the new realities of the East in the 1920s.[125] As a rule, the French simply translated the term. Doisnel de Saint Quentin, the French envoy in Cairo during the war, referred to the "Orient Moyen" in his memoranda.[126]

This did not mean that all those who made use of the term had the same definitions of the region. Despite the dissemination of the concept, no indisputable vision had crystallized. The various institutions and actors dealing with the area had their own view on what the Middle East was. The de Bunsen Committee (1915), one of the first efforts to elaborate a coordinated approach to British policy in the East, adopted a minimalist Middle East, centering on the Ottoman Empire, exclusive of the Balkans, eastern Persia, and Arabia. It issued a set of five maps to depict the various alternatives which it had envisioned.[127] Its members mostly dealt with the Ottoman Empire, thus leaving the "independent Arabs" out of their cartographic focus. By 1918, General George MacDonogh's Middle East extended to Georgia, Azerbaijan, Daghestan, and Transcaucasia.[128] In a meeting of the imperial war cabinet (June 25, 1918), Curzon defined his Middle East as follows: "The East is a loose geographical phrase which requires definition [. . .]. I am going to deal with with a portion of the East, that is frequently called the Middle East. You will see it defined geographically on the map which is in front of the members of the Imperial War Cabinet." The area in question was delimited by two lines. The northern line ran from "the Black Sea" to "the borders of the Chinese Turkestan," following "the Trans-Caspian railway." The southern line ran "from the Eastern coast of the Mediterranean," through "Palestine," "Arabia," "Persia," "to the borders of the Indian Empire." Egypt was not part of Curzon's Middle East.[129] In october 1918, the PID, which provided each inter-allied committee with a matching advisory commission, adopted a territorial organization. It established a Middle East section under Louis Mallet, which covered Turkey, Arabia, Tripoli, Egypt, Abyssinia, the Caucasus, Persia, and Central Asia.[130] These diverging definitions within the British governmental circles were not fortuitous. They reflected internal contradictions concerning the structuring principles which were to be applied to the new territorial breakdown in the East.

The British administrative maze mirrored these disagreements.[131] A succession of overlapping committees, in addition to a wide range of uncoordinated actors, presided over the management of the East during and after the war. The Mesopotamia Administration Committee, established in March 1917 and chaired by Curzon, evolved into the Middle Eastern Committee in August 1917, then into the Eastern Committee in March 1918. By the end of the war, Alfred Milner's Egyptian Administration Committee and Balfour's Persia Committee also dealt with Middle Eastern territories while the Arab Bureau oversaw the Arabian Peninsula. As we have seen earlier, both the IO and the Indian government had a say in Persian and Mesopotamian matters. Bureaucratic fragmentation was in evidence. Curzon's ascendency was called into question after the war. Attacks from Robert Cecil, Alfred Milner, and Balfour weakened his position. Yet it was more than just a case of the competition between great men: these tensions were the expression of deep-running forces. Victorian as well as Edwardian conceptions of empire clashed with newer (and cheaper) conceptions of British global influence.

By 1920, the FO was advocating the creation of a department dealing with "the countries lying between the western frontiers or outskirts of India and the eastern end of the Mediterranean."[132] The IO concurred, stating that "all Middle Eastern areas [. . .] shall be placed under a single controlling agency."[133] The lack of coordination in the management of the British Empire in the Middle East by the postwar treaties was becoming a major drawback. Increasingly, the option of a large department within the CO or the FO with responsibility for the Middle East came to be considered the only realistic solution. Mark Sykes had already advocated such a solution in July 1918.[134] The map enclosed in his report divided his projected Middle East into five areas: Egypt (A), Sudan (B), Syria-Palestine and the western part of Arabia including Aden (C), the Nejd and the Arabian shore of the Persian Gulf (D), and a rather sketchy area comprising Mesopotamia, Kurdistan, and Armenia (E).[135] Sykes's recommendations had little effect at the time. In January and February 1921, a consensus between the interested parties resulted in the creation of an all-encompassing structure under the CO's supervision headed by an assistant undersecretary in the person of Winston Churchill.[136] The department, which at first did not supervise British policy in Egypt and which was centered on the mandates (Iraq, Palestine, Transjordan), eventually oversaw a larger Middle East that was inclusive of Egypt, Arabia, and Persia. The new institutions reconciled diverging views. Churchill was careful to seek the assistance of officials from a variety of backgrounds. Hubert Young from the FO joined the Middle East Deparment. John Shuckburg, from the Mesopotamia Department of the IO, was also transferred to enable him to work under Churchill.[137] T. E. Lawrence was recruited as political adviser.[138] R. Meinertzhagen became the department military expert.[139]

Soon enough, updated maps portrayed this new Middle East that had been redrawn under British and French auspices. Bartholomew's 1922 *Map of the Middle East*, which acknowledged the most recent political boundaries, quickly found a place in the CO map library.[140] The postwar Middle East was introduced as the latest and necessary addition to the British Empire. The overstretched and existing structure was complemented by the new acquisitions. A gap in the imperial network was filled. The interconnections between the different parts of the empire, from Vancouver to Wellington, were all-encompassing. What Vaughan Cornish (1862–1948), a prominent 20th-century geographer, called the "Axis of Empire," was straight and uninterrupted. Nothing could substantiate the point better than a map: Cornish enclosed a very convincing one in his books.[141]

Though still in its infancy, the department represented a relatively innovative approach to imperial issues. Traditional presumptions, such as the necessity to protect India and Suez, or the British elite's Victorian-grounded Zionism, still presided over the reasoning in Whitehall. New constraints nonetheless came to shape the perception of Middle Eastern issues. The assertion, within the context of an emerging international public opinion, of Wilsonian principles including that of self-determination after the war, restricted the options that were available to London. Raj-like and Cromer-like repressions of the revolts that inflamed the Middle East from the early 1920s proved ineffectual. The late 19th-century and early 20th-century imperial methods proved to be somewhat less effective than before. In Great Britain, the extension of the franchise turned Labour into a full-blown opponent possessing its own expertise. Leonard Woolf, for instance, wrote elaborate criticisms of the British Empire's involvement in the East. His *Empire & Commerce in Africa* enclosed various maps illustrating his attacks against British tutelage over the continent and the Red Sea shores in particular.[142] Financial difficulties also dictated Churchill's policies. Growing budgetary difficulties in the United Kingdom necessitated spending cuts.[143] Informal control and indirect rule were privileged as a way of saving the British taxpayer the cost of this new Middle Eastern empire. Everything was done to ensure that the costs of the region's policing would fall on the Middle Easterners themselves. Air control was favored over more customary military solutions, because it was much cheaper. The days of the Victorian "Second Empire" had passed.

CONCLUSION

It is impossible in one chapter to do justice to the complexity of the series of events which led to the postwar settlements in the Middle East. This section has not dwelt at any length on the ineluctable artificiality of the borders established after 1918. I have rather sought to show how cartographic and geographical visions participated in the shaping of Britain's understanding of the region both during and after the war. The role of maps should not, however, be overestimated. Information and knowledge in general were just one of the many variables of the peace negotiations. The clash between British and French interests was certainly a more decisive factor. Yet maps and geographies did play no small part in the establishment of British supremacy in the Middle East in the late 1910s and the early 1920s. More than ever, they had become both technocratic tools for decision makers and vehicles for propaganda and justifications. They substantiated the rational gaze that was to prevail over international statecraft after the war.

The British Middle East was, at least in part, the product of this renovation of Western visions. It was eventually located within what Leopold Amery named the "Southern British world" to become the last prized addition to the British Empire.[144] Yet the overhaul was incomplete. Strategic and imperial thinking presided over its invention, but Victorian educations and Edwardian mirages still lingered in the background. Disregard for religious distinctions (in particular between the Shia and the Sunnis), Anglo-Indian mid- and late Victorian conceptions of indirect rule, as well as overconfidence in the virtues of airpower to control supposedly passive populations, were not the least significant of these assumptions shaped by decades of British involvement in the area. It took very little time for facts to show the limits of this new frame for strategic planning; the rise of Mustafa Kemal in Turkey and revolts in Palestine, Egypt, and Iraq quickly proved decision makers wrong.

The ineluctable imposition of Europe-based territorial sovereignty on the Middle East participated as well in fashioning the region. The demarcation of boundaries and the invention of new states gave an ex post facto coherence to decades of loosely connected attempts to grasp the region through accounts, maps, and statistics. The accumulation of knowledge was normalized and even naturalized into the semblance of a system. These new classifications, imagined spaces, and prospective maps profoundly challenged traditional views on places. They participated in unleashing forces that transformed the way local populations and powers made sense of their own position. By 1921, Palestine was not a metaphorical space anymore. From Syria, Armenia, Mesopotamia, and Egypt, the lines on the maps of the peace conferences by Western powers had been contradicted by a tidal wave of arguments, at once cartographic, historical, and ethnographic. The very rhetoric that had supported Britain's imperial involvement in the area was turning against its creator, precisely when mapping and mapmaking had seemingly started to ordain the Orient. The Middle East's own modernity, one that was consistently overlooked by most surveyors, mapmakers, and experts, would soon contravene London's interests.

GENERAL CONCLUSION

This has been a book about the fashioning of the Middle East in a British imperial context. Maps played a crucial role in its development and articulation, and we have explored the various different ways in which the mapping process was instrumental in the elaboration of a mental space which then went on to play a significant role in the geopolitics of a region which is now at the heart of current global strategic preoccupations. The book's title might appear to suggest that there was even a causal link between the British framing of a Middle East and the currently fraught Middle Eastern situation of today. Nothing, however, could be further from the truth. This has been a book which has not been about the determining of guilt, but the delineation of context. That context has been an all-important point at issue throughout its pages. It has turned out to be pluralist—involving many different participants, geographical locations, and scientific disciplines. The process was emphatically not a case of the majestic imposition of an imperial vision. It was as much bottom-up as top-down, rooted in local expertise and engagement as bureaucratic official practices and procedures. It was as much driven by private initiatives, and by individual interests and ambitions, as by grand strategy and overall political purpose.

By choosing to focus on the elaboration of a British notion of the Middle East, we might be in danger of ignoring the important non-British component in that pluralist context. Was it not the case that the French played as instrumental a part as the British in the wake of the First World War in the consolidation of a notion of the Middle East? Did not the Turkish state itself recycle the categories and rhetoric of Western 19th-century geographies and thereby mediate a notion of the Middle East more broadly to the 20th century? That this was manifestly the case underscores an important general point about the adoption and reutilization of the Middle East by local players. For Western mapping was often appropriated by so-called Orientals in order to foster their own nation-building endeavors. The mapping process accounts for the existence of an ongoing collaboration between the explored and the explorer. At first sight, cartography might seem to be the perfect embodiment of Western dominance over the East. In reality, however, it was seldom the outcome of unilaterally imposed cultural monoliths. British imperial expansion, in its cultural as in its political dimension, was a game involving a large number of players. Local knowledge pervaded the British cartographic and geographical ordering of the East. That knowledge was acquired in conditions typified as much by contestation as by collaboration. Indigenous resistance characterized this asymmetrical coproduction of information as much as local collusion. The few surviving "countermaps" are a testimony to that reality. The patterns of such collaboration and confrontation were inscribed in preexisting power relationships. For the British and French were by no means the first to mobilize the mapping process to subdue local populations. Ottoman, Egyptian, and Persian overlordship predated the arrival of Europeans in the area. British officers engaged in surveying often attested to how local inhabitants conspicuously compared their activities and techniques with those of Ottoman surveyors and their instruments. Specific local groups (the Assyro-Chaldeans are a notable example) instrumentalized British surveying agents in their long-standing distrust of Ottoman rule.[1] The process of British mapping in the East was not created ex nihilo. Nor was it working with a clean sheet.

The processes of mapping the Middle East that have been analyzed in this book had a major impact upon it. They fundamentally "dislocated" the region. That dislocation needs to be understood in terms of a profound disruption, and not simply as an exercise in territorial reorganization. This study of British maps reflects how entire areas were transposed from their traditional position, first culturally, but then physically, by the impact of the ever more intense grasp of European knowledge. The most outward and visible sign of Britain's mapping and surveying enterprise in the East in the 19th century was border delimitation. It fabricated frontiers which were, and still are, fundamentally extraneous to some of the people living in the region. Calling that region the "Middle East" was simply the most evident large-scale manifestation of that dislocation. When British mapmakers designated the area a "Middle East," were they not creating a simulacrum (to use the terminology of the French sociologist Jean Baudrillard) summoning into existence something which had no prior existential reality?[2] In that sense, was not this "Middle East" hyper-real, never a real place, never acknowledged as such by its inhabitants? What this book has shown is that the region was mapped and delineated as such, well before that label of "Middle East" became a common frame of reference for policy makers. More important still, the phrase itself was never really adopted by local people at all. True, the phrase in its Arabic translation serves as the title of the popular newspaper . . . but this was founded in London as recently as 1978. The "Middle East" that we have documented in the works of British experts in a variety of domains in the early years of the 20th century was, in short, nothing less than a chimera.

It was a chimera whose existence, however, was the result of a constant interplay between extraneous conceptualizations and indigenous convictions. Post-1918 pan-Arabism was ineluctably intertwined with British policies that were framed within this geographical conception. Local responses to boundary making, place-naming, and geophysical surveying in turn changed their own notions of what was "real" in the world as they knew it around them. Indigenous geographies, in short, could not ignore the encroachments of British and French intellectual conceptualization, any more than European knowledge could afford to ignore local inputs. The invention of the "Middle East" is therefore testimony to something more substantial, subtle and pervasive than a mere chimera. It is the integral product of a dialectical movement between representation and reality. An innovative geographical delineation, it undoubtedly was incapable of representing the characteristics of a region that was fundamentally pluralistic. In that sense, it had to remain a "mirage," incapable of representing the reality which it purported to encompass.[3] Yet it became an instrument in a wider process of disciplining that space and, therefore, shaped what happened to that space. The distinction between Syria and Palestine, two overlapping terms in the mid-19th century, the invention of the "Balkans," and the unification of the lands between the Mediterranean and India under one single denomination are just three linked examples of one and the same phenomenon. That phenomenon is the dialectic between cultural evolutions and geopolitical realities which transformed the "Near and Middle East" in this period. A similar dialectic was at work in the overlapping scientific, military, and political agendas of most British surveyors in the area. It is of immense significance that this dialectic process did not come to an end with the collapse of the British Empire. The late 20th century inherited divisions in Asia that were the legacy of the British imperial vocabulary. Those who first coined the phrase "the Middle East" would most likely be astounded at what their creation had turned into, and not least by that turn of events in which the Cold War re-created the rationale for a transitional East between the Communist world and "the West."

I have demonstrated that the postwar Middle East was not an "empty map," as often suggested. Multiple layers of geographical assumptions, some of them to be traced back to the mid-19th century, shaped what the region was to be in the eyes of both decision makers and the general public. Post-1918 comprehensions of the area, as institutionalized in Churchill's Middle East Department, were built on a substantial Victorian and Edwardian legacy, subsumed into modern modes of imperial governance. Mapmaking and surveys testify to the existence of these differing facets of British imaginations. Their analysis reveals the conflicting cultural representations, coupled with the varying levels of interest in time, space, and social groups. Imperial centers in India, Cairo, Tehran, and London competed in drawing the contours of British views on the Middle East. The geographical and topographical material I have examined in this book confirms the dysfunctional and sometimes improvised nature of British control in the region. The imperial structure, like other large colonial edifices, was a disparate projection of a multitude of visions, ideas, desires, careers, and hopes whose aggregate never translated into a coherent entity except in the speeches and books of the imperialists and their opponents. The empire of the Raj in the Persian Gulf had, until the early 20th century, its own tempo and logic. That, in turn, differed from that of the British administration in Egypt, which developed its own independent outlook. The postwar Middle East was a creature of the decentralized empire of the 19th century. It reveals the ambiguity in the imperial project. The diversity of Orients devised from within the British sphere was not only a matter of intraimperial contradictions. It was also the expression of the complexities of a rapidly evolving society, in which imperial propaganda, enhanced geographical literacy, and the emergence of a mass reading public all played their part in understanding how maps came to play a central role in empire. This work has striven to demonstrate how the study of mapmaking has crucially to be situated with the social history of a culture. Britain's figuration of the Middle East was dependent on numerous strands that were socially diverse and rooted in its own cultural evolutions. It cannot, therefore, be properly understood without taking into account the market for maps, the formative influence of school cartography, and the metropolitan observer's access to geographical information

While the British geographical imagination was fundamentally plural, it was shaped by a convergent culture of primacy, which in its turn stemmed from a wide-ranging commercial, financial, and geopolitical supremacy. Apart from a few exceptions, British agents exploring and mapping the East looked at the Orient from what they believed to be a civilizational higher ground. The impression thus given by maps can be mistaken. Their imperial gaze belonged in the domain of wishful thinking. So the creation of the British East was not merely about a profound "dislocation" of a region. It was also about an equally deep-rooted imperial "mislocation." The agents at work in the creation of the British Middle East were at a loss, anxious that their surveys and maps were inaccurate, subject to internal doubts about the extent to which they were orchestrating their own illusions, rather than representing the real world around them. The gap between the convictions of domination harbored by British agents and their actual ability to influence the course of events was wide. Campaigns in the East during World War I give ample evidence that the British "imperial gaze" was at best through a glass darkly. The creation of the East by the West was also a facade legitimizing the careers and adventures of the numerous Eastern travelers of the imperial age. What the dissection of the cartographic understanding of the Middle East underscores is that the projected objects repeatedly escaped the grip of the surveyors, geographers, and cartographers.

Early 20th-century British mapping of the Orient was, measured in technical terms and in terms of the geographical accretion of knowledge, a success nonetheless. By 1918, most blanks on the maps of the area had been filled. They were legitimate instruments through which to devise what seemed rational policies. The massive increase in map acquisitions in Whitehall during and after the war testifies to this. Despite the mistrust on the part of some of the main British leaders, cartography, along with statistics, became a crucial cognitive accessory of international statecraft with the peace settlements. This had two consequences. First, it gave 19th-century British topographies of the East an authority they were not supposed to have in the first place. Sometimes, old British maps even resurfaced to serve the interests of unexpected parties, as did an 1886 WO map of the Persian Gulf which was used by Iran in the early 1990s to prove that Abu Musa and the Tunbs islands were Iranian. Second, the traditional instruments used in 19th-century Britain to become familiar with the region—some of the most significant being the travelogues or exotic engravings of the period—came to be replaced by new mediation tools. The era of adventure was drawing to an end after the war. Thanks to the journalistic map among other things, a geopoliticization and militarization of the gaze was taking hold, both in decision-making circles and in the general public. Various modes of scientific and graphic figurations superseded the written word of the travel account. Visions of oil and of clashing ethnic groups supplanted the romantic accounts of exploration. More accurate maps fostered illusions of control, as new visual and mathematical rhetoric was developed to describe the East.

In that regard, the growing autonomy and authority of cartographic information in the 19th and early 20th century is another crucial point. While most early 19th-century maps were illustrative of travelogues, accounts, and reports, they progressively transformed into autonomous discourses, sometimes central to the argument. The graphic visualization of space was an increasingly significant organizing principle of the British worldview. The relationship with places through cartography, travel imagery, and other forms of visual culture reshaped interpretative frameworks. That is true both for decision makers and for the reading public in general. The part played by the various dimensions of imperialism and informal influence was instrumental in this process. The elaboration of maps, from fieldwork, and surveys to data compilation and engraving, participated in the rationalization and legitimation of authority on a territory in an age of expansion of the British world system. The cartographic discourse was therefore a decisive element in the process of the strategic reconfiguration of Britain's role in the "Middle East." The geographical framing of the Eastern Question, rooted in 18th- and 19th-century assumptions, was replaced by a regional entity constructed by experts.

The invention of the Middle East is a test case by which we may understand how topography, cartography and geography gained epistemological credibility on the threshold of the 20th century. Graphic conceptualization of the area acquired a new meaning. The popularization and commodification of cartography as well as exploration in general developed a greater familiarity with the world in general, and the Middle East in particular. The somewhat aristocratic conception of the East slowly gave way to more democratic and accessible entity. The Middle East lost its specific and elite aura as it became more readily penetrable, both intellectually and physically. As a consequence, the possibility of Western intervention in the East was accompanied by its trivialization as a notion. More accurate, detailed, and sophisticated cartography, as well as the advent of other representational media capable of widespread mechanical reproduction, such as photography, reflects the wider decline of the historic Western distrust for visual representation. This is well illustrated by the shifting uses of maps by the political and military elite I analyzed in this work. George Villiers's mid-Victorian suspicions were obsolete by the 1900s.[4] To the early 20th-century eye, cartography was not an eidolon, a simulacrum, anymore, but a trusted instrument to shape reality.

A cultural understanding of British influence in the age of imperialism must therefore, necessarily, be rooted in its precise and distinct historical periods. That is because, as in the case of British images of the East, there were enormous changes over time. We have registered some of these in the course of this book—from the slow decline of the biblical perspective of the East to the emergence of commercial geographies and the consequential divergence from a traditionally located image of a stagnant Orient, for example. Such cultural shifts have to be measured within the appropriate context and chronological framework. From a cultural perspective, the conventional framework and chronology of the British Empire is too restrictive. There was a broader context and background to Britain's international outlook, of which the popular expressions of imperial concern as regards the Middle East were merely one inseparable part. This book opened with an analysis of the relative position of Arabia, Palestine, and Turkey, as documented in the pages of the journals of the RGS. That was a way of reinforcing the importance of his broader context when it comes to understanding the complex mental maps which accompanied the actors and agents in the British "age of imperialism."

There has long been a tendency to write the history of British imperialism within a comforting and unitary triumphalist narrative of exploration and progress from which strands of a more narrowly nationalist British historiography have drawn sustenance. The last decades witnessed deepening criticism of "the fiction of English/Anglo global imperial exceptionalism."[5] Christopher Bayly has underlined the dependence of British India's information order on local information. Disruption and the inconsistencies that separated professed ideals from brute facts on the ground are the focus on Mrinalini Sinha's study of colonial masculinities.[6] Antoinette Burton recently underlined the significance of "narratives of trouble" and native agency.[7] This book contributes to this ongoing discussion. It has sought to show that the prehistory of the idea of a British "Middle East" is rooted in confusion, ignorance, and misunderstanding. Its construction was the work of many agents, in diverse locations, operating with different agendas and aiming at widely divergent goals. Its far-reaching consequences, which have outlasted the British Empire and the contexts which gave it birth, have been to reify and instrumentalize geographical entities which, in turn, have served to complicate the geopolitical realities of the Middle East of the present day. This is, in short a "worried history," which makes no claims to an absolute truth, still less to offering immediate lessons for the present.[8] It does, however, aim to demonstrate that our contemporary and seemingly settled notions of place need to be denaturalized, for they are based on historically determined cultural constructs. In the light of the area's recent past, unsettling certainties is hopefully useful.

ACKNOWLEDGMENTS

This research has greatly benefited from the help of family, friends, and colleagues. Particular thanks go to Cornelius Crowley, Mark Greengrass, Pierre Simoni, Alice Byrne, Sandrine Parageau, Clémence Pioche, and Fabrice Bensimon for their help with the manuscript. Alain Pioche was kind enough to help me with statistical analysis. I should also like to thank John M. MacKenzie for his advice in the early stages of this project. Mary E. Laur, my editor, and the anonymous reviewers played an invaluable part in the creation of this book. I am also very grateful for Erik Carlson's help. Thanks also to Paris Ouest Nanterre la Défense University for its generous support. This book is dedicated to my family.

NOTES

INTRODUCTION

1. See Denis Cosgrove, *Geography and Vision: Seeing, Imagining and Representing the World* (London: I. B. Tauris, 2008).
2. Martin W. Lewis and Kären E. Wigen, *The Myth of Continents: A Critique of Metageography* (Berkeley: University of California Press, 1997).
3. On regional constructs see Anssi Paasi, "Place and Region: Looking through the Prism of Scale," *Progress in Human Geography* 28, no. 4 (2004).
4. Roger Adelson, *London and the Invention of the Middle East: Money, Power, and War, 1902–1922* (New Haven, CT: Yale University Press, 1994); Michael E. Bonine, *Is There a Middle East? The Evolution of a Geopolitical Concept* (Stanford, CA: Stanford University Press, 2012); Karen Culcasi, "Cartographic Constructions of the Middle East" (PhD diss., Syracuse University, 2008); Zachary Lockman, *Contending Visions of the Middle East: The History and Politics of Orientalism* (Cambridge: Cambridge University Press, 2004).
5. Marc Bloch, *Les Caractères Originaux de l'Histoire Rurale Française*, vol. 1 (Paris: Les Belles Lettres, 1931), introduction.
6. David N. Livingstone, *The Geographical Tradition: Episodes in the History of a Contested Enterprise* (Oxford: Blackwell, 1993); Matthew H. Edney, "Cartography without 'Progress': Reinterpreting the Nature and Historical Development of Mapmaking," in *The Map Reader* (John Wiley & Sons, 2011).
7. James Wyld, *Countries lying between Turkey & Birmah, Comprising Asia Minor, Persia, India,*

Egypt & Arabia and including the Black, Caspian & Red Seas, scale 1:7,375,000, 101 × 77 cm (London: J. Wyld, 1839). For a similar projection, see also J. Wyld, *The Persian gulf: Persia with the adjacent countries of Russia, India and Turkey,* 67 × 50 cm (London: J. Wyld, ca. 1850).

8. Jean-Baptiste Bourguignon d'Anville, *Première Partie de la Carte d'Asie,* engraved by Guillaume-Nicolas Delahaye, 79 × 82 cm, Bibliothèque Nationale de France, GE C-6137.
9. Lewis and Wigen, *The Myth of Continents,* chap. 2.
10. Christopher Hagerman, *Britain's Imperial Muse: The Classics, Imperialism, and the Indian Empire, 1784–1914* (Basingstoke: Palgrave Macmillan, 2013).
11. John Carne, *Recollections of Travels in the East: Forming a Continuation of the Letters from the East* (London: H. Colburn and R. Bentley, 1830).
12. Francis Bedford, *Photographic Pictures Made by Mr. Francis Bedford During the Tour in the East* (London: Day & Son, 1863).
13. Henry de Worms Pirbright, *England's Policy in the East* (London: Chapman and Hall, 1877).
14. Baker was one of the heroic figures of the Great Game, see Valentine Baker, *Clouds in the East: Travels and Adventures on the Perso-Turkoman Frontier* (London: Chatto and Windus, 1876).
15. George Curzon, *Problems of the Far East, Japan—Korea—China* (Westminster: A. Constable, 1894), XI.
16. Lewis and Wigen, *The Myth of Continents,* 57.
17. Thomas Scheffler, "'Fertile Crescent,' 'Orient,' 'Middle East': The Changing Mental Maps of Southwest Asia," *European Review of History* 10, no. 2 (2003).
18. Rob Kitchin, Justin Gleeson, and Martin Dodge, "Unfolding Mapping Practices: A New Epistemology for Cartography," *Transactions of the Institute of British Geographers* 38, no. 3 (2013).
19. Martin Dodge, Rob Kitchin, and C. R. Perkins, *Rethinking Maps: New Frontiers in Cartographic Theory,* Routledge Studies in Human Geography 28 (Routledge, London, 2009).
20. In this work, I will therefore use "British map of the East" and "British map of the Middle East" in the singular form to refer to the aggregate of polyphonic imaginations which participated in the elaboration of the area.
21. John Pickles, *A History of Spaces: Cartographic Reason, Mapping, and the Geo-coded World* (London: Routledge, 2004).
22. Derek Gregory, *Geographical Imaginations* (Cambridge, MA: Blackwell, 1994), 7; Edward W. Said, *Orientalism* (New York: Pantheon Books, 1978).
23. J. B. Harley, P. Laxton, and J. H. Andrews, *The New Nature of Maps: Essays in the History of Cartography* (Baltimore: Johns Hopkins University Press, 2002), 53.
24. Veronica Della Dora, "Performative Atlases: Memory, Materiality, and (Co-)Authorship," *Cartographica* 44, no. 4 (2009).
25. F. Driver, "Geography's Empire: Histories of Geographical Knowledge," *Environment and Planning D: Society and Space* 10, no. 1 (1992).
26. D. Graham Burnett, *Masters of All They Surveyed: Exploration, Geography, and a British El Dorado* (Chicago: University of Chicago Press, 2000); John J. Moscrop, *Measuring Jerusalem: The Palestine Exploration Fund and British Interests in the Holy Land* (London: Leicester University Press, 1999); Matthew H. Edney, *Mapping an Empire: The Geographical Construction of British India, 1765–1843* (Chicago: University of Chicago Press, 1997); Haim Goren, *Dead Sea Level: Science, Exploration and Imperial Interests in the Near East* (London: I. B. Tauris, 2010). On France, see Hélène Blais, *Mirages de la Carte l'Invention de l'Algérie Coloniale* (Paris: Fayard, 2014).
27. Driver, "Geography's Empire."
28. Kathleen Stewart Howe, "Mapping a Sacred Geography: Photographic Surveys by

the Royal Engineers in the Holy Land, 1864–1868," in *Picturing Place: Photography and the Geographical Imagination*, ed. Joan Schwartz and James Ryan (London: I. B. Tauris, 2003).

29. See the chronology established by Felix Driver, *Geography Militant: Cultures of Exploration and Empire* (London: Wiley, 2001); see also Matthew Edney, "The Irony of Imperial Mapping," in *The Imperial Map: Cartography and the Mastery of Empire*, ed. J. R. Akerman (Chicago: University of Chicago Press, 2009), 11–45; Morag Bell, Richard Butlin, and Michael Heffernan, *Geography and Imperialism, 1820–1940* (Manchester: Manchester University Press, 1995).
30. Edney, *Mapping an Empire*, 1.
31. John Darwin, *The Empire Project: The Rise and Fall of the British World-System, 1830–1970* (Cambridge: Cambridge University Press, 2009), 3.
32. This is how Herman Merivale described British dominance in Asia in 1871; see Bernard Porter, *The Absent-Minded Imperialists: Empire, Society, and Culture in Britain* (Oxford: Oxford University Press, 2004), 9.
33. For a more specific distinction between imperialism and colonialism, see C. Hall, *Cultures of Empire: Colonizers in Britain and the Empire in the Nineteenth and Twentieth Centuries; A Reader* (New York: Routledge, 2000), 5.
34. Antoinette M. Burton, *After the Imperial Turn: Thinking with and through the Nation* (Durham, NC: Duke University Press, 2003), 1–2.
35. J. M. MacKenzie, *Orientalism: History, Theory and the Arts* (Manchester: Manchester University Press, 1995); Andrew S. Thompson, *The Empire Strikes Back? The Impact of Imperialism on Britain from the Mid-Nineteenth Century* (Harlow: Pearson Longman, 2005).
36. C. A. Bayly, *Empire and Information: Intelligence Gathering and Social Communication in India, 1780–1870* (New Delhi: Cambridge University Press, 2007), 245–46.
37. For a similar approach, see M. F. Jacobs, *Imagining the Middle East: The Building of an American Foreign Policy, 1918–1967* (Chapel Hill: University of North Carolina Press, 2011).
38. On these "anomalies of Empire," see L. Benton, *A Search for Sovereignty: Law and Geography in European Empires, 1400–1900* (Cambridge: Cambridge University Press, 2009), 1–2.
39. Benedict Anderson, *Imagined Communities* (New York: Verso, 1982).
40. Hélène Blais, *Mirages de la Carte l'Invention de l'Algérie Coloniale* (Paris: Fayard, 2014).
41. See David N. Livingstone and Charles W. J. Withers, *Geographies of Nineteenth-Century Science* (Chicago: University of Chicago Press, 2011).
42. Pierre Bourdieu, *Homo Academicus* (Paris: Editions de Minuit, 1984).
43. The phrase "transitional East" is borrowed from Lewis and Wigen, *The Myth of Continents*.
44. Data from JSTOR, where both the *Journal* and the *Proceedings* of the RGS are available in a digitized version, were first exported to Excel. Each article was then located within the geographical categories defined by the RGS itself in its 1880 index in order not to impose anachronistic geographical divisions. I chose to refer to a wide "Middle East" in my statistical analyses. It comprises the whole Ottoman Empire, Persia, Egypt, and Sudan.
45. The graph is based on the maps listed in FO 925 (National Archives). Most descriptions for the Foreign Office map library are those provided by its map librarians when they listed the maps for their own purposes. The dating is also provided by the librarians of the 19th and early 20th centuries. It indicates when the documents were actually purchased or created for actual use, not their dates of creation. These two defects are actually assets from our point of view: they provide a direct insight into the way cartography became part of the decision-making process from the mid-19th century to the early 20th century.
46. Homi K. Bhabha, *The Location of Culture* (London: Routledge, 1994).
47. Pickles, *A History of Spaces*.

48. John Darwin, *Unfinished Empire: The Global Expansion of Britain* (London: Lane, 2012).
49. I will use the term "Anglo-Indian" to refer to British officials in India; see Margot Finn, "Anglo-Indian Lives in the Later Eighteenth and Early Nineteenth Centuries," *Journal for Eighteenth-Century Studies* 33, no. 1 (2010). For a similar use of the term "Anglo-Egyptian" see Lanver Mak, *The British in Egypt: Community, Crime and Crises, 1882–1922* (London: I. B. Tauris, 2011).
50. Adelson, *London and the Invention of the Middle East*.

CHAPTER 1

1. See also A. Crispin Jewett, *Intelligence Revealed: Maps, Plans and Views at Horse Guards and the War Office, 1800–1880* (London: British Library, 2011), app. 6. On Best Jervis's Indian career, see Edney, *Mapping an Empire*, 269–89.
2. The Ferro Prime Meridian was used in most European maps from the 17th century on before the Greenwich Meridian became the international standard in 1884.
3. GPS coordinates for the cathedral's bell tower are N 44 36.643, E 33 31.408. Thomas Best Jervis's longitude coordinates were E 33 31.3 according to the information on the ninth sheet of his map.
4. John Phillips, "Anniversary Address of the President," *Quarterly Journal of the Geological Society of London* 14 (1858): lix.
5. Thomas B. Jervis, *Military Topographical Map of the Krima Peninsula*, sheet 6, ca. 1:165 000 (London, 1855). Bibliothèque Nationale de France (BNF) (Cartes et Plans GE DD-5938 (A)).
6. Phillips, "Anniversary Address of the President," lix.
7. Stephen Wade, *Spies in the Empire: Victorian Military Intelligence* (London: Anthem Press, 2007), 25. See also Stephen M. Harris, *British Military Intelligence in the Crimean War, 1854–1856* (London: Frank Cass, 1999).
8. WO 1/382, February 23, 1856.
9. William H. Russell, *The War: From the Death of Lord Raglan to the Evacuation of the Crimea* (London: Routledge, 1856), 474.
10. Peter J. Jackson and Jennifer L. Siegel, *Intelligence and Statecraft: The Use and Limits of Intelligence in International Society* (Westport, CT: Praeger, 2005), 95.
11. See, for example, David George Hogarth, *The Penetration of Arabia: A Record of the Development of Western Knowledge Concerning the Arabian Peninsula* (London: Lawrence and Bullen, 1904).
12. Denis Wood, "The Fine Line between Mapping and Map Making," *Cartographica: The International Journal for Geographic Information and Geovisualization* 30, no. 4 (1993).
13. WO 78/1000/34; MPHH 1/5 (maps and plans extracted from records of the War Office, now held by the National Archives, Kew).
14. Except for Thomas Spratt's report ("An Inquiry into the soundness of M Lesseps," 1858), MPK 1/436/2 (maps and plans extracted from records of the Foreign Office, now held by the National Archives, Kew).
15. James R. Akerman, *Cartographies of Travel and Navigation* (Chicago: University of Chicago Press, 2006), 95.
16. Charles Dillon, *Beïrout Bay*, scale 1:1,250,000, 46 × 61 cm, Hydrographic Office, surveyed in 1842, published in 1844.
17. Dov Gavish, *The Survey of Palestine under the British Mandate, 1920–1948* (London: Taylor & Francis, 2004), 6–7. See *Beirut Bay with a view; Saïda with a view; Sûr with a view; Acre with a view; Haifa bay with a view; Yafa with a view*.

18. Andrew N. Porter, *The Oxford History of the British Empire*, vol. 3, *The Nineteenth Century* (Oxford: Oxford University Press, 1999), 113.
19. Edward Dunsterville, ed., *Admiralty Catalogue of Charts, Plans, Views, and Sailing Directions* (London: Eyre and Spottiswoode, 1868), 53.
20. See Edward Dunsterville, ed., *Admiralty Catalogue of Charts, Plans, Views, and Sailing Directions* (London: Eyre and Spottiswoode, 1860).
21. *Admiralty Catalogue of Charts, Plans, Views, and Sailing Directions* (London: Published by order of the Lords Commissioners of the Admiralty, 1866).
22. Roderick Murchison, "Address to the Royal Geographical Society of London," *Proceedings of the Royal Geographical Society of London* 2, no. 5 (1857): 265.
23. Jeremy Black, *British Diplomats and Diplomacy, 1688–1800* (Exeter: University of Exeter Press, 2001), 121–22.
24. James Brant, "Journey through a Part of Armenia and Asia Minor, in the Year 1835," *Journal of the Royal Geographical Society of London* 6 (1836).
25. See FO 925/3059 or MPK 1/482/6.
26. R. F. Thomson and H. K. Schomberg, "Journey through the Mountainous Districts North of the Elbúrz, and Ascent of Demavend, in Persia," *Proceedings of the Royal Geographical Society of London* 3, no. 1 (1858).
27. Robert A. Stafford, *Scientist of Empire: Sir Roderick Murchison, Scientific Exploration and Victorian Imperialism* (Cambridge: Cambridge University Press, 1989), 101–2.
28. Henry Poole, "Report of a Journey in Palestine," *Journal of the Royal Geographical Society of London* 26 (1856): 69.
29. Delimitation and demarcation were used indiscriminately before the late 19th century. By the 1890s, delimitation described the definition of the border on paper, demarcation the actual work on the ground.
30. See FO 366/295: Turco-Persian Boundary Commission: Accounts and Correspondence, National Archives.
31. See his manuscript at the Geological Society of London (GB 378 LDGSL/69). W. K. Loftus, "On the Geology of Portions of the Turko-Persian Frontier, and of the Districts Adjoining," *Quarterly Journal of the Geological Society* 11, no. 1–2 (1855).
32. Sabri Ates, *The Ottoman-Iranian Borderlands: Making a Boundary, 1843–1914* (New York: Cambridge University Press, 2013), 105.
33. Firoozeh Kashani-Sabet, *Frontier Fictions: Shaping the Iranian Nation, 1804–1946* (Princeton, NJ: Princeton University Press, 1999), 29.
34. Antony Anghie, *Imperialism, Sovereignty and the Making of International Law* (Cambridge: Cambridge University Press, 2007), chap. 2; Jennifer Pitts, "Boundaries of Victorian international law," in *Victorian Visions of Global Order: Empire and International Relations in Nineteenth-Century Political Thought*, ed. Duncan Bell (Cambridge: Cambridge University Press, 2007), 67–88.
35. Evelyn Ashley, *The Life and Correspondence of Henry John Temple*, 2 vols. (London: Bentley, 1879), 2:337–38.
36. Edney, *Mapping an Empire*.
37. Thomas R. Hughes, ed., *Historical and Other Information Connected with the Province of Oman, Muskat, Bahrein, and Other Places in the Persian Gulf* (Bombay: Printed for Government at the Education Society's Press, 1856).
38. Charles Cruttenden, "Journal of an excursion to Sanaa the capital of the Yemen," *Transactions of the Bombay Geographical Society* 26 (1839).
39. James Raymond Wellsted, *Travels in Arabia* (London: John Murray, 1838).
40. Robert Gavin, *Aden under British Rule, 1839–1967* (London: C. Hurst, 1975), 36–38.

41. Charles Rathbone Low, *History of the Indian Navy: 1613–1863* (London: R. Bentley and Son, 1877), 2:385.
42. Frederic J. Goldsmid, *Telegraph and Travel: A Narrative of the Formation and Development of Telegraphic Communication between England and India* (London: Macmillan, 1874).
43. Arnold Talbot Wilson, *The Persian Gulf: An Historical Sketch from the Earliest Times to the Beginning of the Twentieth Century* (Oxford: Clarendon Press, 1928), 278.
44. The equivalent of more than a million pounds in 2014.
45. Francis Rawdon Chesney, *The Expedition for the Survey of the Rivers Euphrates and Tigris, Carried on by Order of the British Government, in the Years 1835, 1836, and 1837* (London: Longman, Brown, Green & Longmans, 1850).
46. Clements R. Markham, *A Memoir on the Indian Surveys* (London: W. H. Allen, 1878), 48–49.
47. See Haim Goren, *Dead Sea Level*, chap. 9. The Lynch brothers were active promoters of British trade in the East. Thomas Kerr Lynch (1818–91) published a travel narrative in which he insisted how the "trade of Egypt has lately been developed." See T. K. Lynch, *A Visit to the Suez Canal* (London: Day, 1866), 31.
48. J. B. Bewsher, "On Part of Mesopotamia Contained between Sheriat-El-Beytha, on the Tigris, and Tel Ibrahim," *Journal of the Royal Geographical Society of London* 37 (1867).
49. James Outram, *Lieut.-General Sir James Outram's Persian Campaign in 1857* (London: Smith, Elder, 1860).
50. Guillemette Crouzet, "A Slave-Trade Jurisdiction: Attempt against the Slave Trade and the Making of a Space of Rights in the Arabo-Persian Gulf, the Indian Ocean and the Red Sea," in *Legal Histories of the British Empire: Laws, Engagements and Legacies*, ed. Shaunnagh Dorsett and John McLaren (Abingdon: Routledge, 2014), 213–35.
51. *Catalogue of the Books and Maps in the Library of the Geological Society of London* (London: R. and J. E. Taylor, 1846); *Supplemental Catalogue of the Books, Maps, Sections and Drawings in the Library of the Geological Society of London* (London: Taylor and Francis, 1856).
52. Felix Jones, "Topography of Nineveh, Illustrative of the Maps of the Chief Cities of Assyria; and the General Geography of the Country Intermediate between the Tigris and the Upper Zab," *Journal of the Royal Asiatic Society of Great Britain and Ireland* 15 (1855).
53. Driver, *Geography Militant*, 24–26.
54. Francis Herbert, "The Royal Geographical Society's Membership, the Map Trade, and Geographical Publishing in Britain 1830 to ca. 1930: An Introductory Essay with Listing of Some 250 Fellows in Related Professions," *Imago Mundi* 35, no. 1 (1983).
55. Stafford, *Scientist of Empire: Sir Roderick Murchison*, 105.
56. George Augustus Wallin, "Narrative of a Journey from Cairo to Medina and Mecca, by Suez, Arabá, Tawilá, Al-Jauf, Jubbé, Háil, and Nejd, in 1845," *Journal of the Royal Geographical Society of London* 24 (1854). W. G. Palgrave, "Observations Made in Central, Eastern, and Southern Arabia During a Journey through That Country in 1862 and 1863," *Journal of the Royal Geographical Society of London* 34 (1864). Lewis Pelly, "A Visit to the Wahabee Capital, Central Arabia," *Journal of the Royal Geographical Society of London* 35 (1865).
57. Driver, *Geography Militant*.
58. William Gifford Palgrave, *Narrative of a Year's Journey through Central and Eastern Arabia (1862–1863)* (London: Macmillan, 1865), 166.
59. William Makepeace Thackeray, *Vanity Fair: A Novel without a Hero* (London: Bradbury and Evans, 1848), 456–57. Ellipses in square brackets represent material that has been omitted from the quotation.
60. Benjamin Disraeli, *Tancred, or, the New Crusade* (London: Henry Colburn, 1847).
61. See Peter Cochran, *Byron and Orientalism* (Newcastle: Cambridge Scholars Press, 2006).

62. On the concept of romantic exoticism, see Maxime Rodinson, *La Fascination de l'Islam* (Paris: F. Maspero, 1980).
63. Richard F. Burton, "Narrative of a Trip to Harar," *Journal of the Royal Geographical Society of London* 25 (1855): 138: "Perhaps my adventures and a short description of a city hitherto unvisited by Europeans may not be unacceptable to a Society which, though essentially scientific, does not withhold encouragement from the pioneer of discovery, reduced by hard necessity to use nature's instruments—his eyes and ears."
64. Richard Francis Burton, *Personal Narrative of a Pilgrimage to El-Medinah and Meccah* (London: Longman, Brown, Green and Longmans, 1855), 1.
65. John MacGregor, *The Rob Roy on the Jordan, Nile, Red Sea & Gennesareth, &C.: A Canoe Cruise in Palestine and Egypt, and the Waters of Damascus* (London: J. Murray, 1869).
66. On British missionaries in Palestine, see Yaron Perry, *British Mission to the Jews in Nineteenth-Century Palestine* (London: Frank Cass Publishers, 2003). On missions in Mesopotamia and Kurdistan, see John Joseph, *The Modern Assyrians of the Middle East: Encounters with Western Christian Missions, Archaeologists, and Colonial Powers* (Leiden: Brill, 2000).
67. William Jowett, *Christian Researches in Syria and the Holy Land, in 1823 and 1824: In Furtherance of the Objects of the Church Missionary Society* (London: L. B. Seeley: J. Hatchard, 1825).
68. Geoffrey Roper, "George Percy Badger (1815–1888)," *British Journal of Middle Eastern Studies* 11, no. 2 (1984).
69. George Percy Badger, *A Visit to the Isthmus of Suez Canal Works* (London: Smith, Elder, 1862).
70. See George Percy Badger, *The Nestorians and Their Rituals: With the Narrative of a Mission to Mesopotamia and Coordistan in 1842–1844* (London: Masters, 1850).
71. In George Percy Badger, *History of the Imâms and Seyyids of 'Omân* (London: Hakluyt Society, 1871).
72. Cambridge University Library, Add.2909–10.
73. "Route survey" and "traverse survey" often had the same meaning in 19th-century technical literature. I will use these terms interchangeably.
74. Lewis Pelly, *Report No. 67 of 1863 by Pelly on the tribes, trade and resources of the Gulf littoral*, IOR/MSS Eur F 126/48.
75. See draft journal of Lewis Pelly's journey from Kuwait to Riyadh and back IOR/MSS Eur F126/57, in Qatar Digital Library, http://www.qdl.qa/pelly's-unprecedented-trip-riyadh, accessed February 24, 2015.
76. Pelly, "A Visit to the Wahabee Capital," 169–70.
77. Ibid., 171.
78. Henry Raper and FitzRoy Robert, "Hints to Travellers," *Journal of the Royal Geographical Society of London* 24 (1854).
79. Plane tabling is a surveying method that combines data collection and map drawing. Plane tables needs several accessories, such as a compass and drawing equipment.
80. Gérald Arboit, "Un Jésuite au Service de la France du Second Empire: Les Missions du Père Cohen dans le Désert de Syrie et d'Arabie," *Revue d'Histoire Ecclésiastique* 96, no. 1–2 (2010). See, for example, Cohen's letter to Schefer, August 31, 1861, f. 35–38, Turkish Ministry of Foreign Affairs Archives, Memoirs and Documents.
81. Palgrave, *Narrative of a Year's Journey through Central and Eastern Arabia*.
82. "Palgrave's Eastern and Central Arabia," *British Quarterly Review* 42 (July–October 1865): 329.
83. Hogarth, *The Penetration of Arabia*, 249.
84. Richard Francis Burton, *Narrative of a Pilgrimage to Meccah and Medinah* (London: W. Mullan, 1879), preface.
85. David George Hogarth, *The Life of Charles M. Doughty* (London: H. Milford, 1928), 151.

86. Charles F. Beautemps-Beaupré, Alexander Dalrymple, and Matthew Flinders, *An Introduction to the Practice of Nautical Surveying and the Construction of Sea-Charts* (London: R. H. Laurie, 1823).
87. John F. W. Herschel, *A Manual of Scientific Enquiry: Prepared for the Use of Her Majesty's Navy: And Adapted for Travellers in General* (London: J. Murray, 1849).
88. Jeremy Black, *The British Seaborne Empire* (New Haven, CT: Yale University Press, 2004).
89. Eitan Bar-Yosef, *The Holy Land in English Culture, 1799–1917: Palestine and the Question of Orientalism* (Oxford: Clarendon Press, Oxford University Press, 2005).
90. Low, *History of the Indian Navy*, 404.
91. Ibid., 409.
92. Ibid., 406.
93. Ordnance Survey, *Map of the Turco-Persian Frontier made by Russian and English Officers in the years from 1849 to 1855*, scale 1:253,440 (Southampton, 1873).
94. James Felix Jones, *Vestiges of Assyria: An Ichnographic Sketch of the remains of ancient Nineveh with the enceinte of the modern Mosul*, scale 1:12,150, 3 sheets (London: J. Walker, 1855), Bibliothèque Nationale de France, GE C-10631.
95. James Felix Jones, *Notes on the Topography of Nineveh and the Other Cities of Assyria: And on the General Geography of the Country between the Tigris and the Upper Zab* (London: Harrison, 1855), 66.
96. Low, *History of the Indian Navy*, 409.
97. Ibid.
98. See James Felix Jones, *Memoirs of Baghdad, Kurdistan and Turkish Arabia, 1857: Selections from the Records of the Bombay Government, No. Xliii.—New Series* (Farnham Common, Slough, UK: Archive Editions, 1998).
99. Charles Van De Velde, *Narrative of a Journey through Syria and Palestine in 1851 and 1852*, vol. 1 (London: W. Blackwood, 1854), 423–24.
100. Low, *History of the Indian Navy*, 419.
101. See the map of Mesopotamia by Collingwood in Bewsher, "On Part of Mesopotamia."
102. Low, *History of the Indian Navy*, 416.
103. Neil Safier, *Measuring the New World: Enlightenment Science and South America* (Chicago: University of Chicago Press, 2008).
104. John Irwine Whitty, *Proposed Water Supply and Sewerage for Jerusalem: With Description of Its Present State and Former Resources* (London: W. J. Johnson, 1863).
105. Charles W. Wilson, *Ordnance Survey of Jerusalem* (London: E. Stanford, 1864), 1.
106. Charles W. Wilson, H. Palmer, and James H. Spencer, *Ordnance Survey of the Peninsula of Sinai*, 3 vols. (London: Eyre and Spottiswoode, 1869).
107. Howe, "Mapping a Sacred Geography."
108. Kathleen Wilson, *A New Imperial History: Culture, Identity and Modernity in Britain and the Empire, 1660–1840* (Cambridge: Cambridge University Press, 2004), 172.
109. Moscrop, *Measuring Jerusalem*, 72.
110. William Thomson, *Sermons Preached in Lincoln's Inn Chapel* (London: John Murray, 1861), 128–29.
111. Shakespeare revived the myth in his *Henry IV*.
112. Arthur Penrhyn Stanley, *Sermons Preached before His Royal Highness the Prince of Wales: During His Tour in the East in the Spring of 1862* (London: Spottiswoode, 1862), 27.
113. Verses from "And did those feet in ancient time," in William Blake, *Milton: A Poem* (London: W. Blake, 1804), preface.
114. Rachel Hewitt, *Map of a Nation: A Biography of the Ordnance Survey* (London: Granta, 2010), 302.

115. See Vincent Lemire, *La Soif de Jérusalem: Essai d'Hydrohistoire (1840–1948)* (Paris: Publ. de la Sorbonne, 2011).
116. Bar-Yosef, *The Holy Land in English Culture*, chap. 3.
117. See Denis Woodward, *Five Centuries of Map Printing* (Chicago: University of Chicago Press, 1975).
118. Cyril C. Graham, "Explorations in the Desert East of the Haurán and in the Ancient Land of Bashan," *Journal of the Royal Geographical Society of London* 28 (1858).
119. See also Innes M. Keighren, Charles W. J. Withers, and Bill Bell, *Travels into Print: Exploration, Writing, and Publishing with John Murray, 1773–1859* (Chicago, University of Chicago Press, 2015), 161–66.
120. MacGregor, *The Rob Roy on the Jordan, Nile, Red Sea & Gennesareth*, 168.
121. Eliot Warburton, *The Crescent and the Cross: Or, Romance and Realities of Eastern Travel* (London: H. Colburn, 1845), 507–8.
122. Jean-Baptiste B. d'Anville and William Herbert, *A Geographical Illustration of the Map of India* (London: H. Gregory, 1759), 66. See also Lucy P. Chester, "The Mapping of Empire: French and British Cartographies of India in the Late-Eighteenth Century," *Portuguese Studies* 16 (2000): 257–58.
123. Henry B. Tristram, *The Land of Israel: A Journal of Travels in Palestine, Undertaken with Special References to Its Physical Character* (London: Society for Promoting Christian Knowledge, 1865), 275
124. David Buisseret, *Monarchs, Ministers and Maps: The Emergence of Cartography as a Tool of Government in Early Modern Europe* (Chicago: University of Chicago press, 1992).
125. John Gilbert, *The Coalition Ministry, 1854*, pencil, pen, and ink and wash, 1855, 17½ in. × 27 in., National Portrait Gallery, NPG 1125.
126. PRO (Domestic Records of the Public Record Office) 30/22/11D/7, Lord John Russell Papers, Clarendon to Russell, 7 May 1854.
127. WO 32/7291.
128. Anthony C. Cooke, *Catalogue of the Maps, Plans and Views in the Topographical and Statistical Depôt of the War Office* (London: Topographical and Statistical Depôt, 1859).
129. See Geraldine Beech, *Maps and Plans in the Public Record Office: Europe and Turkey* (London: Stationery Office, 1998).
130. Trelawney Saunders, *An Introduction to the Survey of Western Palestine: Its Waterways, Plains & Highlands* (London: R. Bentley and Son, 1881).
131. Trelawney Saunders, Claude R. Conder, and Horatio H. Kitchener, "Map of Western Palestine" (London: Stanford, 1882); Trelawney Saunders, "The Holy Land at Successive Periods" (London: John Murray, 1874).
132. Low, *History of the Indian Navy*, 425.
133. *A Catalogue of Maps of the British Possessions in India and Other Parts of Asia* (London: Allen, 1870), 1.
134. From the poem "The Solitude of Alexander Selkirk" (1782) by William Cowper (1731–1800).

CHAPTER 2

1. Lewis and Wigen, *The Myth of Continents*.
2. On critical toponymy, see Naftali Kadmon, *Toponymy: The Lore, Laws, and Language of Geographical Names* (New York: Vantage Press, 2000); Lawrence D. Berg and Jani Vuolteenaho, eds., *Critical Toponymies: The Contested Politics of Place Naming* (Farnham:

Ashgate Pub., 2009); Mark S. Monmonier, *From Squaw Tit to Whorehouse Meadow: How Maps Name, Claim, and Inflame* (Chicago: University of Chicago Press, 2006).
3. Carl Ritter, Heinrich Gage, and William Leonard Boegekamp, *Geographical Studies* (Boston: Gould and Lincoln, 1863), 332.
4. Warburton, *The Crescent and the Cross*; John Ruskin, *The Stones of Venice* (London: Dent, 1853), 111.
5. Bonine, *Is There a Middle East?*, 19. See H. Kiepert's map of the Near East in H. Kiepert, *Neuer Handatlas* (Berlin: Reimer, 1871), *Vorder-Asien*, map no. 28. Online version available at http://www.qdl.qa/en/archive/81055/vdc_100023321679.0x000002.
6. "Communications with the Far East," *Fraser's Magazine* 54 (1856).
7. "The Germanic Confederation and the Austrian Empire," *Quarterly Review* 84 (1849): 236
8. See Edward Stanford, *The Eastern question in Europe and Asia*, 42 × 68 cm (London: Stanford, 1886). Online version available at http://www.loc.gov/item/2013593034/.
9. See *The Eastern Question-Map of Turkey and the surrounding countries*, *Graphic,* July 1, 1876, 5. The journalistic map shows the Balkans and "Turkey in Asia."
10. Larry Wolff, *Inventing Eastern Europe: The Map of Civilization on the Mind of the* Enlightenment (Stanford, CA: Stanford University Press, 1994).
11. Maps of "Turkey in Europe" and "Turkey in Asia" in Alexander Keith Johnston, *The Royal Atlas of Modern Geography* (Edinburgh: William Blackwood, 1868), plates 1 and 2.
12. Anon., *A Gazetteer of the World, or, Dictionary of Geographical Knowledge*, 7 vols., vol. 5 (Edinburgh: Fullarton, 1856), 362.
13. The name Istanbul comes from the Greek *eis tín polin*, "to the city." English 19th-century writings used various spellings, such as "Stambool," "Istambol," or "Istanbul." In the 19th century, some city officials were called *Istanbul Efendisis*, "judges of Istanbul." However, the Ottoman government used both "Istanbul" and "Kostantiniyye" in its documents (see Alain Servantie, *Le Voyage à Istanbul: Byzance, Constantinople, Istanbul: Voyage à La Ville aux Mille et un Noms Du Moyen Age Au XXe Siècle* [Brussels: Editions Complexe, 2003]), 4. "Istanbul" was officially adopted as the only name for the capital only in 1930.
14. George Fisk, *A Pastor's Memorial of Egypt, the Red Sea, the Wilderness of Sin and Paran, Mount Sinai, Jerusalem, and Other Principal Localities of the Holy Land* (London: Seeley, 1844), 413.
15. John Arrowsmith, *The London Atlas of Universal Geography, Exhibiting the Physical and Political Divisions of the Various Countries of the World*, vol. 2 (London: J Arrowsmith, 1842), map no. 28.
16. "Vocabulary for the Smyrna Station," *United Service Magazine*, 1849, no. 3: 443.
17. See Robert S. Nelson, *Hagia Sophia, 1850–1950: Holy Wisdom Modern Monument* (Chicago: University of Chicago Press, 2004).
18. "Geographical Nomenclature," *British Educator*, 1856, 120–21.
19. Thomas Milner, *The Gallery of Geography: A Pictorial a Descriptive Tour of the World* (London: William Mackenzie, 1863), 586.
20. Francis Beaufort, *Karamania, or, a Brief Description of the South Coast of Asia-Minor and of the Remains of Antiquity* (London: R. Hunter, 1818), 1.
21. Stratford Canning, *The Eastern Question* (London: J. Murray, 1881), 11.
22. John Murray, *A Hand-Book for Travellers in the Ionian Islands, Greece, Turkey, Asia Minor, and Constantinople* (London: J. Murray, 1840).
23. *Handbook for Travellers in Greece* (London: John Murray, 1854). John Murray, *A Handbook for Travellers in Turkey* (London: J. Murray, 1854).
24. Edward Gibbon, *The History of the Decline and Fall of the Roman Empire*, vol. 1 (Basil: Tourneisen, 1787), 33.

25. Genesis 15:18–21.
26. Herodotus, *The History of Herodotus* (London: John Murray, 1858), 25n9.
27. See James Gilbert, *A Map of Palestine or the Holy Land, with Egypt and the Wilderness of Mount Sinai*, engraved by J. & C. Walker (London: s.n., 1849).
28. Arthur Penrhyn Stanley, *Sinaï and Palestine in Connection with Their History* (London: J. Murray, 1856), 5.
29. "General Index to the Proceedings of the Royal Geographical Society, 1855–1878," *Proceedings of the Royal Geographical Society of London* 22 (1877): vii.
30. See Yitzhak Gil-Har, "Egypt's North-Eastern Boundary in Sinai," *Middle Eastern Studies* 29, no. 1 (1993).
31. See Edward Hertslet, *The Map of Africa by Treaty*, vol. 1 (London: Harrison, 1894), 341–42.
32. A copy was recovered in 1934. The map was very inaccurate. See Gideon Biger, *The Boundaries of Modern Palestine, 1840–1947* (London: Routledge Curzon, 2004), 26.
33. Georg W. F. Hegel, *Lectures on the Philosophy of History* (London: H. G. Bohn, 1857).
34. See Heinrich Kiepert, *A Manual of Ancient Geography*, trans. George Augustin Macmillan (London: Macmillan, 1881), 32.
35. "The Races of the Old World," *Ecletic Magazine* 62 (1864): 188.
36. James Gilbert Rennell, *A Treatise on the Comparative Geography of Western Asia in Two Volumes with a Complete Index: Accompanied with an Atlas of Maps* (London: Printed for C. J. G. & F. Rivington, 1831), ix.
37. Walter McLeod, *A Class-Atlas of Physical Geography: Comprising 20 Maps and 10 Sections and Diagrams, with Notes on the Maps* (London: Longman, Brown, Green, Longmans & Roberts, 1856), map 20.
38. Alexander Keith Johnston, *A School Atlas of Physical Geography, Illustrating, in a Series of Original Designs, the Elementary Facts of Geology, Hydrology, Meteorology and Natural History* (Edinburgh: W. Blackwood and Sons, 1852), pl.17.
39. David Page, *Advanced Text-Book of Physical Geography* (Edinburgh: W. Blackwood and Sons, 1864), 240.
40. Edward Gibbon, *The History of the Decline and Fall of the Roman Empire*, vol. 5 (London: W. Strahan; T. Cadell, 1776–88), chap. 50; Edward Lane, *An Account of the Manners and Customs of the Modern Egyptians* (London: Knight, 1836), 312.
41. Wyld, *Map of the Countries lying between Turkey & Birmah*.
42. See the map compiled by James Wyld for James Barber, *The Overland Guide-Book; a Complete Vademecum for the Overland Traveller, to India Via Egypt* (London: W. H. Allen, 1850).
43. See Goldsmid, *Telegraph and Travel*, 321.
44. John Bartholomew, *The British Colonial Pocket Atlas, a Complete Series of Map Illustrating the Geography of the British Empire* (London: J. Walker, 1887), map no. 13.
45. See volume 8 on *South Western Asia* in Ernst G. Ravenstein's and Augustus H. Keane's translation of Elisée Reclus's works, Elisée Reclus, Ernst Georg Ravenstein, and Augustus H. Keane, *The Universal Geography* (London: J. S. Virtue, 1878).
46. David G. Hogarth, "Problems in Exploration, I: Western Asia," *Geographical Journal* 32, no. 6 (1908).
47. Walter McLeod, *The Geography of Palestine or the Holy Land: Including Phoenicia and Philistia; with a Description of the Towns and Places in Asia Minor Visited by the Apostles* (London: Longman, 1858), 2.
48. See Nadia Abu El-Haj, *Facts on the Ground: Archaeological Practice and Territorial Self-Fashioning in Israeli Society* (Chicago: University of Chicago Press, 2001), 82–90.
49. Edmund Evans, *Pictorial Map of Palestine*, wood engraving on paper (London: James Nisbet, [mid- to late 19th century]), Victoria and Albert Museum, access: E.945:48-1976.

50. Edwin James Aiken, *Scriptural Geography: Portraying the Holy Land* (London: I. B. Tauris, 2010), 19.
51. David Friedrich Strauss, *The Life of Jesus*, trans. George Eliot (London: Chapman, 1846), 1.
52. Jeremy Black, *Maps and History: Constructing Images of the Past* (New Haven, CT: Yale University Press, 2000), 33.
53. See chap. 1.
54. George Armstrong, Charles W. Wilson, and Claude R. Conder, *Names and Places in the Old and New Testament and Apocrypha: With Their Modern Identifications* (London: A. P. Watt, 1889).
55. Edward Henry Palmer, *The Desert of the Exodus* (Cambridge: Deighton, Bell, 1871), 15.
56. Sophie Perthus and Jutta Faehndrich, "Visualizing the Map-Making Process: Studying 19th Century Holy Land Cartography with Mapanalyst," *e-Perimetron* 8, no. 2 (2013).
57. Charles Warren, "On the Reconnaissance of a New or Partially Known Country," *Proceedings of the Royal Geographical Society of London* 19, no. 2 (1874): 168.
58. Palmer, *The Desert of the Exodus*, 11.
59. Henry Laurens, *La Question De Palestine*, vol. 1 (Paris: Fayard, 1999), 71–74.
60. See chap. 6 on Percy Molesworth Sykes's references.
61. Edward Weller, *Kingdoms of the Successors of Alexander: After the Battle of Ipsus, B.C. 301*, 1:17,000,000 (London: Weller, ca 1860). Available online at http://www.loc.gov/item/2013593022/.
62. Raphael Samuel, *Theatres of Memory*, vol. 1 (London: Verso, 1994).
63. Johann Büssow, *Hamidian Palestine Politics and Society in the District of Jerusalem 1872–1908* (Leiden: Brill, 2011), 52–53.
64. Ian Manners and Pinar Emiralioğlu, *European Cartographers and the Ottoman World, 1500–1750: Maps from the Collection of O. J. Sopranos* (Chicago: Oriental Institute of the University of Chicago, 2010), 52.

CHAPTER 3

1. John W. Dower, *A School Atlas of Modern Geography: Containing Forty Maps, Prepared from the Best Authorities and Including the Latest Discoveries with a Copius Index* (London: Ward & Lock, 1854).
2. Heinrich Kiepert, *General Karte des Türkischen Reiches in Europa und Asien nebst Ungarn, Südrussland, den Kaukasischen Ländern und West-Persien*, scale 1:3,000,000, 88 × 123 cm (Berlin: Kiepert, 1855).
3. Stefanie Markovits, *The Crimean War in the British Imagination* (Cambridge: Cambridge University Press, 2013).
4. Karl Marx, *The Eastern Question, a Reprint of Letters Written 1853–1856 Dealing with the Events of the Crimean War* (London: S. Sonnenschein 1897), 25.
5. See *Illustrated London News*, August 14, 1855; *Catalogue of the Oriental & Turkish Museum, St. George's Gallery, Hyde Park Corner, Piccadilly* (London: W. J. Golbourn, 1854).
6. Ulrich Keller, *The Ultimate Spectacle: A Visual History of the Crimean War* (Amsterdam: Gordon and Breach, 2001), 64; Albert Smith, *Hand-Book to the Grand Moving Diorama of Constantinople* (London: s.n., 1854).
7. Andrew Maclure (lithographer) and Edward Stanford, *Stanford's Bird's-eye view of the Seat of War in the Crimea*, scale not given, 56.8 × 76 cm (London: Stanford, ca. 1854), National Maritime Museum, Greenwich, London, accession: PAH6343.

8. James Wyld, *The Crimea by James Wyld*, ca. 1:500,000, 68 cm × 50.5 cm 2nd ed. (London, 1854), National Maritime Museum, Greenwich, London, accession G297:59/7.
9. "Consul Petherick's Expedition from Khartúm on the Nile, to Cooperate with That under Captains Speke and Grant from Zanzibar on the East Coast: Subscription List, 1861," *Proceedings of the Royal Geographical Society of London* 5, no. 1 (1860).
10. Bartholomew Archives, National Library of Scotland (NLS): Acc.10222/Reference Maps/28.
11. William Chambers, "Publishing and Its Allied Trades in Edinburgh," *Transactions of the National Association for the Promotion of Social Science*, 1864, 882–83.
12. Walter W. Ristow, "Lithography and Maps, 1796–1850," in Woodward, *Five Centuries of Map Printing*, 77.
13. In the intaglio process, the image is incised on a copper or zinc plate.
14. Chromolithography allowed for color prints. A separate plate was engraved for each color.
15. Henry James, *Photo-Zincography* (Southampton: Printed by Forbes & Bennett, High Street, 1860). The real inventors were Captain Alexander de Courcy Scott and Lance-Corporal Rider working under him.
16. Karen Severud Cook, "The Historical Role of Photomechanical Techniques in Map Production," *Cartography and Geographic Information Science* 29, no. 3 (2002).
17. "Notices of books," *Bookseller*, 1861, 339.
18. A digitized copy Faden's original print is available online at David Rumsey Historical Map Collection (www.davidrumsey.com, no. 2104.040).
19. John M. Duncan, *Report of the Trial at the Instance of the Atlas Company of Scotland against A. Fullarton and Company, Publishers, Edinburgh and London Tried before a Jury at Edinburgh on the 27th, 28th, and 29th Days of July 1853* (Edinburgh: Blackwood, 1853).
20. Bartholomew Archives, NLS, Acc.10222/Business Record/12.
21. Édouard Taitbout de Marigny, *Pilote de la Mer Noire et de la Mer d'Azov* (Constantinople: A. Koromila and P. Caspalli, 1850), preface; James Wyld, *Geographical & Hydrographical Notes to Accompany Mr. Wyld's Maps of the Ottoman Empire and the Black Sea* (London: James Wyld, 1854).
22. Lynda Nead, *Victorian Babylon: People, Streets and Images in Nineteenth-Century London* (New Haven, CT: Yale University Press, 2000), 26.
23. Edward Stanford, *Stanford's Library Map of Asia*. 1:6,960,600, 89.8 × 156.8 cm (Edinburgh, A. K. Johnston: 1864).
24. B. R. Davies and C. Walker, *The Harrow Atlas of Classical Geography* (London: Stanford, 1857); SDUK, *Maps of the Society for the Diffusion of Useful Knowledge* (London: Chapman and Hall, 1844); *The Cyclopaedian: Or Atlas of General Maps* (London: Edward Stanford, 1857).
25. George B. Greenough, "Address to the Royal Geographical Society of London," *Journal of the Royal Geographical Society of London* 11 (1841).
26. Clare Pettitt, *Patent Inventions: Intellectual Property and the Victorian Novel* (Oxford: Oxford University Press, 2004), 46.
27. James R. Planché, *Mr. Buckstone's Voyage Round the Globe (in Leicester Square): A New and Original Cosmographical, Visionary Extravaganza, and Dramatic Review, in One Act and Four Quarters* (London: Thomas Hailes Lacy, 1854).
28. See Denis Cosgrove, *Apollo's Eye: A Cartographic Genealogy of Earth in the Western Imagination* (Baltimore: Johns Hopkins University Press, 2001), chap. 8.
29. James Wyld, *Notes to Accompany Mr. Wyld's Model of the Earth* (London: J. Wyld, 1851), 125.

30. Ibid., 72.
31. Richard D. Altick, *The Shows of London* (Cambridge, MA: Belknap Press, 1978).
32. Eliza Cook, "A Voyage Round the Globe," *Eliza Cook's Journal* 5 (1851).
33. Bar-Yosef, *The Holy Land in English Culture*.
34. John M. MacKenzie, *Popular Imperialism and the Military: 1850–1950* (Manchester: Manchester University Press, 1992), 186–87.
35. *Catalogue of the Educational Division of the South Kensington Museum* (London: Spottiswoode, 1867), 325.
36. Henry Mayhew and George Cruikshank, *The World's Show, 1851, or, the Adventures of Mr. And Mrs. Sandboys, Their Son and Daughter, Who Came up to London to Enjoy Themselves and to See the Great Exhibition* (New York: Stringer and Townsend, 1851), 132.
37. Henry Evans Smith, *The Crystal Palace Game, A Voyage Around The World*, ca. 1:85,000,000, 45.5 × 64.7 cm (London, 1851). Online version available at http://catalogue.nla.gov.au/Record/2600611.
38. Richard D. Altick, *The English Common Reader: A Social History of the Mass Reading Public, 1800–1900* (Chicago: University of Chicago Press, 1976).
39. See how John Pearman, a policeman, used *Cassel's Popular Educator* geography sections in Carolyn Steedman and John Pearman, *The Radical Soldier's Tale: John Pearman, 1819–1908* (London: Routledge, 1988), 267.
40. Michael Heffernan, "The Cartography of the Fourth Estate: Mapping the New Imperialism in British and French Newspapers, 1875–1925," in Akerman, *The Imperial Map*, 261–300.
41. *Bird's-Eye view of the Crimean peninsula*, Illustrated Times, July 31, 1855, 164–65.
42. *Chelmsford Chronicle*, January 12, 1855, 5.
43. Bedford Pim, "Remarks on the Isthmus of Suez, with Special Reference to the Proposed Canal," *Proceedings of the Royal Geographical Society of London* 3, no. 4 (1858).
44. See chap. 1.
45. The same was true in the United States; see Mark S. Monmonier, *Maps with the News: The Development of American Journalistic Cartography* (Chicago: University of Chicago Press, 1989).
46. "Newspaper Maps and Maps of Newspapers," *Chamber's Journal* 16 (1861): 390–92.
47. Edward Weller, *Map of Syria Showing all the Ancient Sites hitherto Identified in Palestine*, ca. 1:800,000, 43 × 31 cm (London: 1858).
48. Online version available at http://www.wdl.org/en/item/12885/.
49. Jonathan Rose, *The Intellectual Life of the British Working Classes* (New Haven, CT: Yale University Press, 2001), 341.
50. Thomas Huxley, "A Liberal Education and where to find it," *Every Saturday*, April 11, 1868, 468.
51. Bristol Record Office, J. T. Francombe's notebook, 1859, accession: 40477/F/2.
52. Walter McLeod, *The Geography of Palestine or the Holy Land*.
53. Colin Shrosbree, *Public Schools and Private Education: The Clarendon Commission, 1861–64, and the Public Schools Acts* (Manchester: Manchester University Press, 1988); Christopher Stray, *The Clarendon Report: English Public Schools in the Nineteenth Century* (Bristol: Thoemmes, 2004).
54. House of Commons Debates (hereafter HC Deb), May 6, 1864, vol. 175, col. 115.
55. Aaron Arrowsmith, *A Compendium of Ancient and Modern Geography, for the Use of Eton School* (London: E. Williams, 1839).
56. Aaron Arrowsmith and C. G. Nicolay, *A Compendium of Ancient and Modern Geography, for the Use of Eton School* (London: E. P. Williams, 1856).

57. See Rex Walford, *Geography in British Schools, 1850–2000: Making a World of Difference* (London: Woburn Press, 2001), 43.
58. *Minutes of the Committee of Council on Education*, vol. 2 (London: Stationery Office, 1850), 74.
59. *Minutes of the Committee of Council on Education, with Appendices* (London: Stationery Office, 1852), 277.
60. Ibid., 280–81.
61. Ibid., 215.
62. Wiltshire and Swindon Archives, accession: 1585/146.
63. Daniel Varisco, "When Did the Holy Land Stop Being Holy? Surveying the Middle East as Sacred Geography," in Bonine, *Is There a Middle East?*, 119–38.
64. *Minutes of the Committee of Council on Education*, vol. 2 (1850), 754.
65. *Report, for the Year 1864, by Her Majesty's Inspector, Charles E. Wilson, on the Free Church Normal School, Glasgow,* in *Report for the Committee of Council on Education, 1864–1865* (London: Eyre and Spottiswoode, 1865), 428.
66. *Minutes of the Committee of Council on Education Correspondence, Financial Statements, Etc., and Reports by Her Majesty's Inspectors of Schools* (London: The Committee, 1847), 100.
67. British and Society Foreign School, *A Hand-Book to the Borough Road Schools: Explanatory of the Methods of Instruction Adopted by the British and Foreign School Society* (London: Printed for the Society, 1854), 79–81.
68. George Henry Taylor, *Notes and Lessons on the Geography and History of Palestine* (London: Longman, 1851), xi.
69. See the outline map of the Holy Land in ibid., 5.
70. James Cornwell, *Map Book for Beginners* (London: Simpkin, Marshall, 1861).
71. Rose, *The Intellectual Life of the British Working Classes*, 350–52.
72. *Reports of the Commissioners of Inquiry into the State of Education in Wales* (London: W. Clowes and Sons, 1848), 457.
73. Ibid., 9.
74. British and Foreign School Society, "Annual Meeting," *The Educational Record, with the Proceedings at Large of the British and Foreign School Society* 5 (1863), 178.
75. Joyce Goodman, Gary McCulloch, and William Richardson, *Social Change in the History of British Education* (London: Routledge, 2008), 78.
76. *Report of the Commissioners*, vol. 7 (London: Eyre et Spottiswoode, 1868), 509.
77. Ibid., 436.
78. Richard Hiley, *Progressive Geography, Adapted to the Junior Classes in Classical and Commercial Schools* (London: Longman, 1843). The book went through several editions in the 1850s and 1860s.
79. Ibid., 108.
80. Francis Sitwell, *Four Centuries of Special Geography* (Vancouver: University of British Columbia Press, 1993), 18.
81. *Minutes of the Committee of Council on Education, with Appendices*, 305.
82. Jane McDermid, *The Schooling of Girls in Britain and Ireland, 1800–1900* (New York: Routledge, 2012), 117.
83. *Minutes of the Committee of Council on Education Correspondence*, 99.
84. McDermid, *The Schooling of Girls in Britain and Ireland, 1800–1900*, 117.
85. Henry White, *Guide to the Civil Service Examinations* (Westminster: P. S. King, 1856), 61.
86. For this categorization see Susan Schulten, *Mapping the Nation: History and Cartography in Nineteenth-Century America* (Chicago: University of Chicago Press, 2012).

CHAPTER 4

1. Andrekos Varnava, *British Imperialism in Cyprus, 1878–1915: The Inconsequential Possession* (Manchester: Manchester University Press, 2009).
2. James Rye and Horace G. Groser, *Kitchener in His Own Words* (London: T. F. Unwin, 1917), 52.
3. MFQ 1/862/5, Admiralty chart of the Mediterranean Archipelago (July 1877) illustrating a report by Sir Lintorn Simmons (MFQ is a series including maps and plans from records of various departments held at the Public Record Office, now held by the National Archives, Kew).
4. FO 358/2, *Simmons Papers*, Papers and memoranda on Eastern Affairs (1876–80).
5. Dwight E. Lee, "A Memorandum Concerning Cyprus, 1878," *Journal of Modern History* 3, no. 2 (1931).
6. HC Deb, August 2, 1878, vol. 242, col. 1009.
7. Alice Denny, John Gallagher, and Ronald Robinson, *Africa and the Victorians* (London: Macmillan, 1961), 21.
8. On the consequences of this late 19th-century panoptic vision of the world for representation see Timothy Mitchell, *Colonizing Egypt* (Cambridge: Cambridge University Press, 1988), 7–8.
9. See Driver, *Geography Militant*; and Joseph Conrad, "Geography and Some Explorers," in *Last Essays* (New York: Doubleday, 1926), 1–31.
10. See Andreas Stylianou and Judith A. Stylianou, *The History of the Cartography of Cyprus* (Nicosia: Cyprus Research Centre, 1980); Rodney W. Shirley, *Kitchener's Survey of Cyprus 1878–1883: The First Full Triangulated Survey and Mapping of the Island* (Nicosia: Bank of Cyprus Cultural Foundation, 2001).
11. George Arthur, *Life of Lord Kitchener* (London: Macmillan, 1920), 40–41.
12. FO 881/4217, A. Gordon, *Annual Cost of Force of Police and Pioneers in Cyprus for the year 1880–81*, 1.
13. Arthur, *Life of Lord Kitchener*, 36.
14. Michael Given, "Maps, Fields, and Boundary Cairns: Demarcation and Resistance in Colonial Cyprus," *International Journal of Historical Archaeology* 6, no. 1 (2002): 24.
15. James C. Scott, *Weapons of the Weak: Everyday Forms of Peasant Resistance* (New Haven, CT: Yale University Press, 1985).
16. Thomas A. Hull, "The Unsurveyed World, 1874," *Journal of Royal United Services Institution Journal* 19, no. 80 (1875): 70.
17. Quoted in Ronald Hyam, *Understanding the British Empire* (Cambridge: Cambridge University Press, 2010), 75.
18. Thomas G. Otte, *The Foreign Office Mind: The Making of British Foreign Policy, 1865–1914* (Cambridge: Cambridge University Press, 2011), 92.
19. Ray Jones, *The Nineteenth-Century Foreign Office: An Administrative History* (London: Weidenfeld and Nicolson, 1971), 63.
20. See FO 95/505, *Establishment of the Foreign Office*, May 10, 1889.
21. Zara S. Steiner, *The Foreign Office and Foreign Policy, 1898–1914* (Cambridge: Cambridge University Press, 1969), 8; Adolphus William Ward and G. P. Gooch, *The Cambridge History of British Foreign Policy 1783–1919 vol.3* (Cambridge: Cambridge University Press, 1923), 610.
22. Edward Hertslet, *The Map of Europe by Treaty; Showing the Various Political and Territorial Changes Which Have Taken Place since the General Peace of 1814* (London: Butterworths, 1875); Hertslet, *The Map of Africa by Treaty*, vol. 1 (London: Harrison, 1894).

23. FO 881/2432, *Memorandum on the Turkish Claim to Sovereignty over the Eastern shores of Arabia*, March 1874.
24. Zara Steiner, "On Writing International History: Chaps, Maps and Much More," *International Affairs (Royal Institute of International Affairs 1944–)* 73, no. 3 (1997).
25. Robert W. Seton-Watson, *Disraeli, Gladstone, and the Eastern Question: A Study in Diplomacy and Party Politics* (London: Macmillan, 1935), 454.
26. Russian General Staff, Wilson, C.W., earlier than 1875, Russian military map of Persia, in three sheets, map copied by Wilson with notes and indications in English, note attached: "My dear Mr. Hertslet, I forward you herewith for the Foreign Office, a copy of part of a Russian military map of Persia which was lent to me by Khanikof. The new information has been incorporated in major part of this map of Persia prepared for the Indian Office."
27. FO 881/5339, F. L. Bertie, *Relations with Tribes in vicinity of Aden and on South Coast of Arabia: Proposals of Government of India*, October 2, 1886.
28. Arnold P. Kaminsky, *The India Office, 1880–1910* (Westport, CT: Greenwood Press, 1985), 29.
29. India Office Records, X and Y series.
30. C. P. Lucas, *Introduction to a Historical Geography of the British Colonies* (Oxford: Clarendon Press, 1887), 120.
31. See CO 700, but there are very few references for the MO in this series.
32. Chewton Atchley, *Catalogue of the Maps, Plans and Charts in the Library of the Colonial Office* (London: s.n., 1910).
33. Charles I. Hamilton, *The Making of the Modern Admiralty: British Naval Policy-Making 1805–1927* (Cambridge: Cambridge University Press, 2011), 217–18.
34. Charles Golding Constable and Arthur William Stiffe, *The Persian Gulf Pilot* (London: Hydrographic Office, Admirality, 1883); Hydrographic Department, *The Gulf of Aden Pilot* (London: J. D. Potter, 1882).
35. FO 925/2853, *Persian Gulf, Approaches to Shatt al Arab and Bahmishir Rivers*, May 1891; MPI 1/717/1, *Shat-Al-Arab and the Bahmanshir River*, Marine Survey Office Bombay, 1888.
36. Hydrographic Department, *Sailing Directions for the Dardanelles, Sea of Marmora, and the Bosphorus* (1877).
37. *Catalogue of Admiralty Charts, Plans, and Sailing Directions* (London: Darling & Son, 1898), 61–64.
38. James Louis Hevia, *The Imperial Security State: British Colonial Knowledge and Empire-Building in Asia* (Cambridge: Cambridge University Pres, 2012), 55–59.
39. A. Crispin Jewitt, *Maps for Empire: The First 2000 Numbered War Office Maps 1881–1905* (London: British library, 1992).
40. See FO 925/2683, copy of map furnished by Russian Minister in Tehran to Persian Government for negotiation of Akhal-Khorassan boundary question, December 1881.
41. L/P&S/18/C69/6.
42. MFQ 1/869/1–3.
43. MPHH 1/114/2.
44. MFQ 1/869/4–5.
45. See Angela Burdett, *Iran in the Persian Gulf: 1820–1966*, vol. 1 (Slough: Archive ed., 2000), 856.
46. Charles Napier Robinson, *Celebrities of the Army* (London: G. Newnes, 1900), 58.
47. MFQ 1/28.
48. Otte, *The Foreign Office Mind*.
49. William S. Hamer, *The British Army; Civil-Military Relations, 1885–1905* (Oxford: Clarendon, 1970), 59.
50. K. G. Robertson, *British and American Approaches to Intelligence* (New York: St. Martin's Press, 1987), 260.

51. Jeremy Black, *The Power of Knowledge: How Information and Technology Made the Modern World* (New Haven, CT: Yale University Press, 2014), 309.
52. Moscrop, *Measuring Jerusalem*, 202.
53. Survey of Western Palestine: working papers relating to the Conder and Kitchener survey of Palestine, 1872–1877, British Library, Western Manuscripts, Add MS 69848: 1872–1877.
54. Claude R. Conder, *Tent Work in Palestine. A Record of Discovery and Adventure* (London: R. Bentley & Son, 1878), 96.
55. Ibid., 280.
56. *La Terre Sainte: Journal des Lieux Saints*, no. 63 (February 18, 1878): 653–54.
57. The preliminary maps are visible in the British Library Manuscript Collection (working papers relating to the Conder and Kitchener survey of Palestine, 1872–77, accession: Add MS 69848: 1872–1877).
58. As the name indicates, zincography consisted in etching a zinc plate using acid. It was particularly useful to print large-scale maps.
59. See the *Map of Western Palestine in 26 sheets from surveys conducted for the Committee of the Palestine Exploration Fund* (1880). Available online at http://nla.gov.au/nla.map-rm1949.
60. See *The Tyropœon Valley*, in Claude Conder and Charles Warren, *The Western Survey of Palestine* (London: PEF, 1884), 287.
61. Claude R. Conder, *The Survey of Western Palestine: Jerusalem* (London: Committee of the Palestine Exploration Fund, 1884), 286.
62. Howe, "Mapping a Sacred Geography."
63. William Mitchell Ramsay, *Impressions of Turkey During Twelve Years' Wanderings* (New York: Hodder & Stoughton, 1897), viii.
64. Lowis D'Aguilar Jackson, *Aid to Survey-Practice for Reference in Surveying, Levelling, and Setting-Out* (London: Lockwood, 1880); H. L. Thuillier and R. Smyth, *A Manual of Surveying for India* (Calcutta: Thacker, 1875).
65. Charles E. D. Black, *A Memoir on the Indian Surveys, 1875–1890* (London: E. A. Arnold, 1891), 216.
66. Lowis D. Jackson, *Aid to Survey-Practice for Reference in Surveying, Levelling, and Setting-Out*, 299–300.
67. Ibid., 295.
68. Kapil Raj, *Relocating Modern Science: Circulation and the Constitution of Knowledge in South Asia and Europe, 1650–1900* (Basingstoke: Palgrave Macmillan, 2006).
69. Kaminsky, *The India Office, 1880–1910*.
70. L/R/5/1: F. C. Danvers, "Memorandum of a Central Registry Department for the India Office," March 30, 1878.
71. Paul R. Brumpton, *Security and Progress: Lord Salisbury at the India Office* (Westport, CT: Greenwood Press, 2002), 189–90.
72. Valentine Baker, "The Military Geography of Central Asia," *Journal of the Royal United Services Institution* 79 (1874).
73. Baker, *Clouds in the East*, v.
74. Edward B. Hamley, "The Strategical Conditions of Our Indian North-West Frontier," *Journal of the Royal United Services Institution* 22, no. 98 (1878).
75. See WO 33/32, Reports and Memoranda 1878.
76. Valentine Baker and Lt W. J. Gill, *A Map of the North-Eastern Frontiers of Persia Embracing also Khiva and the Intervening Deserts* (London: E. Stanford: 1878), 1 inch = 32 miles. National Archives, FO 925/2192.

77. James Onley, "The Raj Reconsidered: British India's Informal Empire and Spheres of Influence in Asia and Africa," *Asian Affairs* 40, no. 1 (2009).
78. The name "Arabic Gulf" was widely used in 19th-century literature, see J. R. M'culloch, *A Dictionary Geographical, Statistical, and Historical of the Various Countries, Places, and Principal Natural Objects in the World*, vol. 1 (Longmans, 1866), 151.
79. HC Deb, August 22, 1883, vol. 283, col. 1688. Arnold wrote several travelogues on the Orient; see Robert A. Arnold, *Through Persia by Caravan* (London: Tinsley, 1877).
80. *London Gazette*, May 13, 1873, 2395.
81. Samuel B. Miles and Werner Muntzinger, "Excursion through the Fudhli Country," *Transactions of the Bombay Geographical Society* 19 (1874).
82. Samuel B. Miles, "Account of an Excursion into the Interior of Southern Arabia," *Proceedings of the Royal Geographical Society of London* 15, no. 5 (1870); S. B. Miles and Werner Munzinger, "Account of an Excursion into the Interior of Southern Arabia," *Journal of the Royal Geographical Society of London* 41 (1871).
83. Samuel B. Miles, "Journey from Gwadur to Karachi," *Journal of the Royal Geographical Society of London* 44 (1874).
84. Brian Marshall, "The Journeys of Samuel Barrett Miles in Oman, between 1875 and 1885," *Journal of Oman Studies* 10 (1989).
85. Samuel B. Miles, "Across the Green Mountain of Oman," *Geographical Journal* 18 (1901): 465–98; Miles, "On the Border of the Great Desert: A Journey in Oman," *Geographical Journal* 36, no. 2 (1910); Miles, "On the Border of the Great Desert: A Journey in Oman (Continued)," *Geographical Journal* 36, no. 4 (1910); Miles, *The Countries and Tribes of the Persian Gulf* (London: Harrison and Sons, 1919).
86. See his correspondence with Austen H. Layard (Layard Papers, vol. CCXXX, confidential print relating to Turkey, Aug., 1878–May, 1880, British Library Manuscripts Collection, accession: Add MS 39160).
87. S. B. Miles, *Sketch Map of Wadi Tyin in Oman* (London: George Philip & Son, 1896); Miles, *Map of Oman; from route traverses and other information by Lieut. Col. S. B. Miles* (London: Royal Geographical Society, 1910).
88. Miles, "Journey from Gwadur to Karachi," 173.
89. Miles, "On the Border of the Great Desert: A Journey in Oman (Continued)," 410.
90. Miles, "On the Border of the Great Desert: A Journey in Oman," 162.
91. Ibid., 160.
92. Miles and Munzinger, "Account of an Excursion into the Interior of Southern Arabia," 240.
93. Edward C. Ross, *Map of Oman, showing Distribution of the Principal Tribes*, 1:1,000,000, 61 × 89 cm (Bombay: Bombay Geographical Society: 1874). Online version available at http://www.qdl.qa/en/archive/81055/vdc_100023399223.0x000001.
94. Edward C. Ross, *Parts of Arabia and Persia, Survey of India*, 1 inch = 32 miles, compiled in the Office of the Trigonometrical Branch, Dehradun, 1883. See also BL IOR W/LPS/21/B13, M. J. Gerard, *A Map of Mesopotamia and Persia, surveyed and compiled between January 1882 and August 1886* (map), 1 inch = 4 miles. Survey carried out by Iman Sharif Khan Bahadur.
95. Yigal Sheffy, *British Military Intelligence in the Palestine Campaign, 1914–1918* (London: F. Cass, 1998).
96. Henry G. Lyons, *The Cadastral Survey of Egypt, 1892–1907* (Cairo: Nat. Print. Dept., 1908), 218.
97. Robert L. Tignor, "The 'Indianization' of the Egyptian Administration under British Rule," *American Historical Review* 68, no. 3 (1963).

98. Robert Hanbury Brown, *The Fayûm and Lake Moeris* (London: Stanford, 1892), prefatory note.
99. William Willcocks, *Egyptian Irrigation* (London: E. & F. N. Spon, 1889).
100. David Gange, *Dialogues with the Dead: Egyptology in British Culture and Religion, 1822–1922* (Oxford: Oxford University Press, 2013), 305.
101. Mak, *The British in Egypt*, 132.
102. *Wanganui Chronicle* 31, no. 11284 (June 28, 1888): 3.
103. Edward McQueen Gray, *Government Reclamation Work in Foreign Countries* (Washington: Govt. Print. Off., 1909), 54.
104. Richard Francis Burton, *The Gold-Mines of Midian and the Ruined Midianite Cities* (London: C. Kegan Paul, 1878).
105. Ibid., 377.
106. Charles Edward Stewart, *Report on the Petroleum Districts Situated on the Red Sea Coast* (Cairo: National Print. Office, 1888), 24.
107. Alfred Milner, *England in Egypt* (London: Arnold, 1892).
108. Derek Gregory, "Scripting Egypt," in *Writes of Passage: Reading Travel Writing*, by James S. Duncan and Derek Gregory (London: Routledge, 1999), chap. 6.
109. Charles Montagu Doughty, *Travels in Arabia Deserta*, vol. 1 (Cambridge: Cambridge University Press, 1888), preface.
110. See the advertisement for Negretti & Zambra optical instruments in Thomas Cook's *Handbook for London* (London: Cook, 1878), 108.
111. John Knox Laughton, *Hints to Travellers Scientific and General*, 5th ed. (London: RGS, 1883).
112. Charles W. J. Withers, Diarmid Finnegan, and Rebekah Higgitt, "Geography's Other Histories? Geography and Science in the British Association for the Advancement of Science, 1831–c.1933," *Transactions of the Institute of British Geographers* 31, no. 4 (2006).
113. On this chronology, see Felix Driver, "Distance and Disturbance: Travel, Exploration and Knowledge in the Nineteenth Century," *Transactions of the Royal Historical Society* 14 (2004).
114. Anne Blunt and Wilfrid Scawen Blunt, *A Pilgrimage to Nejd, the Cradle of the Arab Race: A Visit to the Court of the Arab Emir, and "Our Persian Campaign"* (London: J. Murray, 1881).
115. Ali Behdad, *Belated Travelers: Orientalism in the Age of Colonial Dissolution* (Durham, NC: Duke University Press, 1994), 98.
116. Geoffrey Nash, *From Empire to Orient: Travellers to the Middle East, 1830–1926* (London: I. B. Tauris, 2005), 215.
117. See Wilfrid Scawen Blunt, *The Future of Islam* (London: Kegan Paul, Trench, 1882).
118. Blunt and Blunt, *A Pilgrimage to Nejd*, 14.
119. See Anne Blunt's Journals, December 20, 1878. The notebooks are in the British Library manuscript collection, Wentworth Bequest, vol. CCLXXX. Five Arabic notebooks. Add MS 75282.
120. Anne Blunt's Journals, March 29, 1879.
121. "New Books," *Proceedings of the Royal Geographical Society and Monthly Record of Geography* 3, no. 3 (1881).
122. Billie Melman, *Women's Orients: English Women and the Middle East, 1718–1918: Sexuality, Religion, and Work* (Ann Arbor: University of Michigan Press, 1992), 145; Reina Lewis, *Gendering Orientalism: Race, Femininity, and Representation* (New York: Routledge, 1996).
123. Peter J. Morus and Iwan Rhys Bowler, *Making Modern Science: A Historical Survey* (Chicago: University of Chicago Press, 2005), 495.
124. Lovelace Papers, September 1836, Oxford, Bodleian Library.
125. Quoted in Avril Maddrell, *Complex Locations Women's Geographical Work in the UK, 1850–1970* (Chichester: Wiley-Blackwell, 2009), 32.

126. Morag Bell and Cheryl McEwan, "The Admission of Women Fellows to the Royal Geographical Society, 1892–1914: The Controversy and the Outcome," *Geographical Journal* 162 (1996).
127. Anne Blunt's Journals, December 8, 1879.
128. Aneroid barometer used by Doughty in Arabia, 1876–78 (RGS, 700635); Aneroid barometer used by General Gordon in the Sudan, on his survey of the Nile, 1874–76 (RGS, 700474).
129. Bruno Latour, *La Science en Action* (Paris: Editions La Découverte, 1989).
130. Charles W. J. Withers, "Science, Scientific Instruments and Questions of Method in Nineteenth-Century British Geography," *Transactions of the Institute of British Geographers* 38, no. 1 (2013).
131. Ruth Gay, "Charles Doughty—Man and Book," *American Scholar* 50, no. 4 (1981).
132. Charles Montagu Doughty, *Travels in Arabia Deserta*, vol. 2 (Cambridge: Cambridge University Press, 1888), 327.
133. Ibid., 333–34.
134. Charles M. Doughty, "Reisen in Arabien," *Globus* 39 (1882): 8.
135. Charles Montagu Doughty, *Documents Épigraphiques Recueillis dans le Nord de l'Arabie* (Paris: Imprimerie nationale, 1884).
136. Charles M. Doughty, "Travels in North-Western Arabia and Nejd," *Proceedings of the Royal Geographical Society and Monthly Record of Geography* 6, no. 7 (1884).
137. See Stephen E. Tabachnick, *Explorations in Doughty's Arabia Deserta* (Athens: University of Georgia Press, 1987).
138. Hogarth, *The Life of Charles M. Doughty*, 158.
139. Doughty, *Travels in Arabia Deserta*, vol. 2, 624. See also Hogarth, *The Life of Charles M. Doughty*, 157.
140. Conrad, "Geography and Some Explorers."
141. See Pirouz Mugtahidzada, *The Small Players of the Great Game: The Settlement of Iran's Eastern Borderlands and the Creation of Afghanistan* (London: Routledge Curzon, 2003).
142. Among the many examples of maps compiled by the Indian Survey see IOR/X/3126/3/1, David Bower, *Route Map of the Special Mission to Seistan and Mekran, from Ispahan to Gwadur* (1872); IOR/W/L/PS/21/D25, *Skeleton Map to Illustrate Captain Jennings' Diary of his Journey through Western Baluchistan, Eastern Persia, Sarhad and Sistan, in 1884–1885* (1886); IOR/W/L/PS/21/D14, Colonel Charles M. Macgregor and Captain Lockwood, *Map to Illustrate a Reconnaissance across the Desert of Baluchistan in 1877* (1878).
143. See chap. 1. See also Frederic J. Goldsmid and Oliver B. C. St. John, *Eastern Persia: an Account of the Journeys of the Persian Boundary Commission, 1870–71–72* (London: Macmillan, 1876).
144. Ibid. See *Goldsmid Papers*, MSS Eur F 134 1836–1899, British Library, Asia, Pacific and Africa Collections.
145. Soli Shahvar, "Communications, Qajar Irredentism and the Strategies of British India: The Makran Coast Telegraph and British Policy of Containing Persia in the East (Baluchistan), Part II," *Iranian Studies* 39, no. 4 (2006).
146. Peter John Brobst, "Sir Frederic Goldsmid and the Containment of Persia, 1863–73," *Middle Eastern Studies* 33, no. 2 (1997).
147. Burdett, *Iran in the Persian Gulf: 1820–1966*, 840.
148. Derek J. Waller, *The Pundits: British Exploration of Tibet and Central Asia* (Lexington: University Press of Kentucky, 1990), 158.
149. Black, *A Memoir on the Indian Surveys, 1875–1890*, 135.
150. Thomas H. Holdich, *The Indian Borderland, 1880–1900* (London: Methuen, 1901), 318.
151. Black, *A Memoir on the Indian Surveys, 1875–1890*, 99.

152. Ibid.
153. Ibid., 177.
154. Holdich, *The Indian Borderland, 1880–1900*, 168–69.
155. See Mugtahidzada, *The Small Players of the Great Game*, 69–70.
156. See in the RGS Map Room: Mohammad-Reza Mohandes, *Persia*, approx. 1:480,000, 72.5 × 44.5 cm (Tehran: s.n., 1883); *Map of Fars/Naqsha-ye Fars*, 1 inch = 9 miles (Tehran: s.n., 1895); *Plan of Tabriz from a survey by the pupils of the Tabriz Military College*, 1:506,880 (Tabriz: 1894).
157. Nancy Lee Peluso, "Whose Woods Are These? Counter-mapping Forest Territories in Kalimantan, Indonesia," *Antipode* 27, no. 4 (1995); Denis Wood, *Rethinking the Power of Maps* (New York: Guilford Press, 2010).
158. Turkish General Staff, *Map of Mesopotamia*, 1:250,000, 44.5 × 53.3 cm (Constantinople: Turkish General Staff, 1910s), RGS Map Room (mr Iraq Div.9) (1).
159. The treaty not only ascribed sovereignties in the Balkans but also dealt with the Turco-Persian border as well as the holy places in Palestine.
160. Doughty, *Travels in Arabia Deserta*, vol. 2, 398.

CHAPTER 5

1. Anon., *Birds-eye View of the Battle of Tamanieh, 13th March 1884* (London: G. W. Bacon, 1884); *Bird's-eye View of the Battle of El-Teb, February 29th 1884* (London: G. W. Bacon, 1884); *Bacon's War Map of Egypt, Including the Soudan, Abyssinia Etc.* (London: G. W. Bacon, 1884).
2. W. and A. K. Johnston, *W. & A. K. Johnston's Large Scale War Map of the Eastern Soudan*, scale 1:1,100,000, 48 × 74 cm (Edinburgh: W. & A. K. Johnston, 1885).
3. G. Philip and Son, *Philip's Special Large Scale War Map of the Soudan, Extending to Soakim on the Red Sea, with Enlarged Plan of Khartum*, scale 1:1,013,760, 75 × 54 cm (London: G. Philip and Son, 1885), Bibliothèque Nationale de France, GE C-539.
4. "Fac-Simile of a Map Drawn by General C. G. Gordon, at Khartum, March 17, 1874 [sic], of His Route from Suakin to Berber and Khartum" (London: E. Stanford, 1885).
5. Available online at http://cartotecadigital.icc.cat/cdm/ref/collection/africa/id/568.
6. On this notion see Jonathan Crary, *Techniques of the Observer: On Vision and Modernity in the Nineteenth Century* (Cambridge, MA: MIT Press, 1990).
7. "New Maps," *Scottish Geographical Magazine* 3, no. 10 (1887).
8. See Bartholomew Archives, NLS, Acc.10222/Reference Maps/Drawer.
9. Bartholomew, *The British Colonial Pocket Atlas*, map no. 13.
10. John Scott Keltie, *The World-Wide Atlas of Modern Geography, Political and Physical, Containing One Hundred and Twelve Plates and Complete Index* (Edinburgh: Johnston, 1892), map no. 53.
11. Karen S. Pearson, "The Nineteenth-Century Colour Revolution: Maps in Geographical Journals," *Imago Mundi Imago Mundi* 32, no. 1 (1980).
12. See Alexander K. Johnston, *Syria and the Isthmus of Suez* (map no. 5), in *Six Penny Atlas of Physical Geography*, by A. K. and W. Johnston (Edinburgh: Johnston, 1870).
13. George G. Chisholm, *Handbook of Commercial Geography* (London: Longmans, 1889).
14. Trevor J. Barnes, "'In the Beginning Was Economic Geography': a Science Studies Approach to Disciplinary History," *Progress in Human Geography* 25, no. 4 (2001).
15. John G. Bartholomew, *Atlas of Commercial Geography Illustrating the General Facts of Physical, Political, Economic and Statistical Geography* (Cambridge: Cambridge University Press, 1889).
16. Arthur Silva White, *Britannic Confederation* (London: George Philip & Son, 1892).

17. Arthur Silva White, "On the Comparative Value of African Lands," *Scottish Geographical Magazine Scottish Geographical Magazine* 7, no. 4 (1891). Digitized copy available at http://imagesearch.library.illinois.edu/u?/africanmaps,1409.
18. The map is available online at http://prints.bl.uk/art/406682/Africa.
19. Arthur Silva White, *The Development of Africa* (London: Philip, 1892).
20. Arthur Silva White, *The Expansion of Egypt: Under Anglo-Egyptian Condominium* (London: Methuen, 1899).
21. See Black, *Maps and History*, 109.
22. FO 881/5339, Francis L. Bertie, *Proposals of the Government of India respecting relations with the Tribes in the vicinity of Aden and on the south coast of Arabia*, October 2, 1886.
23. Frederick M. Hunter and Charles W. H. Sealey, *An Account of the Arab Tribes in the Vicinity of Aden* (Bombay: Government Central Press, 1886).
24. John G. Bartholomew, *The Royal Shilling Atlas; a Complete Series of Maps with Index* (London: T. Nelson, 1892).
25. Israel Zangwill, *Without Prejudice* (London: T. Fisher Unwin, 1896).
26. "The Seat of War," *Leeds Mercury*, July 11, 1876, 8.
27. "War Map Indicating Sir Garnet Wolseley's Line of March from Ismailia to Cairo," *Penny Illustrated Paper*, September 2, 1882, 145.
28. William T. Stead, *The Maiden Tribute of Modern Babylon: Reprinted from the "Pall Mall Gazette"* (London: F. A. Roberts, 1885).
29. Stéphanie Prévost, "W. T. Stead and the Eastern Question (1875–1911); or, How to Rouse England and Why?," *Interdisciplinary Studies in the Long Nineteenth Century* 16 (2013).
30. William T. Stead, *Map of the Armenian Massacres*, *Review of Reviews*, December 1896, 506.
31. "The Cholera Map of Egypt," *Pall Mall Gazette*, July 27, 1883, 8.
32. See Peter Vinten-Johansen, *Cholera, Chloroform, and the Science of Medicine: A Life of John Snow* (Oxford: Oxford University Press, 2003).
33. See chap. 3.
34. Elisabeth Crawford, Terry Shinn, and Sverker Sörlin, *Denationalizing Science: The Contexts of International Scientific Practice* (Dordrecht: Kluwer Academic Publishers, 1993), 1.
35. "The Russian Approach to India," *Graphic*, May 8, 1884, 3.
36. On this notion see Porter, *The Absent-Minded Imperialists*.
37. Adelson, *London and the Invention of the Middle East*, 17; Michelle Elizabeth Tusan, *Smyrna's Ashes: Humanitarianism, Genocide, and the Birth of the Middle East* (Berkeley: University of California Press, 2012). See also Stéphanie Prévost, "The Reception of the Eastern Question in Britain and Its Impact on British Political Culture (1875–1898)" (diss., Tours, 2010).
38. The map, entitled "Isochronic Passage Chart for Travellers," was created by Francis Galton and first published in 1881 (see *Proceedings of the Royal Geographical Society and Monthly Record of Geography* 3, no. 11 (November 1881): 701.
39. See George Bradshaw, *Bradshaw's through Route Overland Guide to India, and Colonial Handbook; or, Manual for Travellers* (London: W. J. Adams, 1884).
40. *British Journal of Photography* 30 (August 17, 1883), 478: "Thus, for example, the recent occurrences in Egypt created a demand for Lecture sets of that country."
41. H. Arnold Bemrose, "A Tracing Paper Screen," *Nature* 31, no. 801 (1885): 409.
42. "Photographic Industries," *British Journal of Photography*, July 28, 1893, 480.
43. Samuel Manning, *The Land of the Pharaohs: Egypt and Sinai: Illustrated by Pen and Pencil* (London: Religious Tract Society, 1875), 81.
44. See the French translation of York's handbook, *Enseignement par les Projections Lumineuses: Lectures Traduites de l'anglais pour Accompagner les Collections de Photographies sur Verre de York and Son*, vol. 9, *L'Égypte*; and vol. 10, *Voyage en Terre Sainte* (Paris: A. Molteni, 1893).

45. William C. Darrah, *The World of Stereographs* (Gettysburg, PA: Darrah, 1977). See also Jib Fowles, "Stereography and the Standardization of Vision," *Journal of American Culture* 17, no. 2 (1994).
46. Francis Galton, "On Stereoscopic Maps, Taken from Models of Mountainous Countries," *Proceedings of the Royal Geographical Society of London* 9, no. 3 (1864).
47. Henry F. Brion and Revd. Edmund McClure, *Photo Relief Map of Asia*, published by the Society for Promoting Christian Knowledge, scale approx. 1:38,000,000, 46 cm × 38 cm (1885), National Maritime Museum, Greenwich, Object ID G294:1/5.
48. Royal Geographical Society, *Yearbook and Record*, 1899, 28.
49. George Armstrong, *Raised-Map of Palestine*, London, Palestine Exploration Fund, scale 3/8 inch = 1 mile, GA1321.7.A7.
50. "Notices and News," *Palestine Exploration Fund Quarterly Statement* 27–28 (1895): 91.
51. Francis L. Denman, *Notes of Lectures on the Raised Map of Palestine* (London: London Society for Promoting Christianity Amongst the Jews, 1907).
52. Thompson, *The Empire Strikes Back?*
53. Thomas Hardy, *Jude the Obscure* (London: Osgood, McIlvaine, 1895).
54. James P. Maginnis, "Raised Models," *Geographical Teacher* 5, no. 5 (1910); Hugh McLeod, *Religion and Society in England, 1850–1914* (New York: St. Martin's Press, 1996).
55. John Scott Keltie, *Geographical Education—Report to the Council of the Royal Geographical Society* (London: J. Murray, 1886).
56. See Douglas W. Freshfield, "The Place of Geography in Education," *Proceedings of the Royal Geographical Society and Monthly Record of Geography* 8, no. 11 (1886); Ernst G. Ravenstein, "On the Aims and Methods of Geographical Education," *Proceedings of the Royal Geographical Society and Monthly Record of Geography* 8, no. 2 (1886).
57. "Proceedings of the Geographical Section of the British Association: Birmingham Meeting, 1886," *Proceedings of the Royal Geographical Society and Monthly Record of Geography* 8, no. 10 (1886); "Presentation of the Royal and Other Awards," *Journal of the Royal Geographical Society of London* 46 (1876).
58. "Presentation of the Royal and Other Awards."
59. "The Exhibition of Appliances Used in Geographical Education," *Proceedings of the Royal Geographical Society and Monthly Record of Geography* 8, no. 1 (1886): 53.
60. Halford J. Mackinder, "Modern Geography, German and English," *Geographical Journal* 6, no. 4 (1895): 375.
61. Halford J. Mackinder, "On the Scope and Methods of Geography," *Proceedings of the Royal Geographical Society and Monthly Record of Geography* 9, no. 3 (1887).
62. William F. Connell, *The Educational Thought and Influence of Matthew Arnold* (London: Routledge, 1998), 222.
63. Standards Day School Code of 1894, in *Report of the Committee of Council on Education, 1894* (London: Eyre and Spottiswoode, 1894), 347–48.
64. Walford, *Geography in British Schools, 1850–2000*, 51.
65. *Report of the Committee of Council on Education, 1894*, xxv.
66. *Report of the Committee of Council on Education (England and Wales), 1872* (London: H.M. Stationery Office, 1872), 87.
67. *Report of the Committee of Council on Education with Appendix, 1871* (London: Eyre and Spottiswoode, 1871), 208.
68. *Instruction of Lower Standards in Schools for older Scholars*, no. 332, Education Department, January 6, 1894.
69. *Report of the Committee of Council on Education (England and Wales), 1872*, 40.
70. *Report of the Committee of Council on Education with Appendix, 1871*, cxliii.

71. See William G. V. Balchin, *The Geographical Association: The First Hundred Years 1893–1993* (Sheffield: Geographical Association, 1993).
72. Michael Wise, "The Campaign for Geography in Education: The Work of the Geographical Association 1893–1993," *Geography* 78, no. 2 (1993).
73. Halford J. MacKinder, "Geographical Education: The Year's Progress at Oxford," *Proceedings of the Royal Geographical Society and Monthly Record of Geography* 11, no. 8 (1889).
74. Ibid.
75. "On the Instruction at Present Supplied in This Country, in Practical Astronomy, Navigation, Route Surveying, and Mapping," *Proceedings of the Royal Geographical Society and Monthly Record of Geography* 4, no. 5 (1882).
76. *Report of the Committee of Council on Education with Appendix, 1871*, cxix.
77. *Report of the Committee of Council on Education, 1894*, 342.
78. Ibid., 40.
79. Ibid., 140.
80. Julie McDougall, "The Publishing History and Development of School Atlases and British Geography, c. 1870–c. 1930" (PhD diss., University of Edinburgh, 2012).
81. Jeffrey Richards, *Imperialism and Juvenile Literature* (Manchester: Manchester University Press, 1989).
82. George A. Henty, *For the Temple: a Tale of the Fall of Jerusalem* (London: Blackie, 1887).
83. George A. Henty, *A Chapter of Adventures: Or, through the Bombardment of Alexandria* (London: Blackie, 1891).
84. George A. Henty, *The Dash for Khartoum: A Tale of the Nile Expedition* (London: Blackie 1892); Henty, *With Kitchener in the Soudan: A Story of Atbara and Omdurman* (London: Blackie, 1903).
85. George A. Henty, *For Name and Fame, or, to Cabul with Roberts* (London: Blackie, 1900).
86. Teresa Ploszajska, *Geographical Education, Empire and Citizenship: Geographical Teaching and Learning in English Schools, 1870–1944* (England: Historical Geography Research Group, 1999), 299.
87. John M. D. Meiklejohn, *Fifth Geographical Reader, Standard VI: Asia, Africa, America, and Oceania* (London: Blackwood, 1884), 79–80.
88. Darwin, *Unfinished Empire*, 100.

CHAPTER 6

1. Middle East Centre, Saint Anthony's College, Oxford (accession GB165-0276).
2. Percy Molesworth Sykes, *Ten Thousand Miles in Persia or Eight Years in Irán* (London: Murray, 1902); P. Molesworth Sykes, D. G. Hogarth, and M. Longworth Dames, "Anthropological Notes on Southern Persia," *Journal of the Anthropological Institute of Great Britain and Ireland* 32 (1902); P. Molesworth Sykes, "A Fourth Journey in Persia, 1897–1901," *Geographical Journal* 19, no. 2 (1902); P. Molesworth Sykes, "A Fifth Journey in Persia," *Geographical Journal* 28, no. 5 (1906); P. Molesworth Sykes, *The Glory of the Shia World: The Tale of Pilgrimage, Tr. & Ed. from a Persian Manuscript* (London: Macmillan, 1910); P. Molesworth Sykes, *A History of Persia* (London: Macmillan, 1915).
3. Antony Wynn, *Persia in the Great Game: Sir Percy Sykes, Explorer, Consul, Soldier, Spy* (London: Murray, 2003), 176.
4. Steiner, *The Foreign Office and Foreign Policy, 1898–1914*; F. H. Hinsley, *British Foreign Policy under Sir Edward Grey* (Cambridge: Cambridge University Press, 1977), 17.

5. See, for example, the maps of Persia mentioned in HC Deb, February 13, 1908, vol. 184, cols. 200–201.
6. HC Deb, January 22, 1902, vol. 101, cols. 574–628.
7. HC Deb, February 21, 1912, vol. 34, cols. 628–95.
8. CAB 37/44/15.
9. FO 925/3194.
10. Keith A. Hamilton, *Bertie of Thame: Edwardian Ambassador* (Woodbridge: Royal Historical Society, 1990), 6. See FO 881/6412, FO 881/6413, FO 881/6408.
11. See Bertie F. L., FO 881/5339, with map showing distribution of tribes.
12. Thomas G. Fergusson, *British Military Intelligence, 1870–1914: The Development of a Modern Intelligence Organization* (Frederick, MD: University Publications of America, 1983), 231.
13. Keith Jeffery, *Field Marshal Sir Henry Wilson: A Political Soldier* (Oxford: Oxford University Press, 2006), 87.
14. WO 106/1468; WO 106/1468; WO 106/216.
15. IOR/W/L/PS/21.
16. *Auckland Star* 31, no. 119 (May 21, 1900): 4.
17. Maria O'Shea, *Trapped between the Map and Reality: Geography and Perceptions of Kurdistan* (London: Routledge, 2003), 103–6.
18. Brumpton, *Security and Progress*, 35.
19. CAB 7/2–4.
20. Donald C. Gordon, "The Colonial Defence Committee and Imperial Collaboration: 1885–1904," *Political Science Quarterly* 77, no. 4 (1962).
21. Kenneth Bourne, *Britain and the Balance of Power in North America, 1815–1908* (Berkeley: University of California Press, 1967), 313.
22. John P. MacKintosh, "The Role of the Committee of Imperial Defence before 1914," *English Historical Review* 77, no. 304 (1962).
23. CAB 11/75, *Defence scheme: Cairo*, 1911. See also CAB 38/2/1, *The effect on naval strategic position in the Mediterranean of a Russian occupation of Constantinople*, February 7, 1903; and CAB 38/3/43, *Russian intentions in the Middle East*, June 6, 1903.
24. A. W. Mabbs, *List of Papers of the Committee of Imperial Defence to 1914* (London: H.M. Stationery Off., 1964).
25. Wade, *Spies in the Empire*, 192. See also Royal Commission on the War in South Africa, *Minutes of Evidence Taken before the Royal Commission on the War in South Africa*, vol. 1 (London: Printed for H.M. Stationery Off., by Wyman and Sons, 1903).
26. Gavish, *The Survey of Palestine under the British Mandate, 1920–1948*, 236.
27. Lyons, *The Cadastral Survey of Egypt, 1892–1907*.
28. *Egyptian surveyors at work*, in Lyons, *The Cadastral Survey of Egypt, 1892–1907*, plate XIII.
29. Timothy Mitchell, *Rule of Experts: Egypt, Techno-Politics, Modernity* (Berkeley: University of California Press, 2002).
30. Ernest M. Dowson, "The Cadastral Survey of Egypt, 1879–1907," *Survey Review* 8, no. 58 (1945).
31. A triangulation of a high order avoids excessive errors by measuring a very long baseline of several miles and large triangles with sides reaching a hundred miles. Only then may accurate stations be established all over a territory. This type of triangulation is the only one that ensures the accuracy of small-scale maps. A datum is a point, line, or surface used for reference by the surveyors. The Helwân observatory was a horizontal datum. One could find the coordinates of another point in Egypt by referring to this station. A vertical datum, to calculate elevation, is also needed. The mean sea level observed in the Mediterranean served this purpose. The effects of the curvature of the earth must be

fully computed for a truly geodetic survey and not just a plane survey, which ignores the actual figure of the earth. The earth is not a perfect sphere, it is oblate. This fact may be neglected when the surveyed area is small, but it implies a growing margin of error as the territory under consideration expands. It becomes an issue on large projections. The shape of a reference ellipsoid is used to compute very accurate coordinates. It is a mathematical approximation of the true shape of the globe. Various ellipsoids were calculated in the 19th century. The Clarke 1866 ellipsoid was used until 1906 in Egypt. The Egyptian survey started to use the Helmert ellipsoid in 1907.

32. James H. Cole, *Geodesy in Egypt* (Bûlâq: Govt. Press, 1944).
33. Thomas H. Holdich, "African Boundaries, and the Application of Indian Systems of Geographical Survey to Africa," *Proceedings of the Royal Geographical Society and Monthly Record of Geography* 13, no. 10 (1891).
34. Lyons, *The Cadastral Survey of Egypt, 1892–1907*, 327.
35. See Samir W. Raafat, *Maadi 1904–1962: Society and History in a Cairo Suburb* (Cairo: Palm Press, 1994).
36. See André Bernand, *Pan du Désert* (Leiden: Brill, 1977), 116–17.
37. Thomas Barron and William F. Hume, *The Topography and Geology of the Eastern Desert of Egypt, Central Portion* (Cairo: National Printing Dept., 1902).
38. Cairo Survey Dept., *Geological Map of Egypt*, scale 1:2,000,000, 61 × 60 cm (1910), sheet no. 1.
39. William F. Hume, *The Distribution of Iron Ores in Egypt* (Cairo: National Printing Department, 1909).
40. Ralph S. G. Stokes, *Mines and Minerals of the British Empire, Being a Description of the Historical, Physical, & Industrial Features of the Principal Centres of Mineral Production in the British Dominions beyond the Seas* (London: E. Arnold, 1908), 369.
41. Arthur Llewellyn, *Report on a Mining Concession in the Egyptian Sudan* (London: Egypt & Sudan Mining Syndicate, 1903).
42. Edwin Sharpe, *Field-Marshal Lord Kitchener: His Life and Work for the Empire* (London: Gresham Publ., 1916), 51.
43. This is partly documented in his correspondence (PRO 30/57).
44. Ronald Storrs, *The Memoirs of Sir Ronald Storrs* (New York: Putnam, 1937), chap. 6.
45. Thomas E. Lawrence and Leonard Woolley, *The Wilderness of Zin (Archaeological Report)* (London: Palestine Exploration Fund, 1914).
46. Moscrop, *Measuring Jerusalem*, 209.
47. Henry McMahon, *An Account of the Entry of H.M. Habibullah Khan Amir of Afghanistan into Freemasonry* (London: Favil Press, 1936).
48. Bruce Westrate, *The Arab Bureau: British Policy in the Middle East, 1916–1920* (University Park: Pennsylvania State University Press, 1992).
49. F. Fraser Hunter, "Reminiscences of the Map of Arabia and the Persian Gulf," *Geographical Journal* 54, no. 6 (1919): 357.
50. John Gordon Lorimer, *Gazetteer of the Persian Gulf, Oman and Central Arabia*, vol. 2, *Geographical and Statistical* (Gerrards Cross: Archive Ed., 1986).
51. Hunter, "Reminiscences of the Map of Arabia and the Persian Gulf," 360.
52. Ibid.
53. George C. Gore, *On the Projection for a Map of India and Adjacent Countries* (Dehradun: Office of the Tirgonometrical branch, Survey of India, 1903).
54. Arabia Felix and Arabia Petraea were classical divisions of Arabia dating back to antiquity.
55. Priya Satia, *Spies in Arabia: The Great War and the Cultural Foundations of Britain's Covert Empire in the Middle East* (New York: Oxford University Press, 2008), 30.

56. George Leachman, "A Journey in North-Eastern Arabia," *Geographical Journal* 37, no. 3 (1911).
57. George Leachman, "A Journey through Central Arabia," *Geographical Journal* 43, no. 5 (1914).
58. Leachman's portrait is in the West Sussex Record Office (RSR/PH/11/23).
59. RGS archives, GB 0402 WHI.
60. Political and Secret Correspondence with India, L/P & S/7 Letters from India, vol. 48, no. 89, Shakespear to Political Resident, April 8, 1911.
61. Cabinet Papers, CAB 24/1, March 1915.
62. Eric Macro, "George Wyman Bury South Arabian Pioneer," *Proceedings of the Seminar for Arabian Studies* 13 (1983).
63. George Wyman Bury, *The Land of Uz* (London: Macmillan, 1911), dedication.
64. See chaps. 1 and 3.
65. James Canton, "Imperial Eyes: Imperial Spies; British Travel and Espionage in Southern Arabia, 1891–1946," *Journal of Imperial and Commonwealth History* 37, no. 4 (2009).
66. *British Association for the Advancement of Science, Report of the annual meeting*, no. 63 (London: Murray, 1894), 562.
67. See Charles Edward Yate, *Northern Afghanistan: Or, Letters from the Afghan Boundary Commission* (Edinburgh: W. Blackwood & Sons, 1888), 115.
68. *Map of Arabia showing the routes of Mr. J Theodore Bent*, in *Southern Arabia*, by J. Theodore Bent (London: Smith, Elder, 1900).
69. Henry McMahon, "Recent Survey and Exploration in Seistan," *Geographical Journal* 28, no. 3 (1906); Kashani-Sabet, *Frontier Fictions*.
70. Maurice Zimmermann, "L'exploration du Séistan par Sir Henry Mac Mahon," *Annales de Géographie* 16, no. 85 (1907): 82.
71. William Willcocks, *The Restoration of the Ancient Irrigation Works on the Tigris or the Re-Creation of Chaldea, Being a Lecture Delivered at a Meeting of the Khedivial Geographical Society, Cairo, 25th March 1903* (Cairo: National Printing Department, 1903), 85.
72. See Nadia Atia, "A Relic of Its Own Past: Mesopotamia in the British Imagination, 1900–14," *Memory Studies* 3, no. 3 (2010).
73. William Willcocks, "Mesopotamia: Past, Present, and Future," *Geographical Journal* 35, no. 1 (1910).
74. William Willcocks, "The Garden of Eden and Its Restoration," *Geographical Journal* 40, no. 2 (1912): 240.
75. William Willcocks, *From the Garden of Eden to the Crossing of the Jordan* (Cairo: French Institute of Oriental Archaeology, 1918).
76. Henry G. Lyons, "Sir William Willcocks's Survey in Mesopotamia," *Geographical Journal* 40, no. 5 (1912).
77. Willcocks, however, enjoyed considerable support from Archibald Sayce, who was also convinced of Eden's location in southern Mesopotamia. See Sayce et al., "The Garden of Eden and Its Restoration: Discussion," *Geographical Journal* 40, no. 2 (1912).
78. L/P&S 10/576.
79. Mark Sykes, "The Western Bend of the Euphrates," *Geographical Journal* 34, no. 1 (1909).
80. Captain Smythe was sent in 1904 to study the territories that the Bagdadbahn would cross, and Captain Kemp studied the possibilities of creating a port at Khor Abdullah.
81. FO 371/340/12/17743, *Cabinet memorandum* dated May 31, 1907, cited by Stuart Cohen, *British Policy in Mesopotamia, 1903–1914* (London: Ithaca Press, 1976), 361.
82. The War Office produced several maps in order to visualize the line's potential danger, see

GSGS 2246, *Sketch Map to Show Approximately Railways in Asiatic Turkey, with Notes*, scale 1:7,500,000, 48 × 38 cm (1911).
83. Mark Sykes monitored the line; he was also one of the first to write a report for the British on the region's oil resources (FO 195/2164, report from October 15, 1905).
84. Cohen, *British Policy in Mesopotamia*; Marian Kent, *Oil and Empire: British Policy and Mesopotamian Oil, 1900–1920* (London, 1976).
85. Gertrude Lowthian Bell, *Amurath to Amurath* (London: Heinemann, 1911), 87.
86. Letter dated March 10, 1911, Gertrude Bell Archive, Newcastle University Library.
87. Bell, *Amurath to Amurath*, 186.
88. Letter from Arnold Wilson to Colonel Yate, late November 1914, quoted by John Fisher, *Curzon and British Imperialism in the Middle East, 1916–19* (London: Cass, 1998), 113.
89. Toynbee, *Turkey*, 47.
90. Denis Wright, "Curzon and Persia," *Geographical Journal* 153, no. 3 (1987).
91. Andrew S. Goudie, "George Nathaniel Curzon: Superior Geographer," *Geographical Journal* 146, no. 2 (1980).
92. Lovat Fraser, *India under Curzon and After* (London: Heinemann, 1911), 77.
93. George Curzon, "Memorandum on the Society's New Map of Persia," *Proceedings of the Royal Geographical Society and Monthly Record of Geography* 14, no. 2 (1892). William Turner was the actual cartographer.
94. George Curzon, *Persia and the Persian Question* (London: Longmans, 1892).
95. Shiva Balaghi, "Nationalism and Cultural Production in Iran, 1848–1906" (PhD diss., University of Michigan, 2008), 60–61.
96. The map was filed by the Intelligence Department of the WO under the number IDWO 1583.
97. CAB 37/74/15: *British interests on the coast of Arabia, Koweit, Bahrein and El Katr*, January 30, 1905.
98. See British Library, IOR/MSS Eur F111–112, series 68: Lord Curzon (1859–1925): early life: maps in his collection; MSS Eur F111–112, series 15; and MSS Eur F111/397.
99. Middle East Centre (Oxford), GB165-027.
100. *Journal of the Royal Central Asian Society* 39 (1952): 170.
101. Herbert R. Sykes, *Our Recent Progress in Southern Persia, and Its Possibilities* (London: Central Asian Society, 1905).
102. Herbert R Sykes, "Roads to the Gulf," *Times of India,* October 9, 1903, 2.
103. See Mark Polelle, *Raising Cartographic Consciousness: The Social and Foreign Policy Vision of Geopolitics in the Twentieth Century* (Lanham, MD: Lexington Books, 1999).
104. See chap. 1.
105. See the map entitled "Persian Gulf and Adjacent Countries," showing zones defined by the Anglo-Russian Convention of 1907, published by the GSGS (no. 2385), scale 1:4,055,040 (CO 1047/561).
106. *General Maitland's proposal for area to be surveyed by joint Anglo-Ottoman boundary commission in Dthali region* (Calcutta, 1901).
107. See series L/P&S/10/63. See also Col. Wahab, *Sketch of the boundary in west hidaba claimed by the Amri subjects of the Haushabi*, March 15, 1904 (L/P&S/10/66).
108. *Southern Arabia: map illustrating the boundary demarcated by the Anglo-Turkish Commission 1902–1904* (Simla, 1904). See also CO 1047/9, *Shekh Sa'id: 2 maps with MS additions showing Anglo-Turkish boundary*, scale 1:40,000 (Survey of India, 1904–7).
109. HC Deb, April 23, 1901, vol. 92, col. 1068; August 13, 1901, vol. 99, cols. 585–86.
110. See Edward Hertslet, *Map of the Sinai Peninsula, to Illustrate the Turco-Egyptian Frontier Dis-*

pute Agreement of October 1, 1906, scale 1:500,000, 13 × 11 cm, in *The Map of Africa by Treaty* (London, 1909), vol. 3, map no. 44.

111. Ygal Sheffy, "Une Convergence: La Collaboration entre les Services Secrets Français et Britanniques au Levant pendant la Première Guerre Mondiale," in *De Bonaparte à Balfour: La France, l'Europe Occidentale et la Palestine: 1799–1917*, ed. Ran Aaronsohn and Dominique Trimbur (Paris: Centre National de la Recherche Scientifique, 2001), 92.

112. Guy E. Pilgrim, *Geological Sketch Map of the Persian Gulf and the Gulf of Oman with the Adjoining Portions of Persia and Arabia*, scale 1:2,000,000, 96 × 72 cm (Calcutta: Geol. Survey of India, 1907).

113. Guy E. Pilgrim, *The Geology of the Persian Gulf and Adjoining Portions of Persia and Arabia* (Calcutta: Office of the Geological Survey of India, 1908).

114. See Timothy Mitchell, *Carbon Democracy: Political Power in the Age of Oil* (London: Verso, 2011), 577.

115. See Sydney H. North, *Oil Fuel: Its Supply, Composition, and Application* (London: C. Griffin, 1905).

116. See Cohen, *British Policy in Mesopotamia*; Peter J. Cain and Antony G. Hopkins, *British Imperialism, 1688–2000* (New York: Longman, 2001), 349–50.

117. Arnold Wilson, *S.W. Persia: A Political Officer's Diary 1907–1914* (London, 1941), 25–26.

118. Karl Ernest Meyer and Shareen Blair Brysac, *Kingmakers: The Invention of the Modern Middle East* (New York: Norton, 2008), 133.

119. Wilson, *S.W. Persia*, 37.

120. See Marian Kent, *Moguls and Mandarins: Oil, Imperialism and the Middle East in British Foreign Policy, 1900–1940* (London: Frank Cass, 1993), chap. 2.

121. IOR W/L/PS/21/B46/2: Map of the Anglo Persian oil Company's Concessions. IOR/L/PS/10/ 410/ part.1/1831: Final Report of the Slade Commission (April 6, 1914).

122. HC Deb, June 29, 1914, vol. 64, cols. 5–7.

123. Charles Eliot, *Turkey in Europe* (London: F. Cass, 1908), 419.

124. See the maps of the *Balkan States and Asia Minor 1:250 000 (British)* series (WO 301/59 ff.).

125. This program, conceived by Russia and Austro-Hungary in September 1903 and imposed on the Ottomans, aimed to create administrative reforms in order to protect the Macedonian people from potential abuse. The British became involved in this process in 1908 so that they could begin establishing a sphere of influence around the city of Drama. See John Fraser Foster, *Pictures from the Balkans* (London: Cassell, 1912), 160.

126. FO 925/3194: *Map to illustrate notes by Capt Fairholme, on races and religions in Turkey in Europe*, IDWO 1247, 1:3,220,177, *War Office*.

127. "M. Maunsell," *Bulletin de Géographie d'Aix-Marseille* 18 (1893): 305.

128. Francis R. Maunsell, "Kurdistan," *Geographical Journal* 3, no. 2 (1894).

129. Francis R. Maunsell, "Kurdistan," *Geographical Journal* 3, no. 2 (1894): 81.

130. *Geographical Journal* 26, no. 3 (September 1905): 5.

131. P. H. H. Massy, "Exploration in Asiatic Turkey, 1896 to 1903," *Geographical Journal* 26, no. 3 (1905).

132. Francis R. Maunsell, "Eastern Turkey in Asia and Armenia," *Scottish Geographical Magazine* 12, no. 5 (1896); Maunsell, "Central Kurdistan," *Geographical Journal* 18, no. 2 (1901).

133. Francis R. Maunsell, *Eastern Turkey in Asia*, 78 × 56 cm, scale 1:250,000, IDWO, 1901–15, sheet no. 35; see compilation notes.

134. Francis R. Maunsell, *Eastern Turkey in Asia*, 78 × 56 cm, scale 1:250,000, IDWO, 1901–15, sheet no. 13.

135. See Francis R. Maunsell, "Notes to Accompany Lieut.-Colonel Maunsell's Map of Eastern Turkey in Asia," *Geographical Journal* 28, no. 2 (1906).
136. See Mark Sykes, "Journeys in North Mesopotamia," *Geographical Journal* 30, no. 3 (1907); L. Molyneux-Seel, "A Journey in Dersim," ibid.44, no. 1 (1914).
137. Frederic Maunsell, *Eastern Turkey in Asia*, sheet no. 32, *Mosul*, 78 × 56 cm, scale 1:250,000. Online image available at Digital Library for International Research Archive, item 15892, http://www.dlir.org/archive/items/show/15892, accessed September 4, 2014.
138. Curzon, *Problems of the Far East, Japan—Korea—China*, XI.
139. Rafiuddin Ahmad, "The Battle of Omdurman and the Musulman World," *Living Age* 291, no. 2838 (November 26, 1898): 565–71.
140. Jordan Branch, *The Cartographic State: Maps, Territory and the Origins of Sovereignty* (Cambridge: Cambridge University Press, 2013).

CHAPTER 7

1. Bonine, *Is There a Middle East?*, 36.
2. Halford J. Mackinder, "The Geographical Pivot of History," *Geographical Journal* 23, no. 4 (1904): 431.
3. See, for instance, the anonymous review of the map in *School World*, vol. 15 (Macmillan, 1913), 431.
4. Bonine, *Is There a Middle East?*, 26.
5. "Front Matter," *Geographical Teacher* 4, no. 4 (1908).
6. Johann Wolfgang von Goethe, *West-Oestlicher Divan* (Stuttgart: Cotta, 1819), 301: "dem Mittlern Orient, wie wir Persien und seine Umgebungen"; Carl Friedrich von Rumohr, *Italienische Forschungen 3. 3* (Berlin: Nicolai, 1831), 311.
7. Jules Michelet, *Histoire de France au Seizième Siècle*, vol. 8 (Paris: Chamerot, 1855), 489.
8. Clayton R. Koppes, "Captain Mahan, General Gordon, and the Origins of the Term Middle East," *Middle Eastern Studies* 12, no. 1 (1976).
9. Seth Low, "The International Conference of Peace," *North American Review* 169, no. 516 (1899): 626.
10. Only Low's correspondence with Mahan (Seth Low Papers, Columbia University) might give a definite answer on that point.
11. Università commerciale Luigi Bocconi, *Annuario* (Milan: Societa Tipografica Editrice Popolare, 1903), 13: "La questione del Medio Oriente—Il Medio Oriente non è soltanto una espressione geografica, ma anche una entità geografica, politica ed economica costituita da un ben determinato gruppo di Stati."
12. D.L., "The Problem of the Middle East," *The Outlook in Politics, Life, Letters and the Arts* 1 (1898): 15.
13. "Foreign Affairs," *Blackwood's Edinburgh Magazine* 158 (962): 930.
14. Walter E. Houghton et al., eds., *The Wellesley Index to Victorian Periodicals*, vol. 1 (Toronto: University of Toronto Press: 1966), 194n7246.
15. D. Sherman, "Miscellaneous," *Zion's Herald* 53, no. 19 (May 11, 1876): 146. The ongoing digitization of the 19th-century records could reveal earlier uses in the near future.
16. Alfred T. Mahan, *Retrospect & Prospect: Studies in International Relations Naval and Political* (London: Sampson Low, Marston, 1902), 231.
17. A. J. Anthony Morris, *The Scaremongers: The Advocacy of War and Rearmament, 1896–1914* (London: Routledge & Kegan Paul, 1984), 44.

18. Valentine Chirol, *The Middle Eastern Question; or, Some Political Problems of Indian Defence* (London: J. Murray, 1903); *Times*, October 14, 17, 20, 25, and 27, 1902; November 10, 12, 20, and 24, 1902; December 15, 23, 25, 26, and 30, 1902, March 4 and 30, 1903; April 11 and 21, 1903.
19. George Curzon, *The Far Eastern Question* (London: Macmillan, 1896).
20. Adelson, *London and the Invention of the Middle East*.
21. Quoted in Jason Tomes, *Balfour and Foreign Policy: The International Thought of a Conservative Statesman* (Cambridge: Cambridge University Press, 1997), 112.
22. See CAB 16/4: *Military Requirements of the Empire as Affected by Egypt and the Sudan*, 1909; or CAB 16/15: *Proceedings of the Standing Sub-committee of the Committee of Imperial Defence, on the Persian Gulf*, Nov. 1911.
23. John D. Rees, *The Real India* (London: Methuen, 1908).
24. HC Deb, August 4, 1906, vol. 16, cols. 1813; and March 19, 1908, vol. 186, cols. 879.
25. HC Deb, March 13, 1907, vol. 171, cols. 28–29.
26. Angus Hamilton, *Problems of the Middle East* (London: E. Nash, 1909).
27. See chap. 6.
28. P. M. Sykes, "A Fifth Journey in Persia," 435; P. M. Sykes, "A Sixth Journey in Persia," *Geographical Journal* 37, no. 1 (1911): 6.
29. H. R. Sykes, "The Middle East," *Manchester Courier*, May 4, 1903: 8.
30. Arthur C. Yate, *England and Russia Face to Face in Asia: Travels with the Afghan Boundary Commission* (Edinburgh: W. Blackwood and Sons, 1887).
31. "The Proposed Trans-Persian Railway," *Scottish Geographical Magazine Scottish Geographical Magazine* 28, no. 4 (1912): 172.
32. Ármin Vámbéry, *Western Culture in Eastern Lands; a Comparison of the Methods Adopted by England and Russia in the Middle East* (London: J. Murray, 1906).
33. Henry G. Lyons, "The Basin of the Nile," *Geographical Teacher* 5, no. 4 (1910).
34. Arthur W. Jose, *The Growth of the Empire: A Handbook to the History of Greater Britain* (London: Murray, 1909), facing p. 404 available online at http://ebook.lib.hku.hk/CADAL/B31387135/.
35. Henry F. B. Lynch, *Railways in the Middle East* (London: Central Asian Society, 1911).
36. Henry F. B. Lynch, *Armenia, Travels and Studies* (London: Longmans, Green, 1901).
37. This distinction between an academic "Near East" and a geopolitical "Middle East" is still visible today. For the first occurrence of the phrase "Fertile Crescent," see Henry Thatcher Fowler, *A History of the Literature of Ancient Israel: From the Earliest Times to 135 B. C.* (New York: Macmillan, 1912), 2.
38. Halford J. Mackinder, "The Strategical Geography of the near East," *Journal of the Royal Artillery* 39 (1911).
39. David G. Hogarth, *The Nearer East* (London W. Heinemann, 1902).
40. Ibid., 1; see also Andrew J. Herbertson, "Notes on the Teaching of the Geography of the World," *Geographical Teacher* 1, no. 1 (1901); Spencer Wilkinson et al., "The Geographical Pivot of History: Discussion," *Geographical Journal* 23, no. 4 (1904); D. G. Hogarth, "Geographical Conditions Affecting Population in the East Mediterranean Lands," *Geographical Journal* 27, no. 5 (1906).
41. As noted by Roger Adelson, a colonial weekly entitled *The Near East* began to appear from 1908. Bartholomew's 1910 edition of *The Graphic Atlas of the World* (London: J. Walker, 1910) included an original map of the "Near East" which was inspired by Hogarth's work (see map no. 58).
42. There was a concentration of occurrences of the phrase "Middle East" in these titles in

1902. See *Western Times*, January 20, 1902, 2; *Western Times*, August 30, 1902, 2; *Derby Daily Telegraph*, October 16, 1902, 2; *Western Daily Press*, December 6, 1902, 5.
43. *Morning Post*, September 2, 1898, 5.
44. *London Daily News*, September 5, 1898, 5.
45. *Chart of the Beluchistan, Penny Illustrated Paper*, January 22, 1898, 54.
46. Mark Monmonier, "The Rise of Map Use by Elite Newspapers in England, Canada, and the United States," *Imago Mundi* 38 (1986).
47. *Sketch Map of Persia, Times*, September 25, 1907, 3.
48. *War Map of The Balkan Peninsula, Times* October 28, 1912, 21.
49. "The Near and Middle East," *Times*, May 24, 1913, 22.
50. Simon James Potter, *News and the British World: The Emergence of an Imperial Press System, 1876–1922* (Oxford: Oxford University Press, 2003).
51. *Punch*, March 14, 1900, 188.
52. *Punch*, April 18, 1900, 274.
53. John A. Hobson, *The War in South Africa Its Causes and Effects* (London: Nisbet, 1900), 251.
54. Benjamin G. Horniman, "The New Imperial Policy," *Times*, May 18, 1903, 10; Marmaduke Pickhall, "The Outlook in the Near East: For El Islâm," *Nineteenth Century and After* 72 (1912); Edward Granville Browne, *The Persian Revolution of 1905–1909* (Cambridge: Cambridge University Press, 1910).
55. *Illustrated London News*, November 16, 1912, 708.
56. Halford J. Mackinder, *Democratic Ideals and Reality: A Study in the Politics of Reconstruction* (London: Constable, 1919), 20.
57. Heffernan, "The Cartography of the Fourth Estate."
58. http://www.qualidata.essex.ac.uk/edwardians/about/introduction.asp.
59. Advertisement for "Mellin's Puzzle Map of Europe," *Illustrated London News*, November 26, 1898, 793.
60. John Mackenzie, "The Provincial Geographical Societies in Britain, 1884–1914," in Bell, Butlin, and Heffernan, *Geography and Imperialism, 1820–1940*.
61. William and Alexander K. Johnston, *The M. P. Atlas: A Collection of Maps Showing the Commercial and Political Interests of the British Isles* (W. and A. K. Johnston, 1907).
62. "New Maps," *Geographical Journal* 29, no. 2 (1907).
63. Adolf Stieler, *Stieler's Atlas of Modern Geography, Adapted for the Use of the English-Speaking Public*, trans. Darbishire B. V. (Gotha, 1907).
64. Ernest Rhys, *A Literary and Historical Atlas of Asia* (London: J. M. Dent, 1912).
65. *Times Atlas* (London, 1900).
66. Richard Andree, *Richard Andree's Allgemeiner Handatlas* (Bielefeld: Velhagen & Klasing, 1881). See "The Universal Atlas," *Geographical Journal* 1, no. 5 (1893).
67. Richard Scully, *British Images of Germany: Admiration, Antagonism & Ambivalence, 1860–1914* (Basingstoke: Palgrave Macmillan, 2012), 32.
68. John A. Hammerton, *The Harmsworth Universal Atlas and Gazetteer: 500 Maps and Diagrams in Colour, with Commercial Statistics and Gazetteer Index of 105,000 Names* (London: Amalgamated Press, 1907).
69. John G. Bartholomew, *The XXth Century Citizen's Atlas of the World: Containing 156 Pages of Maps and Plans with an Index, a Gazetteer, and Geographical Statistics* (London: George Newnes, 1902).
70. Ibid., iv.
71. See Edward G. Browne's preface to Mark Sykes, *Dar-Ul-Islam; a Record of a Journey through Ten of the Asiatic Provinces of Turkey* (London: Bickers & Son, 1904), xii.

72. John G. Bartholomew, *Atlas of the World's Commerce; a New Series of Maps with Descriptive Text and Diagrams Showing Products, Imports, Exports, Commercial Conditions and Economic Statistics of the Countries of the World* (London: G. Newnes, 1907); Isaac Pitman and Sons, *Pitman's Commercial Atlas of the World, with an Account of the Trade Productions, Means of Communication, and the Principal Statistics of Every Country of the Globe* (London: Sir Isaac Pitman & Sons, 1914).
73. George Adam Smith, *The Historical Geography of the Holy Land* (London: Hodder, 1894). See Aiken, *Scriptural Geography Portraying the Holy Land*, chap. 5.
74. Edward Woodfin, *Camp and Combat on the Sinai and Palestine Front: The Experience of the British Empire Soldier, 1916–18* (New York: Palgrave Macmillan, 2012), 178.
75. Keith Robbins, *History, Religion, and Identity in Modern Britain* (London: Hambledon Press, 1993), 123.
76. Jeffrey Cox, *The British Missionary Enterprise since 1700* (London: Routledge, 2007), 213.
77. See the incredible "Moslem Box" in Tusan, *Smyrna's Ashes*, 69.
78. See Edmund McClure, *Historical Church Atlas* (London: S.P.C.K., 1897), map XVIII.
79. S. M. Zwemer, *Methods of Mission Work among Moslems* (New York: Revell, 1906), 232.
80. See for example Shakeeb Salih, "The British-Druze Connection and the Druze Rising of 1896 in the Hawran," *Middle Eastern Studies* 13, no. 2 (1977); Necati Alkan, "Fighting for the Nuṣayrī Soul: State, Protestant Missionaries and the ʻalawīs in the Late Ottoman Empire," *Die Welt des Islams* 52, no. 1 (2012).
81. The papers of the mission are held by the Middle East Centre, Saint Antony's College, Oxford (accession GB165-037 and GB165-0161).
82. "Chairs of Geography in British Universities," *Geography* 46, no. 4 (1961); Gary S. Dunbar, *Geography: Discipline, Profession, and Subject since 1870; An International Survey* (Dordrecht: Kluwer Academic Publishers, 2001), 16–17.
83. Erskine Childers, *The Riddle of the Sands* (London: Sidgwick and Jackson, 1903).
84. George Taubman Goldie, "Twenty-Five Years' Geographical Progress," *Geographical Journal* 28, no. 4 (1906).
85. "Geography in British Universities," *Geographical Teacher* 2, no. 1 (1903).
86. "Geographical Courses in the Universities and University Colleges of the United Kingdom in 1904–5," *Geographical Teacher* 2, no. 6 (1904).
87. "Empire Notes," *Journal of the Royal Society of Arts* 62, no. 3208 (1914).
88. W. Maclean Carey, "The Correlation of Instruction in Physics and Geography," *Geographical Teacher* 5, no. 3 (1909).
89. Henry J. Findlay, "The Scope of School Geography," *Scottish Geographical Magazine* 30, no. 3 (1914).
90. *Report of the Board of Education for the year 1910–1911* (London: Eyre and Spottiswoode, 1912), 31.
91. Andrew J. Herbertson, "Recent Regulations and Syllabuses in Geography Affecting Schools," *Geographical Journal* 27, no. 3 (1906).
92. *School World*, June 1902, 229.
93. *School World*, March 1902, 108. On geography and secondary education in Scotland see J. Cossar, "The Teaching of Geography in the Secondary and Higher Grade Schools of Scotland," *Geographical Teacher* 6, no. 6 (1912).
94. "Royal Geographical Society's Syllabuses of Instruction in Geography," *Geographical Teacher* 2, no. 3 (1903).
95. Elizabeth Baigent, "Herbertson, Andrew John (1865–1915)," in *Oxford Dictionary of National Biography* (Oxford: Oxford University Press, 2004); online ed., January 2011, at

https://rprenet.bnf.fr:443/http/www.oxforddnb.com/view/article/40187, accessed 24 June 2014.
96. Andrew J. Herbertson, *Towns, Routes and Political Divisions of SW. Asia*, in *Junior Geography*, by A. J. Herbertson (Oxford: Clarendon Press, 1905), fig. 108.
97. Brian Hudson, "The New Geography and the New Imperialism: 1870–1918," *Antipode* 9, no. 2 (1977).
98. Andrew J. Herbertson, *A Geography of the British Empire* (Oxford: Clarendon Press, 1912).
99. W. P. Welpiox, "The Educational Outlook on Geography," *Geographical Teacher* 7, no. 5 (1914).
100. Bell, Butlin, and Heffernan, *Geography and Imperialism, 1820–1940*, 165–68.
101. Charles P. Lucas, *A Historical Geography of the British Colonies 1: The Mediterranean and Eastern Colonies*. (Oxford: Clarendon Press, 1906).
102. See data published in the *Report of the Board of Education for the year 1910–1911*, 12–30.
103. Ibid., 50–51.
104. *Report of the Board of Education for the year 1906–1907* (London: Eyre and Spottiswoode, 1907), 51.
105. McDougall, "The Publishing History and Development of School Atlases and British Geography, c.1870–c.1930."
106. Wood, "The Fine Line between Mapping and Map Making," 56. On the United States, see Schulten, *Mapping the Nation*.

CHAPTER 8

1. Fo 608/83/3: Memorandum from the Mid-Eastern Committee, March 12, 1919.
2. See Erik Goldstein, *Winning the Peace: British Diplomatic Strategy, Peace Planning, and the Paris Peace Conference, 1916–1920* (Oxford: Clarendon Press, 1991).
3. Biger, *The Boundaries of Modern Palestine, 1840–1947*.
4. Ibid., 111–12.
5. Fo 608/83/3, 241.
6. Fo 608/83/3, 313.
7. Elie Kedourie, *In the Anglo-Arab Labyrinth: The McMahon-Husayn Correspondence and Its Interpretations, 1914–1939* (Cambridge: Cambridge University Press, 1976).
8. Alan Sharp, *The Versailles Settlement: Peacemaking in Paris, 1919* (New York: St. Martin's Press, 1991); Malcolm Yapp, *The Making of the Modern near East, 1792–1923* (London: Longman, 1987).
9. Karen Culcasi, "Disordered Ordering: Mapping the Divisions of the Ottoman Empire," *Cartographica* 49, no. 1 (2014).
10. Crosbie Garstin, *Vagabond Verses* (London: Sidgwick, 1917), 41.
11. Peter Chasseaud and Peter Doyle, *Grasping Gallipoli: Terrain, Maps and Failure at the Dardanelles, 1915* (Staplehurst: Spellmount, 2005).
12. See the traces of the many maps of the area used by British forces in the National Archives and the British Library (BL, Maps 43336.(21.) 1915).
13. Ian S. Hamilton, *Gallipoli Diary*, vol. 1 (London: Edward Arnold, 1920), 25.
14. Ibid., 15.
15. Ibid., 14.
16. Peter Hart, *Gallipoli* (Oxford: Oxford University Press, 2011).
17. Hamilton, *Gallipoli Diary*, 178.

18. http://www.awm.gov.au/exhibitions/gmaps/landing/callaghan/.
19. Fred Waite, *The New Zealanders at Gallipoli* (Auckland: Whitcombe and Tombs, 1921), 72.
20. See British Library: *Plan of S.W. End of Gallipoli Peninsula*, April 25, 1915, scale 1:3,000, surveyed and drawn by Commander H. P. Douglas and Lt Nikolas (Add Ms 82485D).
21. Peter Doyle and Matthew Bennett, *Fields of Battle: Terrain in Military History* (London: Kluwer Academic, 2002), 163.
22. See chap. 4.
23. RGS Collections (accession 543275): *Map of the Anzac position, Gallipoli to illustrate Sir Ian Hamilton's dispatch of December 11th, 1915*, GSGS no. 2805.
24. British Library: MSS Eur C 313/3.
25. Edmund Candler, *The Long Road to Baghdad* (London: Cassell, 1919), 1:110.
26. RGS Collections (accession: 539228/mr Iraq Div.9): *Turkish Map of Mesopotamia*, scale 1:250,000.
27. *A Collection of First World War Military Handbooks of Arabia, 1913–1917, 3, 3* (Gerrards Cross: Archive Editions, 1988).
28. David G. Hogarth, "War and Discovery in Arabia," *Geographical Journal* 55, no. 6 (1920).
29. Candler, *The Long Road to Baghdad*, 117.
30. Oskar Teichman, *The Diary of a Yeomanry M. O. Egypt, Gallipoli, Palestine and Italy, by Captain O. Teichman* (London: T. Fisher Unwin, 1921), 213.
31. See Peter Collier, "The Impact on Topographic Mapping of Developments in Land and Air Survey: 1900–1939," *Cartography and Geographic Information Science* 29, no. 3 (2002).
32. Priya Satia, "The Defense of Inhumanity: Air Control and the British Idea of Arabia," *American Historical Review* 111, no. 1 (2006); Dov Gavish and Gideon Biger, "Innovative Cartography in Palestine 1917–1918," *Cartographic Journal* 22, no. 1 (1985).
33. Hogarth, "War and Discovery in Arabia," 422.
34. Thomas Hamshaw, "Geographical Reconnaissance by Aeroplane Photography, with Special Reference to the Work Done on the Palestine Front," *Geographical Journal* 55, no. 5 (1920).
35. See chap. 1.
36. See Satia, "The Defense of Inhumanity."
37. Michael J. Heffernan, "Geography, Cartography and Military Intelligence: The Royal Geographical Society and the First World War (1996)," *Transactions of the Institute of British Geographers* 21, no. 3 (1996): 504–33; Christopher Andrew and Jeremy Noakes, *Intelligence and International Relations, 1900–1945* (Exeter: University of Exeter, 1987); Michael Heffernan, "Mars and Minerva: Centres of Geographical Calculation in an Age of Total War (Mars Und Minerva: Zentren Geographischer Kalkulation Im Zeitalter Des Totalen Krieges)," *Erdkunde* 54, no. 4 (2000).
38. B. B. Cubitt, "War Work of the Society," *Geographical Journal* 53, no. 5 (1919).
39. Black, *Maps and History*, 102.
40. Heffernan, "The Cartography of the Fourth Estate," 269.
41. On maps published by Wellington House, see Mark A. Wollaeger, *Modernism, Media, and Propaganda British Narrative from 1900 to 1945* (Princeton, NJ: Princeton University Press, 2006), 20.
42. Isaiah Friedman, *The Question of Palestine, 1914–1918: British-Jewish-Arab Relations* (London: Routledge, 1973), 170. See FO/395/139, no. 42320: Lloyd George to Buchan, February 1, 1917.
43. Toynbee, *Turkey*; Toynbee, *Armenian Atrocities: The Murder of a Nation* (London: Hodder and Stoughton, 1915); Toynbee, *The Murderous Tyranny of the Turks* (London: Hodder and Stoughton, 1917).

44. Toynbee, *Armenian Atrocities*, 2–3; Toynbee, *Turkey*, 81–82.
45. http://www.iwm.org.uk/collections/item/object/1060022615.
46. Luke McKernan, "The Supreme Moment of the War: General Allenby's Entry into Jerusalem," *Historical Journal of Film, Radio and Television Historical Journal of Film, Radio and Television* 13, no. 2 (1993).
47. Eitan Bar-Yosef, "The Last Crusade? British Propaganda and the Palestine Campaign, 1917–18," *Journal of Contemporary History* 36, no. 1 (2001).
48. Ibid., 103.
49. National Army Museum, accession 1992-04-141.
50. Imperial War Museum, accession IWM: 1031.
51. Imperial War Museum, accession IWM: IWM 18.
52. Imperial War Museum, accession: IWM 1095.
53. Available online at http://www.criticalpast.com/video/65675026023_battle-of-Ctesiphon_defensive-position_51stdivision_counter-attack and http://www.criticalpast.com/video/65675026026_retreat-to-Kut_halt-position_British-battalion_attack-position.
54. See William Cartwright, "An Investigation of Maps and Cartographic Artefacts of the Gallipoli Campaign 1915, " in *Geospatial Visualisation*, ed. Antoni Moore and Igor Drecki (Berlin: Springer, 2013), 33–35.
55. George Philip, *Philip's Strategical Map of Mesopotamia & Asia Minor*, scale 200 miles to 5 in., 25 × 34 cm (London: G. Philip & Son, 1916). Online version available at http://gallica.bnf.fr/ark:/12148/btv1b53064713v. W. & A. K. Johnston, *War Map of the Middle East: illustrating the campaigns in Persia, Mesopotamia and the Caucasus*, 58 × 88 cm (London, 1916).
56. Akerman, *The Imperial Map*, 270.
57. Alexander Gross, *The Daily Telegraph War Map of Egypt and the Near East*, scale 1:42,240,000, 85 × 54 cm (London: Geographia, 1918). Online version available at http://nla.gov.au/nla.map-vn3100250.
58. See John M. McEwen, "The National Press during the First World War: Ownership and Circulation," *Journal of Contemporary History* 17, no. 3 (1982).
59. *Graphic*, July 24, 1915, 148.
60. *Graphic*, March 25, 1916, 411.
61. James Renton, "Changing Languages of Empire and the Orient: Britain and the Invention of the Middle East, 1917–18," *Historical Journal* 50, no. 3 (September 2007): 645–67.
62. George F. Morrell, "The Restoration of Mesopotamia," *Graphic*, July 14, 1917, 55.
63. George F. Morrell, "More Trials for the Turk," *Graphic*, November 10, 1917, 571.
64. Rosie Kennedy, *The Children's War: Britain, 1914–1918* (Basingstoke: Palgrave Macmillan, 2014), 121.
65. Albert Arthur Cock, *A Syllabus in War Geography and History for Use in Senior Classes in Elementary and Secondary Schools* (London: G. Philip & Son, 1916).
66. Walford, *Geography in British Schools, 1850–2000*, 83.
67. See Sonja Müller, "Toys, Games and Juvenile Literature in Germany and Britain during the First World War: A Comparison, " in *Untold War New Perspectives in First World War Studies*, ed. Heather Jones, Jennifer O'Brien, and Christoph Schmidt-Supprian (Leiden: Brill, 2008), chap. 8.
68. Michael Paris, *Over the Top: The Great War and Juvenile Literature in Britain* (Westport, CT: Praeger, 2004).
69. Frederick S. Brereton, *At Grips with the Turk: A Story of the Dardanelles Campaign* (London: Blackie and Son, 1916); Brereton, *On the Road to Bagdad, a Story of Townshend's Gallant Advance on the Tigris* (London: Blackie and Son, 1917); Brereton, *From the Nile to the Tigris; a Story of Campaigning from Western Egypt to Mesopotamia* (London: Blackie and Son, 1918);

Brereton, *With Allenby in Palestine, a Story of the Latest Crusade* (London: Blackie and Son, 1920).
70. Brereton, *On the Road to Bagdad*, 33.
71. Wallace E. Whitehouse, "Statistical Analysis of a Geography Examination 1," *Scottish Geographical Magazine* 36, no. 1 (1920).
72. Edward Mandell House and Charles Seymour, *What Really Happened at Paris: The Story of the Peace Conference, 1918–1919* (London: Hodder & Stoughton, 1921), 142.
73. Harold George Nicolson, *Peacemaking, 1919* (London: Constable, 1934), 269.
74. Ibid., 337.
75. See Steven Wright and Jonathan Casey, *Mental Maps in the Era of Two World Wars* (Basingstoke: Palgrave Macmillan, 2008), xiii.
76. Naval Intelligence Division, *A Handbook of Syria: Including Palestine* (London: H.M. Stationery Office, 1920).
77. FO 371/4352/f18/PC18.
78. Jeremy W. Crampton, "The Cartographic Calculation of Space: Race Mapping and the Balkans at the Paris Peace Conference of 1919," *Social and Cultural Geography* 7, no. 5 (2006).
79. FO 608/108, "Map III for Table III showing Racial Majorities by Districts," in Intelligence Dept, Naval Staff, *Populations of the Armenian Provinces of Transcaucasia* (June 25, 1919), 127.
80. See O'Shea, *Trapped between the Map and Reality*.
81. *Arab Bulletin*, March 12, 1917.
82. FO 608.96, map no. 2.
83. See also FO 608/96/9: *Ethnographical map of Turkey*, 1919.
84. See Paul Crook, *Darwinism, War and History: The Debate over the Biology of War from the "Origin of the Species" to the First World War* (Cambridge: Cambridge University Press, 1994).
85. Karen Culcasi, "Cartographic Constructions of the Middle East" (Syracuse University, 2008).
86. FO 925/41378.
87. MPK 1/264. The most surprising feature of Sykes's map is probably the revealing figuration of "thinly populated region like parts of Bavaria and the Riviera; fit for European colonization." The British agent's infatuation with the prospect of turning the Middle Eastern wasteland into a land flowing with milk and honey was relatively common, as stated in the previous chapters.
88. MPK 1/397/1 and 2: Maps of the oilfields of Persia and Mesopotamia (NID memorandum, February 26, 1919).
89. CAB 23/43: *Notes of a Meeting of the Prime Ministers of the United Kingdom and of the Overseas Dominions and the British War Cabinet* (August 13, 1918), 146.
90. Henry C. King and Charles R. Crane, *Report of American Section of Inter-Allied Commission on Mandates in Turkey: An Official United States Government Report* (New York: s.n., 1922), vii.
91. FO 608/83/3: Map of a projected Chaldean-Christian state (August 1919), 564.
92. Michael Llewellyn Smith, *Ionian Vision: Greece in Asia Minor, 1919–1922* (New York: St. Martin's Press, 1973).
93. Georgios Soteriades, *An Ethnological Map Illustrating Hellenism in the Balkan Peninsula and Asia Minor* (London: Stanford, 1918).
94. The use of cartography and geography to support the Zionist stance began before 1914. The English Zionist Federation and the London Zionist League were particularly active on that front (see *The Physical and political conditions of Palestine* [London: Rabbinowicz,

1907], 35–40). Aaron Aaronsohn (1876–1919), one of the architects of the NILI, a Zionist spy ring that helped the British Army in Palestine during the war, surveyed potential settlements sites for Jewish immigrants as early as 1901; see Aaron Aaronsohn, *Agricultural and botanical explorations in Palestine* (Washington: Govt. Print. Off., 1910).

95. See Wood, *Rethinking the Power of Maps*; Biger, *The Boundaries of Modern Palestine, 1840–1947*.
96. Polelle, *Raising Cartographic Consciousness*.
97. Mackinder, *Democratic Ideals and Reality*. On the French approach to geography applied to the peace negotiations, see Olivier Lowczyk, *La Fabrique de la Paix: Du Comité d'Études à la Conférence de la Paix, l'Élaboration par la France des Traités de la Première Guerre Mondiale* (Paris: Economica: Institut de Stratégie comparée, 2010).
98. Dimitri Kitsikis, *Le Rôle des Experts à la Conférence de la Paix de 1919* (Ottawa: Editions de l'Université, 1972); B. C. Busch, *Mudros to Lausanne: Britain's Frontier in West Asia, 1918–1923* (State University of New York Press, 1976); Houshang Sabahi, *British Policy in Persia 1918–1925* (London: Cass, 1990); Marian Kent, *The Great Powers and the End of the Ottoman Empire* (London: G. Allen & Unwin, 1984); Fisher, *Curzon and British Imperialism in the Middle East*; Lucian M. Ashworth, "Mapping a New World: Geography and the Interwar Study of International Relations," *International Studies Quarterly* 57, no. 1 (2013); Rory Miller, *Britain, Palestine and Empire: The Mandate Years* (Aldershot: Ashgate Pub., 2010); Christopher N. B. Ross, "Lord Curzon and E. G. Browne Confront the 'Persian Question,'" *Historical Journal* 52, no. 2 (2009).
99. Jacques Ancel, *Peuples et Nations des Balkans* (Paris: Armand Colin, 1926), 1.
100. David Lloyd George, *Memoirs of the Peace Conference*, 1939, 2:722.
101. On Lloyd George's "preference for private rather than official advice," see John F. Naylor, *A Man and an Institution: Sir Maurice Hankey, the Cabinet Secretariat and the Custody of Cabinet Secrecy* (Cambridge: Cambridge University Press, 1984), 75. Lloyd George's philhellenism, which led him to support Greek claims over Smyrna with devastating consequences, was also one of the legacies of his upbringing, see Casey, *Mental Maps in the Era of Two World Wars*, 21.
102. IOR/MSS Eur F 112/267 and 268: Report on Zionism by E. Montagu (October 9, 1917).
103. It must be noted the distinction between experts with an academic or specialized background and policy makers was not always clear cut. The most structured delegation from that point of view was the American one. In both French and British delegations, diplomats were often appointed as experts: the lines were blurred. Moreover, the influence of expertise could vary considerably in relation with the personality of the main negotiators.
104. David Fromkin, *A Peace to End All Peace: Creating the Modern Middle East, 1914–1922* (London: Penguin, 1989); Westrate, *The Arab Bureau*.
105. Fromkin, *A Peace to End All Peace*, 90.
106. British Library, MSS Eur F111–112 Series 68, Curzon Papers.
107. Goldstein, *Winning the Peace*, 171.
108. This has already amply documented in other works. See Peter Sluglett, *Britain in Iraq: Contriving King and Country, 1914–1932* (New York: Columbia University Press, 2007); B. C. Busch, *Britain, India, and the Arabs, 1914–1921* (London: University of California Press, 1971); Toby Dodge, *Inventing Iraq: The Failure of Nation-Building and a History Denied* (New York: Columbia University Press, 2003).
109. Records of the French Armed Forces (SHD Vincennes, accession GR 16 N 3200), Letter from General Bailloud, October 4, 1917: "Pour un certain nombre de fonctionnaires anglais d'Egypte—je ne dis pas d'Angleterre—l'union sacrée a pour bornes le continent."

110. Records of the French Armed Forces (SHD Vincennes, accession GR 16 N 3205), Report by Poidebard, Bassorah, April 14–15, 1918: "Les Officiers anglais qui reviennent de France nous sont très sympathiques, les fonctionnaires et officiers de l'armée des Indes chargés de l'expédition et de l'administration de la Mésopotamie nous demeurent très opposés. Ils mènent contre nous une lutte d'influence politique et commerciale où nous ne sommes guère traités en alliés."
111. See chap. 6.
112. British Library, Add MS 52455.
113. British Library, Add MS 52455, Wilson, dated May 9, 1919.
114. See FO 608/83/3, which includes a report from the IO showing how Wilson supported the creation of a Christian enclave in the Mosul area for his "Nestorian clients."
115. Timothy J. Paris, "British Middle East Policy-Making after the First World War: The Lawrentian and Wilsonian Schools," *Historical Journal* 41, no. 3 (1998).
116. British Library, Add MS 52455, Office of the Civil Commissioner Baghdad, August 14, 1919.
117. British Library, Add MS 52455, ff. 58.
118. See Harold Tollefson, *Policing Islam: The British Occupation of Egypt and the Anglo-Egyptian Struggle over Control of the Police, 1882–1914* (Westport, CT: Greenwood Press, 1999), 40.
119. FO 925/41378: 3.
120. CAB/24/45: Memorandum on the Eastern Committee, March 15, 1918.
121. CAB/24/66.
122. FO 608/161/4.
123. CAB/24/117: *The Situation in the Middle East*, December 16, 1920.
124. CAB/23/5: Minutes of a Meeting of the War Cabinet, March 21, 1918.
125. *L'Humanité*, no. 5788 (February 26, 1920).
126. Records of the French Armed Forces (SHD Vincennes), GR 16 N 3200.
127. CAB 27/1, map no. 5.
128. "Notes on policy in the Middle East," October 28, 1918, General Sir George MacDonogh.
129. Curzon gave a similar description of the Middle East later, see CAB 24/45.
130. FO 371/4352/f18/PC18, *Peace Memoranda*, October 10, 1918.
131. Helmut Mejcher, "British Middle East Policy, 1917–21: The Inter-departmental Level," *Journal of Contemporary History* 8, no. 4 (1973).
132. CAB/24/107/34: Future Administration of the Middle East, June 8, 1920.
133. CAB/24/107/2: *Mesopotamia and the Middle East: Question of future Control*, memorandum by Secretary of State for India, June 1, 1920.
134. MFQ 1/435: Detailed report in FO 800/221: *Proposed Middle East Department*, July 29, 1918.
135. FO 800/221.22: *Rough Draft of Scheme of Organization for Middle Eastern Affairs*, July 29, 1918.
136. CAB 21/186, C.P. 2545, February 7, 1921.
137. CO 323/869/10: *Extension of loan of services to the Colonial Office Middle East Department*.
138. CO 850/269: *Appointment of Lt. Col. T. E. Lawrence*.
139. CO 323/873/17: *Transfer of officers from the Foreign Office and the India Office to the Middle East Department of the Colonial Office*, February–March 1921.
140. CO 1047/584, *Bartholomew's Map of the Middle East Coloured to Show Height of Land*, scale 1:4,000,000, 59 × 92 cm (Edinburgh: Bartholomew, 1921).
141. Vaughan Cornish, *The Axis of the British Empire*, in Vaughan Cornish, *A Geography of Imperial Defence* (London: Sifton, Praed, 1922).
142. Leonard Woolf, *Empire & Commerce in Africa: A Study in Economic Imperialism* (London: Allen & Unwin, 1919), facing pp. 170 and 174.

143. As stated by Churchill, see HC Deb, March 9, 1922, vol. 151, col. 1550.
144. Quoted in Keith Jeffery, *The British Army and the Crisis of Empire, 1918–22* (Manchester: Manchester University Press, 1984), 132.

CONCLUSION

1. See Ruth Kark, "Mamluk and Ottoman Cadastral Surveys and Early Mapping of Landed Properties in Palestine," *Agricultural History* 71, no. 1 (1997). On Eastern Christians, see chap. 1.
2. Jean Baudrillard, *Simulacres et Simulation* (Paris: Éditions Galilée, 1981), 10.
3. See Blais, *Mirages de la Carte l'Invention de l'Algérie Coloniale*.
4. See chap. 1.
5. Antoinette M. Burton, *Empire in Question: Reading, Writing, and Teaching British Imperialism* (Durham, NC: Duke University Press, 2011), 283.
6. Mrinalini Sinha, *Colonial Masculinity: The "Manly Englishman" and the "Effeminate Bengali" in the Late Nineteenth Century* (Manchester: Manchester University Press, 1995), 7.
7. Antoinette Burton, *The Trouble with Empire: Challenges to Modern British Imperialism* (Oxford: Oxford University Press, 2015).
8. See Patrick Boucheron, *L'Entretemps: Conversations sur l'Histoire* (Paris: Verdier, 2012).

NOTES ON METHODOLOGY AND SELECT BIBLIOGRAPHY

I adopted a twofold approach in order to be able to work on both a microhistorical and a macrohistorical level. I applied quantitative analysis to newspapers databases, 19th-century catalogs, RGS publications, and official records. Statistics on map acquisitions by the FO or the WO, data on geographical appellations frequency, and map sales figures thus provide a basis for the study of both knowledge production and its reception. Prominent illustrated papers such as the *Illustrated London News* and the *Graphic* were systematically examined. I chose a qualitative approach to surveys and explorations. Personal papers, diaries, and field notes contain a wealth of information on British agents and local responses. As mentioned in the book, documenting counterdiscourses is a challenge. I found isolated maps and indirect evidence of countercartography in the French National Library records as well as in British reports and memoirs. Finally, local archives, specialist periodicals, and official reports were crucial to get a clear picture of the uses of cartography in schools.

The detail of sources used for this work can be found in the extensive endnotes. Documentation for the study of such a geographically and chronologically wide topic is inexhaustible. I therefore chose to list some of the most relevant primary sources used to research this book and a select bibliography.

1. UNPUBLISHED PRIMARY SOURCES

BIBLIOTHÈQUE NATIONALE DE FRANCE, MAPS AND PLANS

GE C, GE F, GE FF—British Maps from the Roland Bonaparte Collection
SG—Maps from the Société de Géographie

BODLEIAN LIBRARY, OXFORD

E13—Maps of Egypt
D26—Maps of Palestine

BRITISH LIBRARY, MANUSCRIPT COLLECTIONS

Add MSS 55069–70—Percy Molesworth Sykes Papers
Add MSS 53817–54155—Anne Blunt Papers
Add MS 69848: 1872–1877—Survey of Western Palestine
Add MS 52455—Percy Cox and A. T. Wilson Correspondence

BRITISH LIBRARY, MAP COLLECTION

Collection of military maps of the Gallipoli Peninsula; Maps 43336.(21.) 1915

BRITISH LIBRARY, ORIENTAL AND INDIA OFFICE COLLECTION

The Qatar Digital Library
L/PS/7, 10, 11, 18—Political and Secret Department Records
IOR/MSS Eur F 111–112—Curzon Papers
IOR/MSS Eur F 126—Correspondence to and from Lewis Pelly
IOR/L/SUR/6/6—Geographical Department
IOR/W/L/PS/21—Geographical Section

FRENCH MILITARY RECORDS, VINCENNES

GR 16 N 3200, GR 16 N 3205—Reports from Egypt (First World War)

GETTY RESEARCH INSTITUTE, LOS ANGELES

A2.F—Pierre de Gigord Collection

IMPERIAL WAR MUSEUM

First World War propaganda movies: IWM: 18, 1031, 1098

LOCAL ARCHIVES

Sussex Record Office (Chichester), RSR/PH/11/23
Bristol Record Office, 40477/F/2—J. T. Francombe
Wiltshire and Swindon Record Office 1585/146—lesson book

MIDDLE EAST CENTRE ARCHIVE (MECA), ST. ANTONY'S COLLEGE, OXFORD

GB165–0179—Gerard Leachman
GB165–0276—Percy Molesworth Sykes
GB165–274—Herbert Sykes

NATIONAL ARCHIVES, UK

CAB 11, 23, 24, 27, 37
CO 323, 700, 850, 1047
FO 375, 608, 800, 881, 925
MFQ 1/28, 1/435, 1/862/5, 1/869/1–5
MPHH 1/5, 1/114/2
MPK 1/264, 1/397/1 and 2, 1/436/2, 1/482/6
WO 1, 32, 78, 106, 302, 303

NATIONAL ARCHIVES OF INDIA, HISTORICAL MAPS OF THE SURVEY OF INDIA

N.B.: These records were not directly consulted. Analysis of the Survey of India's map collection was based on S. N. Prasad, ed., *Catalogue of the Historical Maps of the Survey of India (1700–1900)* (New Delhi: National Archives of India, 1977).

F.92/16–21—Arabia
F.93/27—Mesopotamia

F.93/33—Persia
F.105/10—Persian Gulf

NATIONAL LIBRARY OF SCOTLAND

Bartholomew Archives

ROYAL GEOGRAPHICAL SOCIETY

Map Collection
RGS Papers: Council, committees, bylaws: 1830–1993
William H. I. Shakespear Papers

ORAL HISTORY SOCIETY

Edwardians On-Line, *The Edwardians: Family Life and Work Experience before 1918*. Colchester, Essex: Qualitative Data Service. Q197 / 07/95—QDD/Thompson1/FLWE.

2. PRINTED SOURCES

2.1 OFFICIAL PAPERS

Cambridge Archive Editions (for instance, Burdett, Angela. *Iran in the Persian Gulf: 1820–1966* [Slough: Archive ed., 2000])
Minutes of the Committee of Council on Education
Report of the Board of Education
Report of the Committee of Council on Education
British Parliamentary Papers

2.2 NEWSPAPERS AND PERIODICALS

Geographical Teacher
Illustrated London News
Journal and the Proceedings of the RGS
Journal of the RUSI
London Times
19th Century British Library Newspapers Database
Quarterly Journal of the Geological Society of London
School World
Scottish Geographical Magazine

2.3 ESSAYS, ATLASES, MEMOIRS, TRAVELOGUES

Anon. *A Gazetteer of the World; or, Dictionary of Geographical Knowledge.* 7 vols. Vol. 5. Edinburgh: Fullarton, 1856.

Arrowsmith, Aaron. *A Compendium of Ancient and Modern Geography, for the Use of Eton School.* London: E. Williams, 1839.

Badger, George Percy. *History of the Imâms and Seyyids of 'Omân.* London: Hakluyt Society, 1871.

Baedeker, Karl. *Egypt: Handbook for Travellers: Part First, Lower Egypt, with the Fayum and the Peninsula of Sinai.* Leipzig: Baedeker, 1885.

Baker, Valentine. *Clouds in the East: Travels and Adventures on the Perso-Turkoman Frontier.* London: Chatto and Windus, 1876.

Bartholomew, J. G. *Atlas of Commercial Geography Illustrating the General Facts of Physical, Political, Economic and Statistical Geography.* Cambridge: University Press, 1889.

———. *Atlas of the World's Commerce; a New Series of Maps with Descriptive Text and Diagrams Showing Products, Imports, Exports, Commercial Conditions and Economic Statistics of the Countries of the World.* London: G. Newnes, 1907.

———. *The Royal Shilling Atlas; a Complete Series of Maps with Index.* London: T. Nelson, 1892.

———. *The XXth Century Citizen's Atlas of the World: Containing 156 Pages of Maps and Plans with an Index, a Gazetteer, and Geographical Statistics.* London: Newnes, 1902.

Black, Charles. *A Memoir on the Indian Surveys, 1875–1890.* London: E. A. Arnold, 1891.

Blunt, Anne, and Wilfrid Scawen Blunt. *A Pilgrimage to Nejd, the Cradle of the Arab Race. A Visit to the Court of the Arab Emir, and "Our Persian Campaign."* London: J. Murray, 1881.

Brereton, F. S. *At Grips with the Turk: A Story of the Dardanelles Campaign.* London: Blackie and Son, 1916.

Burton, Richard Francis. *The Gold-Mines of Midian and the Ruined Midianite Cities: A Fortnight's Tour in Northwestern Arbia.* London: C. Kegan Paul, 1878.

Canning, Stratford. *The Eastern Question.* London: J. Murray, 1881.

Chirol, Valentine. *The Middle Eastern Question; or, Some Political Problems of Indian Defence.* London: J. Murray, 1903.

Chisholm, George G. *Handbook of Commercial Geography.* London: Longmans, 1889.

Cochran, Peter. *Byron and Orientalism.* Newcastle: Cambridge Scholars Press, 2006.

Cock, Albert Arthur. *A Syllabus in War Geography and History for Use in Senior Classes in Elementary and Secondary Schools.* London: G. Philip & Son, 1916.

Conder, Claude R. *Tent Work in Palestine. A Record of Discovery and Adventure.* London: R. Bentley & Son, 1878.

———. *The Survey of Western Palestine. Jerusalem.* London: Committee of the Palestine Exploration Fund, 1884.

Curzon, G. *The Persian Question.* London: Longmans, 1892.

Doughty, Charles Montagu. *Travels in Arabia Deserta.* Cambridge: Cambridge University Press, 1888.

Dower, John W. *A School Atlas of Modern Geography.* London: Ward & Lock, 1854.

Goldsmid, F. J. *Telegraph and Travel: A Narrative of the Formation and Development of Telegraphic Communication between England and India.* London: Macmillan, 1874.

Hamilton, A. *Problems of the Middle East.* London: E. Nash, 1909.

Harmsworth, George Philip. *The Harmsworth Universal Atlas and Gazetteer: 500 Maps and Diagrams in Colour, with Commercial Statistics and Gazetteer Index of 105,000 Names*. London: The Amalgamated Press, 1907.

Henty, G. A. *With Kitchener in the Soudan: A Story of Atbara and Omdurman*. London: Blackie, 1903.

Herbertson, A. J. *A Geography of the British Empire*. Oxford: Clarendon Press, 1912.

Herschel, John F. W. *A Manual of Scientific Enquiry: Prepared for the Use of Her Majesty's Navy: And Adapted for Travellers in General*. London: J. Murray, 1849.

Hertslet, Edward. *The Map of Africa by Treaty*. Vol. 1, London: Harrison, 1894.

Hogarth, David George. *The Penetration of Arabia: A Record of the Development of Western Knowledge Concerning the Arabian Peninsula*. London: Lawrence and Bullen, 1904.

Holdich, Thomas H. *The Indian Borderland, 1880–1900*. London: Methuen, 1901.

Jackson, Lowis D'Aguilar. *Aid to Survey-Practice for Reference in Surveying, Levelling, and Setting-Out*. London: Lockwood, 1880.

Johnston, Alexander Keith. *The Royal Atlas of Modern Geography*. Edinburgh: William Blackwood, 1868.

———. *A School Atlas of Physical Geography, Illustrating, in a Series of Original Designs, the Elementary Facts of Geology, Hydrology, Meterorology and Natural History*. Edinburgh; London: W. Blackwood and Sons, 1852.

Jones, James Felix. *Memoirs of Baghdad, Kurdistan and Turkish Arabia, 1857: Selections from the Records of the Bombay Government, No. Xliii.—New Series*. Farnham Common, Slough, UK: Archive Editions, 1998.

Keltie, John Scott. *Geographical Education.—Report to the Council of the Royal Geographical Society*. London: J. Murray, 1886.

Lynch, H. F. B. *Armenia, Travels and Studies*. London: Longmans, Green, 1901.

Lyons, H. G. *The Cadastral Survey of Egypt, 1892–1907*. Cairo: Nat. Print. Dept., 1908.

MacGregor, John. *The Rob Roy on the Jordan, Nile, Red Sea & Gennesareth, &C.: A Canoe Cruise in Palestine and Egypt, and the Waters of Damascus*. London: J. Murray, 1869.

Mackinder, Halford J. *Democratic Ideals and Reality: A Study in the Politics of Reconstruction*. London: Constable and company, 1919.

Mahan, A. T. *Retrospect & Prospect: Studies in International Relations Naval and Political*. London: Sampson Low, Marston, 1902.

Markham, Clements R. *A Memoir on the Indian Surveys*. London: W. H. Allen, 1878.

Murray, John, ed. *A Handbook for Travellers in Turkey: Describing Constantinople, European Turkey, Asia Minor, Armenia, and Mesopotamia*. London: J. Murray, 1854.

Palgrave, William Gifford. *Narrative of a Year's Journey through Central and Eastern Arabia (1862–1863)*. London: Macmillan, 1865.

SDUK. *The Cyclopaedian: Or Atlas of General Maps*. London: Edward Stanford, 1857.

———. *Maps of the Society for the Diffusion of Useful Knowledge*. London: Chapman and Hall, 1844.

Smith, George Adam. *The Historical Geography of the Holy Land*. London: Hodder, 1894.

Thuillier, H. L., and R. Smyth. *A Manual of Surveying for India*. Calcutta: Thacker, 1875.

Toynbee, Arnold. *Armenian Atrocities: The Murder of a Nation*. London: Hodder and Stoughton, 1915.

Vámbéry, Ármin. *Western Culture in Eastern Lands; a Comparison of the Methods Adopted by England and Russia in the Middle East*. London: J. Murray, 1906.

White, Arthur Silva *The Expansion of Egypt: Under Anglo-Egyptian Condominium*. London: Methuen, 1899.
Willcocks, William. *Egyptian Irrigation*. London: E. & F. N. Spon, 1889.
Wyld, James. *Notes to Accompany Mr. Wyld's Model of the Earth*. London: J. Wyld, 1851.

3. SECONDARY WORKS

My research owes a great deal to Martin W. Lewis and Kären E. Wigen, *The Myth of Continents: A Critique of Metageography* (1997). Matthew H. Edney, *Mapping an Empire: The Geographical Construction of British India: 1765–1843* (1997), was another key reference for this book. In the expanding field of critical cartography, the History of Cartography series (University of Chicago Press), J. B. Harley, P. Laxton, and J. H. Andrews, *The New Nature of Maps: Essays in the History of Cartography* (2002), and Martin Dodge, Rob Kitchin, and C. R. Perkins, *Rethinking Maps: New Frontiers in Cartographic Theory* (2009), are good starting points. Felix Driver, *Geography Militant: Cultures of Exploration and Empire* (2001) is one of the best surveys on the relations between geography and 19th-century imperial expansion. Priya Satia, *Spies in Arabia: The Great War and the Cultural Foundations of Britain's Covert Empire in the Middle East* (2008), provides telling examples of how intelligence collection shaped policies in the area. James Onley, *The Arabian Frontier of the British Raj: Merchants, Rulers, and the British in the Nineteenth Century Gulf* (2007), focuses on non-European collaboration. Derek Gregory, *Geographical Imaginations* (1994), is a brilliant exploration of the relations between space and representations. More detailed examination of the links between empire and culture is to be found in the Studies in Imperialism Series (Manchester University Press). Malcolm Yapp, *The Making of the Modern Near East, 1792–1923* (1987), and Bernard Lewis, *The Shaping of the Modern Middle East* (1994), provide a clear overview on the modern history of the area. Mark Monmonier, *Maps with the News: The Development of American Journalistic Cartography* (1989), and Michael Heffernan's chapter in James R. Akerman, *The Imperial Map: Cartography and the Mastery of Empire* (2008), provide crucial evidence on the birth of mass cartography. Rex Walford, *Geography in British Schools, 1850–2000: Making a World of Difference* (2001), is a useful survey on school geography. To finish with, anyone looking for a comprehensive account of the development of the British Empire may begin with the *Oxford History of the British Empire* and John Darwin, *The Empire Project* (2009).

INDEX

Aaronsohn, Aaron, 235, 314n94
Aberdeen, 4th Earl of (George Hamilton-Gordon), 48
Abdul-Hamid II, 135
Acre, 28, 64
Aden, 12, 24–25, 36, 82, 96, 102–3, 107, 118–19, 146–47, 174, 185–86, 196–98, 201, 262
Admiralty, 19, 21, 62, 70, 99, 102, 103, 174–75, 199, 200, 238; Foreign Intelligence Committee, 103; Hydrographic Office, 36; Naval Intelligence Department, 103, 247, 254. *See also* surveying practices: hydrographic surveys
Adrianople, 201, 215
Afghanistan, 82, 135, 169, 197, 211
Africa: eastern limit of, 52–53, 54 fig. 2.1–2.6, 56, 59–60; exploration of, 110, 130, 174; strategic position within the Empire, 124, 145, 217, 219, 263
African Association, 9
Ahmed, Rafiuddin, 205
Ahwaz, 183
Albania, 58
Aleppo, 112

Alexander the Great, 64, 87, 288n61
Alexandria, 21, 158, 164
Alford, C. J., 178
Ali, Ahmed, 132
Al-Jazira, region, 57, 85
Allenby, Edmund, 111–12, 225, 233, 238, 243, 248, 250
American Palestine Exploration Society, 110
Amery, Leopold, 264
Amman, 111
Anatolia, 64, 106, 112, 126, 189, 204, 235
Andree, Richard, 222
Anglo-Persian Oil Company, 199–200
Anglo-Persian War (1856–57). *See* Persia
Anglo-Turkish Convention (1878). *See* Cyprus: Convention (1878)
Anville, Jean-Baptiste Bourguignon d', 2, 47
ANZAC, 240
Aqaba, Gulf of, 24, 52, 59–60
Arab Bureau, 242, 257, 259–60, 262
Arabia, 2, 10, 24, 25, 27, 28, 33, 34 fig. 1.3, 34–35, 38 fig. 1.4, 41, 48, 50, 52, 53, 61, 62, 64, 65, 78, 80, 88, 102, 107, 112, 114, 115, 117–20, 123,

Arabia (*continued*)
124, 126–30, 129 fig. 4.4, 137, 145–47, 155, 159, 172, 180–88, 181 fig. 6.2, 196–98, 200, 205, 223, 233, 239, 242–43, 246, 253, 257, 259, 261–62, 272

Arabian Sea, 2, 3, 131
Arab nationalism, 196, 235, 255, 260
Ararat, mount, 204
Ardagh, John, 172
Armenia, 29, 56, 203; genocide (1917), 248; mapping of, 20; massacres (1894–1897), 12, 150, 154; at the Paris conference, 8, 70, 105, 236, 253, 262; stereotypes on, 57
Armstrong, George, 157
Arnold, Arthur, 117
Arrowsmith, Aaron, 70, 85, 101
Arrowsmith, John, 47, 57, 76, 142
Asia Minor, 48, 57, 65, 85–86, 102, 105, 112, 194, 203, 220, 249, 252, 256
Aswan Dam, 122–23
Atchley, Chewton, 103
authorship, 72–75, 73 fig. 3.2, 141–43
Aznavour, Serovpe, 69
Azov, Sea of, 74, 105

Babbage, Charles, 126
Babylon, 77
Bacon, George W., 107, 139–41, 140 fig. 5.1, 143, 147, 207–8, 208 fig. 7.1
Badger, George Percy, 29–30, 119, 226
Baedeker, Karl, 155, 181
Bagdadbahn, 180, 190, 199, 221, 304n80
Baghdad, 40, 118, 181–82, 184, 191, 236, 238, 241, 243, 245 fig. 8.4, 246 fig. 8.5, 250
Bahrain, 12
Bailloud, Maurice, 258
Baker, Samuel, 96, 122
Baker, Valentine, 3, 116, 238n14
Baku, 124
Balkans: definition of, 55–56, 201–2, 206, 215, 268; as a geopolitical issue, 58, 95, 105–6, 152, 153 fig. 5.5, 201, 218
Baluchistan, 25, 30, 114–15, 131–32, 153, 197, 213, 218
Bandar Abbas, 194
Baring, Evelyn (Lord Cromer). *See* Cromer, 1st Lord of (Evelyn Baring),
Barker, William Charles, 24
Bartholomew, firm, 27, 61, 142–45, 147 fig. 5.2, 147, 148 fig. 5.3, 163, 165, 222, 224, 231, 267
Bartholomew, John, Jr., 70, 74
Bartholomew, John G. *See* Bartholomew, firm
Basra, 61, 191
Beaufort, Francis, 21, 58

Beautemps-Beaupré, Charles-François, 36
Beersheba, 110, 235
Beirut, 20–21, 255
Belgrade, 215
Bell, Gertrude, 112, 189–91, 233, 236, 259–60
Bent, Mabel, 186
Bent, Theodore, 186
Berlin-Baghdad railway. *See* Bagdadbahn
Bertie, Francis, 102–3, 147, 172
Besant, Walter, 112
Betts, John, 78
Bewsher, James B., 26
biblical culture, influence of, 6, 37, 43, 44, 51, 59, 61, 62–64, 81–87, 111, 128, 141, 144, 157–58, 162, 164, 188–91, 206, 208, 214, 224, 232, 235, 242, 248, 257, 272; Archeology and, 42, 109; Millenarianism and, 41, 43, 226; Scripture geography and, 59, 62–64, 87–89, 225–26
Birjand, 132, 169
Bismarck, Otto von, 100
Blackie, William G., 52, 55 fig. 2.2
Black Sea, 3, 48, 56, 62, 67, 74, 105, 124, 261
Blunt, Anne, 112, 126–28, 130, 161, 183, 190
Blunt, Wilfrid Scawen, 25, 223. *See also* Blunt, Anne
Board of Education, 229–31
Board of Trade, 102, 165, 225
Bookseller, 71
Bosphorus, 56, 124, 240
Bowman, Isaiah, 251
Boy's Own Paper, 163
Brackenbury, Henry, 104, 106
Bradshaw, George, 155
Brant, James, 22
Brereton, Frederick S., 250
Brion, Henry F., 77, 157
British and Foreign School Society, 86–89
British Educator, 57
British Geological Survey, 22
British India Steam Navigation Company, 61
British Quarterly Review, 35, 56
Buchan, John, 247, 249
Buckstone, John Baldwin, 76
Bulgaria, 95, 256; 1878 atrocities in, 141, 150, 154, 220; stereotypes on, 56
Bulgarian atrocities, 141, 150, 154, 220
Bulletin de la Société des études coloniales et maritimes, 261
Burton, Richard, 28, 35, 102, 123, 143, 181, 283n63
Bury, G. Wyman, 185–86, 187 fig. 6.3
Bushire, 25, 37, 118–19, 212
Butler, George, 52, 54 fig. 2.1
Byron, influence of on imaginations. *See* Byronism
Byronism, 6, 27–28, 126

Cairo, 3, 28, 80, 122, 124, 140, 149, 154, 155, 176–77, 189, 198, 219, 226, 236, 242, 246, 260–61, 269
Cairo Conference (1921), 12, 246, 257–58
Cambridge University, 159, 214, 229
Candler, Edmund, 241
Canning, Stratford, 23, 58
Carnarvon Commission (1880), 174
Carne, John, 2
cartography. *See* maps
Caspian Sea, 22, 57, 60, 105, 123
Cassel, Ernest, 199
Cassell's Weekly Dispatch Atlas, 80
Catellani, Enrico, 210
Cattaui, Yacoub, 177
Cecil, Robert, 260
Central Asia, 3, 101, 105, 116, 130, 154, 162, 214, 253, 261
Chah-Rigan, 194
Chamberlain, Joseph, 173, 175
Chambers's Journal, 80
Charteris, Francis, 97
Cheetham, Milan, 235, 236
Chelmsford Chronicle, 79
Chermside, Herbert, 106, 112–13
Chesney, Francis, 25, 119
Childers, Erskine, 227
Chirol, Valentine, 173, 210–12, 211 fig. 7.2, 214, 217, 224, 259
chromolithography, 17, 52, 71, 128, 143–44, 289n14
Churchill, George, 171
Churchill, Henry A., 23
Churchill, Randolph, 115
Churchill, Winston, 211, 238, 246, 257–58, 262–63, 269
Church Missionary Society, 29, 63
Cilicia, 203
Clarendon Commission 1864, 84
Clayton, Gilbert, 179, 198
Clayton, William, 116
Clemenceau, George, 233
Cobden, William, 88
Cock, Albert, 250
Codrington, William John, 17
Collingwood, William, 26, 40–41
Colonial Defence Committee, 174
Colonial Survey Committee, 174
Colvin, Auckland, 121, 177
Committee of Imperial Defence (CID), 174, 185, 200, 212
Conder, Claude Reignier, 109–12
Conrad, Joseph, 98, 129
Constable, Charles Golding, 37
Constable, John, 37
Constantinople Protocol (November 1913), 200

Convention of Commerce between the Ottomans and Britain (1838), 23
Cook, Eliza, 77
Cook, Thomas, 125, 148, 154–55
Cooper, Anthony Ashley. *See* Shaftesbury, 7th Earl of (Anthony Ashley Cooper)
Cornish, Vaughan, 263
Cornwallis, Kinahan, 242
countermaps. *See* maps: local uses of cartography
Cowper, William, 51
Cox, Percy, 183–84, 200, 258–60
Crimea. *See* Crimean War
Crimean War, 10, 11, 15–19, 16 fig. 1.1, 22, 67, 69, 79–81, 105, 240
Croatia, 68
Cromer, 1st Lord of (Evelyn Baring), 175, 257, 263
Crowe, Eyre, 260
Cruttenden, Charles, 24
Ctesiphon, battle of (1915), 248, 313n53
Curzon, George, 3, 12, 81, 157, 181, 191–95, 193 fig. 6.4, 205–6, 211–13, 217, 236, 259, 261–62
Cushing, Thomas, 113
Cyprus, 62, 95–99, 96 fig. 4.1, 103, 137, 145, 179; Convention (1878), 11, 95, 101
Cyprus Convention (1878). *See* Cyprus: Convention (1878)

Daily Herald, 249
Daily Telegraph, 222, 249
Dalhousie, 11th Earl of (Fox Maule-Ramsay), 17, 48
Damascus, 22, 45 fig. 1.5, 127
Dan, 110, 235
D'Arcy, William Knox, 199
Dardanelles, 97. *See also* Gallipoli, battle of (1915)
Davis, Alfred, 78
Dawes, Edwin, 33, 41
Dawson, Alfred, 149
Dead Sea, 42, 86, 144
de Bunsen Committee (1915), 261
Dehradun, 114
Demavend, Mount, 22
Derby Daily Telegraph, 217
Détachement Français de Palestine, 247
Digna, Osman, 140
Dillon, C. H., 21
Disraeli, Benjamin, 11, 27, 57, 95, 102
Djebel Shammar, 161
Doughty, Charles, 35, 127–30, 129 fig. 4.4, 137, 161, 223
Dower, John, 67, 68 fig. 3.1

East. *See* Middle East
Eastern Christians, 8, 29, 66, 150, 220, 253, 316n114

Eastern Mediterranean, 2–3, 12, 21, 64, 96–97, 154, 256
Eastern Question, 10–11, 56, 81, 100, 107, 137, 150, 152, 153–54, 157, 201, 232, 271, 286n8
Edinburgh University, 228
education, cartographical and geographical: academic position, 164, 226, 228, 230, 250; adult education, 162; curricula and examinations, 81–85, 88–89, 159–60, 164, 225, 228–31, 250–51; geographical literacy, philanthropy and, 75; teaching and maps, 83 fig. 3.3, 84 fig. 3.4, 87–88, 160–61, 161 fig. 5.8, 227
Egypt, 2, 4, 5, 21, 30, 59–60, 64, 76, 80, 86, 101, 107, 108 fig. 4.3, 114, 141, 148, 150, 154–56, 156 fig. 5.6, 159, 161, 163–64, 210, 229, 239, 241, 253; 1882 intervention in, 10, 104, 143; British influence in, 11, 12, 106, 112, 115, 120–26, 136, 145, 174–80, 183, 188–90, 205–6, 212, 213, 219, 235–36, 242, 257; Survey of, 244–46, 245 fig. 8.4
Egyptian Hall, 69
Egyptian Institute, 122
El Arish, 52, 59–60
Eliot, Charles, 201
Eliot, George, 63
Erzurum, 20, 22, 97, 203
Erzurum, Treaty of (1847), 23
Esher, Reginald, 227
ethnography, 22, 29, 30, 59, 65, 66, 69, 111, 118–19, 128, 202, 213, 254, 265. *See also* maps: ethnographic
Euphrates, 7, 25, 59, 68, 77, 118, 189, 191, 203, 223, 235
Evans, Edmund, 62
Evans, Henry Smith, 84

Faden, William, 66, 72, 73 fig. 3.2, 289n18
Fairholme, William E., 172, 202
Fayoum, 121–22
Fergusson, James, 43
fieldwork. *See* surveying practices
Findlay, H. J., 228–29
First Balkan War (1912), 220
Fisher, John, 200
Fletcher, Joseph, 87, 89
Foreign Office, 21–23, 98, 115, 196, 235, 262; Eastern Department, 233; map room, 10–11, 11 fig. i.2, 49, 101–3, 106, 132, 172, 202, 219, 253, 279n45; Political Intelligence Department, 252; Propaganda Office, 247; strategic divide with the government of India, 115–16, 118, 185–86, 196–97
Forster, W. E., 88; Forster Act (1870), 85–86, 160
Fowler, Henry, 214
Fox, William R., 107, 108 fig. 4.3

France, influence in the East, 18, 25, 35–36, 228, 233, 246, 252, 256, 258
Francombe, James T., 82–85, 83 fig. 3.3, 84 fig. 3.4
Fraser, Simon. *See* Lovat, 14th Lord of (Simon Fraser)
French-Prussian war (1870), 20, 80

Galilee, Sea of, 47
Gallipoli, battle of (1915), 136, 238–41, 249–50
Galton, Francis, 157
Garstin, Crosbie, 238
Garstin, William, 122
Gascoyne-Cecil, Robert (3rd Marquess of Salisbury). *See* Salisbury, 3rd Marquess of (Robert Gascoyne-Cecil)
Gaza, 239, 243, 244 fig. 8.3
Gaze and Sons, 148
Gazetteer of the Persian Gulf, Oman and Central Arabia, 183
Geils, J., 20
Geographia Map Company, 249
Geographical Association, 162, 228
Geographical Magazine, 145, 213
Geographical Teacher, 251
Geological Society of London, 26
Germany, influence in the East, 9, 12, 96, 145, 180, 188–90, 199, 201, 202, 238, 239, 249. *See also* Bagdadbahn
Gharbia, 121
Gibbon, Edward, 2, 59, 61, 64
Gilbert, John, 48
Gill, William, 116
Gladstone, 12, 102, 117
Glasgow University, 228
Goethe, Johann Wolfgang von, 209, 307n6
Goldsmid, Frederic, 25, 61, 116, 131–32, 186, 193
Good, Frank Mason, 155
Gorchakov, Alexander, 102
Gordon, Charles, 127, 129–41, 150, 154, 157, 218, 248
Gordon, Thomas Edward, 209–10
Gore, George C., 182–83
Gorst, Eldon, 179
Graham, Cyril, 45, 45 fig. 1.5, 46 fig. 1.6
Graham, Gerald, 140
Grand Tour, 2, 58, 90, 154
Grant Duff, Mountstuart E., 84
Graphic, 79, 141, 153, 249
Graves, Thomas, 21
Gray, Francis John, 103
Greece, 55, 58, 64, 68 fig. 3.1, 76, 101, 235, 256
Grey, Edward, 214
Gross, Alexander, 249
Gwadar, 30, 118

Hadramaut, 186, 196
Hague Conference (1899), 210
Ha'il, 127, 183, 223
Haines, Stafford B., 24, 36, 38 fig. 1.4
hajj, 28, 127
Hakluyt Society, 26, 30
Hall, William Henry, 103
Hamilton, Angus, 212
Hamilton, Ian, 240–41
Hamilton-Gordon, George (4th Earl of Aberdeen). See Aberdeen, 4th Earl of (George Hamilton-Gordon)
Hamley, Edward Bruce, 116
Hardy, Thomas, 158
Harmsworth, Alfred, 222
Hauran, 45–46. See also Graham, Cyril
Hedjaz, 60
Helwân, 176–77, 302n31
Henty, George A., 163–64, 250
Herat, 132, 221
Herbert, Aubrey, 172, 202
Herbertson, Andrew J., 214–15, 230
Hermon, Mount, 62
Herodotus, 215
Herschel, John, 36
Hertslet, Edward, 102–3, 132, 293n26
Hicks, William, 150
Hiley, Richard, 89
Hirtzel, Arthur, 233, 259
Hobson, John A., 219–20
Hogarth, David, 81, 112, 189, 214–17, 216 fig. 7.3, 230, 236, 243
Holdich, Thomas Hungerford, 114, 131–33, 177, 197
Home, Robert, 97
Horniman, Benjamin, 220
Howlett, Samuel B., 33
Huber, Charles, 223
Hume, William Fraser, 177–78
Hunter, F. Fraser, 180–83, 181 fig. 6.1
Hussein, Iltifat, 245
Huxley, Thomas Henry, 81, 88

Illustrated London News, 67–69, 68 fig. 3.1, 79, 152, 153 fig. 5.5
imperialism: anti-imperialism, 219–20, 263; constructive, 114–45, 205–6, 213; ignorance and, 99, 158–59, 169–72; Indian subempire, 12, 24, 50, 115, 117–20, 180, 186, 188, 210, 258–60, 269; informal empire, 61, 145, 158, 175; Jingoism and popular imperialism, 141, 219, 227, 230, 248; liberal, 23, 48, 143; strategic planning and, 116–17, 174, 185, 200, 212, 254, 262, 264, 271. See also France, influence in the East; Germany, influence in the East; Middle East: position within the British Empire; Persia: Russian influence in
India: Bombay government, 24–25, 118, 131; British government in, 12, 24–26, 50–51, 117, 130–32, 179, 183, 203, 215, 219; East India Company, 7, 15, 25, 26, 40; Indian Mutiny (1857), 49, 65; Indian Navy, 24–25, 37, 38 fig. 1.4, 50, 80; Indian Political Service, 117–20, 193, 200, 236, 250, 258; protection of the route to, 21, 26, 50, 51, 61, 84, 96, 104; surveying expertise, 18, 37–40, 61, 102, 113–15, 117, 120–21, 173, 176, 188, 198
India Office, 25, 49, 70, 115–16, 173, 175, 200, 233, 259, 262–63; geographical department, 49–50, 103; India Act (1858), 49, 115; Political and Secret Department, 173
Indo-European Telegraph. See telegraphic lines: exploration and
instruments. See surveying practices
International Geographical Association, 152
Iskenderun, 21
Ismail (khedive of Egypt), 122
Ismailia, 149
Ispahan, 183
Italy, colonial competition with, 96
Izmir. See Smyrna

Jackson, Lowis d'Aguilar, 113–14
Jaffa, 21
James, Henry, 20, 49, 71
Jefferys, Thomas, 70
Jervis, Thomas Best, 15–18, 16 fig. 1.1, 20, 48, 69
Johnson, Henry V., 87
Johnston, Alexander Keith and William, 52, 55 fig. 2.3–2.5, 61, 70–71, 74, 99, 143–44, 147, 209, 222, 249, 286n11
Johnston, Thomas Ruddiman, 160, 161 fig. 5.8
John Taylor and Sons, 178
Jones, James Felix, 24, 26, 37, 40, 44, 50, 188, 246
Jordan, river, 46–47, 62, 110
Jose, Arthur Wilberforce, 213–14
Journal of the Royal Geographical Society, 45
Journal of the RUSI. See RUSI (Royal United Service Institute)
Jowett, William, 29
Judea, 86, 179
Jury's Imperial Pictures, 248

Kabir, Amir, 133
Karamania, 58
Karun, river, 194
Keltie, John Scott, 158–59, 185, 251
Kemal, Mustafa, 264

Kentish Gazette, 79
Khartoum, 70, 139–40, 150, 157, 164
Khedivial Geographical Society, 122
Khorasan, 131–33, 169, 170 fig. 6.1
Kiepert, Heinrich, 20, 35, 67, 70, 128
Kitchener, Horatio H., 12, 95–96, 96 fig. 4.1, 109–12, 113, 164, 179–80, 206, 218, 240
Knight, Charles, 75, 79
Křziž, August, 133
Kurdistan, 8, 106, 262; definitions of, 20, 57, 68, 70, 106, 173, 203; stereotypes on, 77
Kut-El-Amara, battle of (1916), 238, 248
Kuwait, 184, 194

Lane, Edward, 61
Lang Anderson, Robert, 122
Lausanne, Treaty of (1923), 254, 257
Lawrence, Thomas E., 179, 190, 233, 235–36, 254, 260, 263
Layard, Austen Henry, 22, 23, 40, 43, 77, 97, 106, 188
Leachman, Gerard, 183–85, 198
Lebanon, 62, 226
Leeke, Henry, 37
Lesseps, Ferdinand de, 20, 80
Levant, 2, 20–21, 28–29, 41, 97, 231, 235
Levantines, stereotypes on, 253–54
L'Humanité, 261
lithography, 71. *See also* chromolithography
Lloyd George, David, 81, 225, 247, 249, 252, 256–57, 259
Lloyd's Weekly Newspaper, 79
Loftus, William Kenneth, 23
London Convention (1840), 60
London Daily News, 218
London School of Economics, 214, 228
Lorimer, John, 181, 190. See also *Gazetteer of the Persian Gulf, Oman and Central Arabia*
Lovat, 14th Lord of (Simon Fraser), 192
Lovelace, Ada, 126
Lovett, Beresford, 131
Low, Seth, 209, 307n10
Lucas, Charles P., 103, 230–31
Lynch, Henry Blosse, 25
Lynch, Henry Finnis Blosse, 213–14
Lynch Brothers steamship lines, 51, 223, 282n1
Lyons, Henry G., 176–77, 180, 188–89, 213

MacDonald, Ramsay, 214
MacDonogh, George, 261
MacGregor, John, 28–29, 47
Mackinder, Halford J., 159, 162, 207, 214–15, 221, 228, 256
Maclean Carey, William, 228

Madame Tussauds, 69
Mahan, Alfred, 209–12, 214, 215, 224, 230, 256, 307n10
Mahomed, Ata, 132
Maitland, Pelham James, 197
Mallet, Louis, 233–36, 252–53, 261
Manchester Courier, 213
Manchester Geographical Society, 159, 162
Manchester Guardian, 241
Manning, Samuel, 155, 156 fig. 5.6
Mansell, A. L., 21
maps: in atlases, 29, 51–52, 54 fig. 2.1–2.6, 56–57, 61, 62, 66, 67–68, 74–75, 79–80, 85, 143–58, 148 fig. 5.3, 163, 222–25, 224 fig. 7.4, 231, 249; boundary making and, 7, 22, 40, 119, 131, 196–98, 200–201; as commodities, 62, 75–78; comparison between British and German cartography, 70, 100, 135, 142, 152, 159, 192, 222, 226–28, 241, 246; compilation of, 22, 29, 70, 72–74, 101–2, 111, 142, 178, 192, 271; difference between mapping and mapmaking, 19, 44–47, 45 fig. 1.5, 46 fig. 1.6; difficulty to read, 47–48, 240, 252; ethnographic, 61, 68–69, 98, 137, 197, 203–5, 215, 216 fig. 7.3, 220, 253, 255 fig. 8.6, 256; journalistic, 67–69, 68 fig. 3.1, 79–81, 149–54, 151 fig. 5.4, 153 fig. 5.5, 164, 217–20, 248–50; local uses of cartography, 134 fig. 4.5, 135 fig. 4.6; market, 74, 141, 142, 147–48, 148 fig. 5.3, 155, 160–61, 222–24, 225, 249, 270; military, 15–17, 21, 100, 102, 103, 113, 120–21, 172, 202, 239–47; and missions, 29–30, 225–26; official uses of, 4, 8, 17, 21, 48, 50, 90, 95–99, 102, 108, 116, 150, 171, 174–75, 185, 192, 195, 204, 209, 251–56, 264, 270; and propaganda, 7, 48, 108 fig. 4.3, 140–41, 145–46, 147 fig. 5.2, 163, 191–95, 212, 219, 227, 230, 248–49; semiotics, 3, 17, 46, 98, 111–12, 128–29, 137, 143–45, 156–57, 161, 204, 208, 239, 253; technical development in map printing, 17, 70–75, 90, 111, 114–15, 141–46, 149, 204, 217, 271. *See also* education, cartographical and geographical: teaching and maps; telegraphic lines: maps and
Marigny, Edouard Taitbout de, 74
Markham, Clements, 49–50, 103, 158, 203
Marmara, Sea of, 103, 152
Marshall, Charles, 69
Marvin, Charles, 101, 153
Marx, Karl, 68
Mashrek, 141
Masjed-Soleyman, 199
Massy, Percy H. H., 203
Maude, Frederick Stanley, 248
Maule-Ramsay, Fox (11th Earl of Dalhousie). *See* Dalhousie, 11th Earl of (Fox Maule-Ramsay)

Maunsell, Francis, 135, 189, 202–4, 242
Maxse, Leopold, 210
Maxwell, Robert James, 105
Mayhew, Henry, 77
McDonald, James, 109
McLeod, Walter, 61–62, 82
McMahon, Arthur Henry, 180, 186, 223
Mecca, 28, 35, 82, 182
Medina, 82, 180
Mehmet II, 57
Meinertzhagen, Richard, 233, 236, 263
Merv, 116, 153
Meshed, 193–94
Mesopotamia, 10, 12, 20, 25, 26, 27, 29, 30, 40, 41, 50, 51, 57, 64–66, 77, 85, 115, 117, 126, 173, 181, 188–91, 199–201, 203–4, 213, 233, 236, 238, 241–43, 245, 248–50, 253–54, 255, 258–60, 262, 264
Michelet, Jules, 209
Michie, Alexander, 210
Middle East: in contrast with the Far East, 3, 55, 201, 205, 211; definitions of, 1–3, 10, 20, 64, 174, 179–80, 190, 206, 207–9, 208 fig. 7.1, 210–11, 211 fig. 7.2; difference between East and West, 2–3, 12, 51–56, 60–61, 64; invention of the concept of, 6–8, 53, 81, 113, 114, 137, 185, 186, 192, 209–11, 212–14, 225–26, 232, 253, 260–61; "Near" and "Middle East", 214–16, 216 fig. 7.3, 218; position within the British Empire, 23, 105 fig. 4.2, 107, 112, 190, 192, 198–99, 201, 202, 203, 205, 212, 220, 223–25, 231, 233–37, 247–49, 257–60, 270–72
Miles, Samuel Barrett, 118–20
Milner, Alfred, 6, 124, 145, 175, 236, 252, 262
Mirza Ibrahim (Persian commissioner), 131
Moeris, Lake, 122
Mohandes, Mohammad-Reza, 133, 134 fig. 4.5, 137, 298n156
Mohi-uddin, 16
Molk, Amir Heshmat Al-, 133
Molyneux Seel, Louis, 202
Montenegro, 58, 95
Moresby, Robert, 24, 37
Morier, Robert, 100
Morning Post, 218
Morrell, George F., 249
Mosul, 29, 40, 77, 199, 233, 235, 254
Mozaffar ad-Din (Shah of Iran), 199
Mukhin, Semën Alexandrovich, 15
Munzinger, Werner, 119
Murchison, Roderick, 26
Murray, Charles, 22
Murray, John, 58, 76
Muscat, 17, 82, 118–19
Musil, Alois, 239

Nablus, 64
Napoleon III, 35
Nasseredin Shah, 133
National Assyrian Council, 256
National Society for Promoting Religious Education, 86
Nejd, 33, 35, 126–27, 143, 198, 215, 262
Nelson, Thomas, 142, 147
Newcastle, 5th Duke of (Henry Pelham-Clinton), 15
Nicolay, Charles Grenfell, 85
Nicolson, Harold, 252
Niebuhr, Carsten, 35
Nile, 12, 59, 60, 70, 87, 104, 107, 111, 121–24, 139, 149, 155, 156 fig. 5.6, 175, 178, 188, 212–13
Nineveh, 40, 77, 87

oil, prospection and uses, 7, 123–24, 178, 198–201, 224, 235, 238, 250, 254, 271
Oman, 30, 37, 118–20, 182
Omdurman, battle of (1898), 12, 179, 205
ordnance survey of the peninsula of Sinai. *See* Sinai: Sinai Survey
Orient. *See* Middle East
Ottoman Empire, 12, 20, 22–23, 27, 49, 52–53, 55–57, 61, 67–69, 68 fig. 3.1, 72, 73 fig.3.2, 74, 82, 85, 95, 104–5, 115, 135–36, 153, 153 fig. 5.5, 201–2, 204, 205, 217, 232, 234, 238, 247, 251, 253, 256, 261; Young Turk Revolution of 1908, 201. *See also* Eastern Question
Outlook (London), 210
Oxford University, 28, 159, 162, 189, 214, 228–30
Oxus, river, 133

Page, David, 61
Palestine, 4, 10 fig. i.1, 22, 42–44, 62–64, 77, 80, 86–88, 112, 144, 148 fig. 5.3, 154, 155, 158 fig. 5.7, 162–63, 208, 226; definitions of, 110, 233–38, 262, 264–65; local toponymy, 65; Palestine campaign (1917), 238, 242–44, 247–49, 256; Syria and, 268. *See also* biblical culture, influence of; education, cartographical and geographical; Palestine Exploration Fund
Palestine Association, 9
Palestine Exploration Fund: creation and underlying assumptions of, 42–44; *Palestine Exploration Fund Quarterly Statement*, 157; religious education and, 77–78, 144, 158 fig. 5.7; Sinai survey (1869), 63–65; surveys between West of Jordan, 110–11; surveys in Palestine (1870s), 104, 109–13, 179, 243; surveys in the Zin desert, 179. *See also* Conder, Claude Reignier; Kitchener, Horatio H.; McDonald, James; Thomson, William; Warren, Charles; Wilson, Charles

INDEX · 333

Palgrave, William G., 27, 34–35, 88, 181
Pall Mall Gazette, 149–50, 151 fig. 5.4, 212, 218. *See also* Stead, William T.
Palmer, Edward, 63, 149
Palmer, Elwin, 177
Palmerston, 3rd Viscount of (Henry John Temple), 12, 20, 48, 143, 194
Palmyra, 46
Panmure, Lord. *See* Dalhousie, 11th Earl of (Fox Maule-Ramsay)
Papadopoulos, Alexander, 235
Paris Peace Conference (1919), 7, 233–38, 234 fig. 8.1, 237 fig. 8.2, 252–60
Paris Treaty with Persia (1857), 23
Pauncefote, Julian, 101
Pelham-Clinton, Henry (5th Duke of Newcastle). *See* Newcastle, 5th Duke of (Henry Pelham-Clinton)
Pelly, Lewis, 27, 33, 34 fig. 1.3, 41, 51
Penny Illustrated Paper, 149, 218
Perim Island, 25, 103
Persia, 2, 3, 4, 5, 7, 10, 12, 16 fig. 1.1, 22–23, 25–26, 27, 28, 30, 40, 48, 53, 60–61, 77, 85, 96, 102, 104, 105, 114, 126, 145, 155, 159, 169–72, 170 fig. 6.1, 193 fig. 6.4, 195 fig. 6.5, 203, 210, 215, 218, 223, 229, 238, 241, 253, 255 fig. 8.6, 261; Anglo-Persian War (1856–57), 26, 37, 51; Indian influence in, 117–20, 173–80, 183–84, 190, 200, 258–59, 269; Persia Committee, 214; Persian Constitutional Revolution, 194, 213, 232; Persian Gulf, 10, 12, 24–26, 33, 50, 62, 68, 103, 115, 181 fig. 6.2, 182–83, 207, 211–13, 259, 262, 267; Russian influence in, 130, 191–95
Persian Gulf. *See* Persia
Persian Question. *See* Persia
Perthes, Justus, 67, 70, 241
Petermann, August, 47, 67, 70, 152
Petherick, John, 70
Philippopolis, 215
Pichon, Stephen, 260
Pickthall, Marmaduke, 220
Pilgrim, Guy E., 198–99
Pim, Bedford, 80
piracy, 24, 117
Planché, James Robinson, 76
Poole, Henry, 22
Porter, Josias, 47
Port Said, 21, 52, 154
Pritchard, James C., 36
Proceedings of the Royal Geographical Society, 128, 279n44
Public Works Department, 49
Punch, 225, 226

pundits, 110, 114, 132, 181–82, 186
Pusey, Edward, 43

Qaliubia, 121

RAF, 346, 260
Rahim, Abdur, 182
Rahman, Abdur (amir of Afghanistan), 133, 135
Ramsay, William Mitchell, 112–13, 189, 214, 235
Rashidi dynasty, 183–84, 239
Rassam, Christian, 29
Ravenstein, E. G., 144–45, 152, 165, 287n45
Rawlinson, George, 59
Rawlinson, Henry, 23, 40, 50, 116
Red Sea, 24–26, 35–36, 52, 59, 60, 61, 82, 101–4, 106, 107, 117, 120, 123–24, 140, 143, 180, 183, 213, 263
Rees, John David, 212
Reimer, Heinrich D., 70
Religious Tract Society, 158, 163
Remzi Pasha (Ottoman commissioner), 197
Rennell, James, 61
Resht, 123
Revue des sciences politiques, 261
Rhodes, Cecil, 219
Rhys, Ernest, 222
Richards, Erle, 254
Ritchie, Ritchmond, 192
Ritter, Carl, 47, 53
Riyadh, 33, 184, 283n75
Rosh Haniqra, 21
Ross, Edward C., 119
Royal Asiatic Society, 26, 128, 172
Royal Engineers, 33–34, 97–98, 109–10, 125; *Papers on subjects connected with the duties of the Corps of Royal Engineers*, 34
Royal Geographical Society: as a center of calculation, 21, 26–27, 29, 46, 50, 51, 128, 137, 158–59, 162, 185, 203, 227–29, 247, 297n128; criticism of, 126; *Hints to Travellers*, 33, 125, 154; membership, 22, 26–28, 30, 42, 61, 78, 80, 84, 85, 116, 118, 144, 152, 157, 192; worldview, 10 fig. i.1. *See also Journal of the Royal Geographical Society*; *Proceedings of the Royal Geographical Society*
Royal Military Academy, 106
Royal Scottish Geographical Society, 144, 159, 162
Rub al Khali, 184
Rumohr, Carl Friedrich von, 209, 307n6
RUSI (Royal United Service Institute), 116
Russell, John, 48
Russell, William H., 18
Russia, influence in Persia. *See* Persia; Russian influence in
Russo-Turkish war (1878), 10, 104, 106, 149

Said, Sayyid Turki bin, 120
Said, Thuwaini bin Said al-, 119
Saint Quentin, Doisnel de, 261
Salisbury, 3rd Marquess of (Robert Gascoyne-Cecil), 86, 97, 100–103, 105, 117, 160–71, 170 fig. 6.1
Salmond, Geoffrey, 243
Sana, 24
San Remo Conference (1920), 225, 257
San Stefano, Treaty of (1878), 95, 102
Sa'ud, Abdulaziz al-, 184–85, 259
Saunders, Trelawney, 49, 70
Schubert, Fëdor Fëdorovich, 15
Scottish Geographical Magazine, 142, 213
Scott-Moncrieff, Colin, 122, 189
Sebah and Joaillier, photographic studio of, 135–36, 135 fig. 4.6
Second Boer war (1899–1902), 107, 174, 231
Seeley, John Robert, 213
Selby, William B., 26, 41
Sèvres, Treaty of (1920), 254, 257
Shaftesbury, 7th Earl of (Anthony Ashley Cooper), 43
Shakespeare, William, 183–85, 198
Sharif, Imam, 132, 186, 295n94
Shatt al Arab, 103, 183, 200
Shuckburg, John, 262
Simmons, John Lintorn A., 97
Sinai, 20, 21, 42, 52, 59–60, 62, 64, 107, 179, 198, 201, 235, 246; Sinai Survey, 44, 63, 74, 109
Sing, Hira, 132
Sivas, 106, 202
Slade, Edmond, 200–201
slavery, 25–26, 104, 120
Smeaton, Donald Mackenzie, 212
Smith, Albert, 69
Smith, Charles, 70
Smith, George Adam, 225
Smyrna, 57, 112, 154, 257
Smyth, Robert, 113
Smyth, William Henry, 21
Society for the Diffusion of Useful Knowledge, 75, 79
Society for the Promotion of Christian Knowledge, 76
Socotra, 103
Soteriades, Georgios, 256
Stanford, Edward, 53, 55 fig. 2.6, 69, 70, 74–75, 98, 128, 141–43, 212
Stanley, Arthur P., 44, 59
Stead, William T., 149–52, 151 fig. 5.4, 165, 218
Stewart, Charles, 123
Stewart, Patrick, 25
Stewart, Robert W., 109

Stieler, Adolf, 222
Storrs, Ronald, 179
Strauss, David Friedrich, 63
Suakin, 140, 143, 164
Subhan, Abdul, 132–33
Sudan, 104, 106–7, 108 fig. 4.3, 111, 122, 124, 140 fig. 5.1, 175–76, 178–80; Madhist rebellion in the, 139–40, 149–50, 151 fig. 5.4, 160, 164. *See also* Omdurman, battle of (1898)
Suez: Canal, 11, 12, 20, 21, 44, 61, 64–65, 95, 97, 103, 104, 109, 124, 140, 144, 145, 173–74, 180, 196, 208, 210, 238, 263; isthmus, 20, 54 fig. 2.1–2.6, 56, 59; port, 24, 28, 52, 80, 140
surveying practices, 22, 28–29, 30–37, 32 fig. 1.2, 109–10, 113–14, 125–30, 176, 183–84, 195 fig. 6.5, 198, 203–4, 283n79; accuracy, 25, 28, 32, 33, 36, 74, 95, 110, 137, 176–77, 184, 242, 302n31; aerial surveys, 241–46, 244 fig. 8.3, 245 fig. 8.4, 246 fig. 8.5; altimetry, 31; cadastral surveys, 7, 124, 176–77, 180, 205, 213; geological surveys, 7, 22, 23, 119, 123–24, 128, 171, 177–78, 194, 198–99, 204; hydrographic surveys, 21, 24, 26, 36–37, 103; local assistance and resistance, 37–41, 48, 63–64, 99–100, 110, 121, 130–36, 182, 198
Sykes, Herbert R., 193–95, 195 fig. 6.5
Sykes, Mark, 190, 202, 204, 223, 248, 254, 260, 262
Sykes, Percy Molesworth, 169–72, 170 fig. 6.1, 190, 193, 213, 239
Sykes-Picot Agreement (1916), 195
Syria, 4, 10, 21, 25, 27, 29, 41, 45 fig. 1.5, 47, 56, 57, 62, 64–65, 80–85, 126, 144, 233, 235, 254, 262, 264, 268

Taba, 52
Taurus, 62, 203–4
Tawfiq (khedive of Egypt), 123, 177
Taylor, George, 87
Teague-Jones, Reginald, 241
Tehran, 25, 106, 131, 133, 259, 269
Tehran Protocol (1911), 198
telegraphic lines: exploration and, 25, 30, 131, 176, 193, 200; maps and, 61, 80, 223. *See also* Goldsmid, Frederic
Temple, Henry John (3rd Viscount of Palmerston). *See* Palmerston, 3rd Viscount of (Henry John Temple)
Thackeray, William Makepeace, 27
Thebes, 87
Thomson, William, 43
Thuillier, Henry L., 113
Thwaites, William, 236
Tigris, 25, 40, 68, 77, 118, 188–89, 203, 223
Times (London), 18, 67, 126, 173, 210, 212, 218, 222, 249

Times of India, 194
Topical Film Company, 248
Torbat, 194
Toynbee, Arnold, 191, 233, 247, 257, 260
Trabzon, 202
Transactions of the Bombay Geographical Society, 119
Trans-Caspian railway, 191, 194, 261
Transcaucasia, 253, 161
Tristram, Henry B., 47
Trucial Coast, 26
Turkey, 4, 48, 97, 101, 104, 190, 220, 253; in Europe and in Asia, 56, 73 fig. 3.2, 147 fig. 5.2, 149, 172, 201–2; mapping of, 49, 57, 68 fig. 3.1, 70, 73–73, 105, 144, 203–4, 223, 255 fig. 8.6. *See also* Ottoman Empire

Urabi, Ahmed, 104, 148, 249

Vámbéry, Arminius, 213
Van, 202, 204
Velde, Charles van de, 40, 47
Victoria Investment Corporation, 178

Wahab, Robert A., 114, 197–98
Wahhabi, 34–35, 41, 127, 239
Walker, Ewery, 128
Wallin, George A., 27
Wallis, John, 78
Walton, Joseph, 172
Warburton, Eliot, 47, 55
War Office: Cinematograph Committee, 248; Directorate of Military Intelligence of the War Office, 172; Geographical Section of the General Staff, 173, 190, 241, 247, 253–54; Intelligence Branch, 104–7; Intelligence Department, 96, 97, 111, 112, 202, 204; Ordnance Survey, 42, 71, 77, 111–12; Topographical and Statistical Department, 15–18, 20

Warren, Charles, 20, 42, 63, 109
Wassmuss, Wilhelm, 239, 241
Weizmann, Chaim, 235, 259
Weller, Edward, 33, 49, 64, 67, 80
Wellsted, James Raymond, 30
Welpiox, W. P., 230
Western Daily Press, 217
Western Times, 217
Westminster Review, 88
Wharton, William, 103
White, Arthur Silva, 144, 145
Whitehouse, Wallace, 251
Willcocks, William, 122–23, 235, 242, 260
Williams, William Fenwick, 23
Wilson, Arnold, 199–200, 236, 258–59
Wilson, Charles, 20, 42, 102, 104, 106, 109, 112, 150, 189
Wilson, Henry, 173, 252
Wilson, Woodrow, 251
Wingate, Reginald, 121, 180, 257
Wolesley, Garnet, 98
Woolf, Leonard, 263
Woolley, Leonard, 179
Worms, Henry de, 3
Wyld, James (the elder), 72–73, 73 fig. 3.2
Wyld, James (the younger), 2, 61, 69, 70, 72–74, 73 fig. 3.2, 76–77, 85, 144

Yacouboff, Lazar, 256
Yate, Arthur Campbell, 213
Yazd, 194
Yazidi, 30
Yemen, 185, 196–98
Young, Hubert, 262

Zagros Mountains, 199
Zionism, 8, 235–36, 256, 263, 314n94
Zwemer, Samuel M., 226